Advances in Science, Technology & Innovation

IEREK Interdisciplinary Series for Sustainable Development

Advances in Science, Technology & Innovation (ASTI) is a series of peer-reviewed books based on important emerging research that redefines the current disciplinary boundaries in science, technology and innovation (STI) in order to develop integrated concepts for sustainable development. It not only discusses the progress made towards securing more resources, allocating smarter solutions, and rebalancing the relationship between nature and people, but also provides in-depth insights from comprehensive research that addresses the **17 sustainable development goals (SDGs)** as set out by the UN for 2030.

The series draws on the best research papers from various IEREK and other international conferences to promote the creation and development of viable solutions for a **sustainable future and a positive societal** transformation with the help of integrated and innovative science-based approaches. Including interdisciplinary contributions, it presents innovative approaches and highlights how they can best support both economic and sustainable development, through better use of data, more effective institutions, and global, local and individual action, for the welfare of all societies.

The series particularly features conceptual and empirical contributions from various interrelated fields of science, technology and innovation, with an emphasis on digital transformation, that focus on providing practical solutions to **ensure food, water and energy security to achieve the SDGs.** It also presents new case studies offering concrete examples of how to resolve sustainable urbanization and environmental issues in different regions of the world.

The series is intended for professionals in research and teaching, consultancies and industry, and government and international organizations. Published in collaboration with IEREK, the Springer ASTI series will acquaint readers with essential new studies in STI for sustainable development.

ASTI series has now been accepted for Scopus (September 2020). All content published in this series will start appearing on the Scopus site in early 2021.

Ante Mandić • Rui Alexandre Castanho •
Uglješa Stankov
Editors

Cultural Sustainable Tourism

Strategic Planning for a Sustainable Development

Third Edition

Springer

Editors
Ante Mandić
Department of Tourism and Economy
Faculty of Economics, Business and Tourism
University of Split
Split, Croatia

Rui Alexandre Castanho
College of Business and Economics
WSB University
Dabrowa Gornicza, Poland

Uglješa Stankov
Department of Geography,
Tourism and Hotel Management
University of Novi Sad
Novi Sad, Serbia

ISSN 2522-8714 ISSN 2522-8722 (electronic)
Advances in Science, Technology & Innovation
IEREK Interdisciplinary Series for Sustainable Development
ISBN 978-3-031-10799-3 ISBN 978-3-031-10800-6 (eBook)
https://doi.org/10.1007/978-3-031-10800-6

Cover illustration: Tourist Board of Split, author: Maksim Bašić/Split from air 2—Copyright Tourist Board of Split; author: l33t digital agency

This Springer imprint is published by the registered company Springer Nature Switzerland AG
The registered company address is: Gewerbestrasse 11, 6330 Cham, Switzerland

Scientific Committee

Acknowledgments

We would like to thank the authors of the research papers that were selected for addition in this book. We would also like to thank the reviewers who contributed with their knowledge and constructive feedback in hopes of ensuring the manuscript is of the best quality possible. A special thanks go to editors of this book for their foresight in organizing this volume and diligence in doing a professional job in editing it. Finally, we would like to express our appreciation to the IEREK team for supporting the publication of the best research papers submitted to the conference.

Contents

Sustainable Environment: Geography and Tourism

Tourists are increasingly seeking authentic experiences, and in the framework of authenticity, cultural and natural heritage have become a significant focus (Duffy, 2009; Mandić & Petrić, 2021; Mandić & Kennell, 2021). According to Timothy and Boyd (2003), for some, a visit to a heritage destination or site provides a sense of pleasant escapism, while for others, it may be a quest for knowledge or a complex amalgam of different social, cultural or leisure oriented-motivations. Successful heritage destinations will be in the "process of continuous reinvention of their imagined pasts" in response to the changing nature of tourism demand (Ashworth, 2009). Acknowledging the challenges associated with the development and management of sustainable and resilient heritage destinations, this edition of the *Cultural Sustainable Tourism* is divided into the three significant parts to reflect on (1) *Sustainable environment: geography and tourism*, (2) *Social and economic impacts*, and finally (3) *Policies and mechanisms for heritage preservation*.

Tourism is dependent upon a healthy and preserved environment; therefore, it should ally itself firmly with the concept of sustainable development (Butler, 2000). Nevertheless, in the last two decades, tourism progress towards this goal has been minimal (Sharpley, 2020). Driven by deregulation, globalisation, and the neoliberal capitalist system, massive tourism growth has become a critical challenge for many tourism destinations as the symptoms associated with over-tourism become more and more pronounced (Peeters et al., 2018; UNWTO, 2018). As a society, we are growing awareness of how a sustainable future of tourism requires the fundamental rethinking of the consumption and the production processes within the entire industry (Higgins-Desbiolles, 2020; Higgins-Desbiolles et al., 2019). This means paying greater attention to the health of ecosystems, economies, and communities while transforming personal patterns of behaviour by committing to more responsible acts and choices (Programme, 2020; Becken & Loehrn, 2022).

The first part of the *Cultural Sustainable Tourism* edition, titled *Sustainable Environment: Geography and Tourism*, is organised into five chapters . Within Chapter "Re-looking Through Ian Mc Harg's Perspective on Landscape Planning: Watershed Management of Cultural Landscapes of Orchha, Madhya Pradesh, India", Ramani employed the holistic approach to the sustainable development of the city's cultural landscape while critically examining water resources management. The comprehensive analysis conducted in Madhya Pradesh, India, using Ian McHarger's framework revealed critical challenges which should be addressed to ensure sustainable development and prevent the degradation of the water resources. La Ghezza (Chapter "Heritage Tourism: The Case Study of I Benedettini (BARI)") discussed the notion of traditional values and modern lifestyles and how they convey the concept of community-based tourism. The study focused on the intersection between communities, economic development and conservation of heritage and delivered several valuable suggestions on how to market such destinations and businesses efficiently. Within Chapter "From the Paradise of Beit Shean Valley to the Contested Landscape of the Valley of Springs; Water Amenities, Environmental Justice and Sustainable Development", Mozes explores the role of the natural heritage in the process of regeneration of Beit Shean Valley in Israel. Waster and natural resources are discussed as agents of social relationships and essential substances. Through the lens of environmental justice, the author proposes an alternative approach for creating a more sustainable future in the region. Martins (Chapter "Religious Tourism During the COVID-19 Period: The Case of Our Lady of Agony Festival, Viana do Castelo, Portugal"), reflects on the impacts of the COVID-19 pandemic on religious tourism development in Viana do Castelo. The analysis confirmed the resilience of this niche tourism form and delivered an extensive discussion of managerial implications. Within Chapter "Visitors' Perception of Tourist Attractions in a Green Protected Area: The Case Study of the Peneda-Gerês National Park", Martins focused on visitor perception of tourist attractions in the Peneda_Geres National park in Portugal. The author employed netnographic approach and conducted an extensive analysis of TripAdvisor review over the three years. The content analysis provided new insights into the visitor experience constituents and factors affecting visitor satisfaction and future behavioural intentions.

Ante Mandić
Department of Tourism and Economy
Faculty of Economics, Business and Tourism
The University of Split
Split, Croatia
e-mail: ante.mandic@efst.hr

References

Ashworth, G. J. (2009). Heritage and economy. *International Encyclopedia of Human Geography*, 104–108. https://doi.org/10.1016/b978-008044910-4.00363-1

Becken, S. & Loehr, J. (2022): Tourism governance and enabling drivers for intensifying climate action, *Journal of Sustainable Tourism*.

Butler, R. W. (2000). Tourism and the environment: A geographical perspective. *Tourism Geographies*, *2*(3), 337–358. https://doi.org/10.1080/14616680050082553

Duffy, M. (2009). Festival and spectacle. *International Encyclopedia of Human Geography*, 91–97. https://doi.org/10.1016/b978-008044910-4.00948-2

Higgins-Desbiolles, F. (2020). The "war over tourism": Challenges to sustainable tourism in the tourism academy after COVID-19. *Journal of Sustainable Tourism*, *0*(0), 1–19. https://doi.org/10.1080/09669582.2020.1803334

Higgins-Desbiolles, F., Carnicelli, S., Krolikowski, C., Wijesinghe, G., & Boluk, K. (2019). Degrowing tourism: Rethinking tourism. *Journal of Sustainable Tourism,27*(12), 1926–1944. https://doi.org/10.1080/09669582.2019.1601732

Mandić, A., & Kennell, J. (2021). Smart governance for heritage tourism destinations: Contextual factors and destination management organisation perspectives. *Tourism Management Perspectives*, *39*(Oct 2020), 100862. https://doi.org/10.1016/j.tmp.2021.100862

Mandić, A., & Petrić, L. (Eds.). (2021). *Mediterranean protected areas in the era of overtourism*. https://doi.org/10.1007/978-3-030-69193-6

Peeters, P., Klijs, J., & Novelli, M. (2018). *Requested by the TRAN committee research for TRAN committee—Overtourism: Impact and possible policy responses*. February 2019.

Programme, O. P. S. T. (2020). *One planet vision for a responsible recovery of the tourism sector*

Sharpley, R. (2020). Tourism, sustainable development and the theoretical divide: 20 years on. *Journal of Sustainable Tourism*, *28* (11), 1932–1946. https://doi.org/10.1080/09669582.2020.1779732

Timothy, D. J. & Boyd, S. W. (2003). *Heritage tourism*. Pearson Education Limited.

UNWTO. (2018). 'Overtourism'?—Understanding and Managing Urban Tourism Growth beyond Perceptions, Executive Summary. *World Tourism Organization (UNWTO); Centre of Expertise Leisure, Tourism & Hospitality; NHTV Breda University of Applied Sciences; and NHL Stenden University of Applied Sciences*. https://doi.org/10.18111/9789284420070

Re-looking Through Ian Mc Harg's Perspective on Landscape Planning: Watershed Management of Cultural Landscapes of Orchha, Madhya Pradesh, India

Sneha Ramani

Abstract

Orchha uplands in central India, once the capital of the Bundela Rajputs from 1531 to 1783, and now in Madhya Pradesh, India, is facing increasing pilgrimage and tourism pressures. The protected monuments of the region are only a part of a complex multi-layered cultural landscape heritage of the setting—the importance of the Orchha is not limited to its architectural heritage, but also encompasses its environmental and landscape heritage. The environmental heritage of Orchha includes—the Orchha Sanctuary, the braided Jamni and Betwa River. During the course, the Betwa and Jamni River carve interesting geological forms like rocky outcrops, rapids, etc., adding to the leisure value of the setting. The Orchha Sanctuary forms a habitat for many species of reptiles, mammals, and birds, making it a must-visit destination for nature lovers. The urban development of Orchha and its adjoining villages, owing to the cities pilgrimage and heritage value, requires a holistic approach to the sustainable development of the city's cultural landscape while critically examining the management of water resources. The idea of the study was to comprehend the tourism pressures of the region and propose a holistic conservation framework (using Ian McHarg's framework, as presented in Design with Nature book) based on managing and protecting the water-dependent cultural landscape. The extent of the region selected for our study aligns with the watershed basin of the rivers Betwa and Jamni. The grounded conjectures of Orchha from site readings, mappings, and interviews of water resources revealed issues about pollution, groundwater degradation, degradation of vegetation, habitat fragmentation, etc. This understanding helped identify the potential areas of intervention to ensure sustainable development and prevent the loss of essential water resources.

Keywords

Watershed management • Cultural landscape • Water resources • Sustainable development • Environmental heritage

1 Introduction to the Study

The region is known for its religious and heritage significance, as it happens to be a stop for the Kawad yatra. The city of Orchha, highlighting the magnificent heritage and architecture of the Bundelas, has temples and monuments surrounded by woody forests. It has monuments like Jahangir Mahal, Chaturuhuj Temple, Ram Raja temple, Raja Mahal, and Lakshmi Narayan Temple. Further, the region is also known for its environmental heritage of Orchha, including—the Orchha Sanctuary, the braided Jamni, and Betwa River. The rivers also add leisure value to the setting as it has interesting geological forms like rocky outcrops, rapids, etc.

The region has tourists visiting the place all year round, with the peak inflow of domestic tourists observed during the month of January and that of foreign tourists observed during October and November. Domestic tourists visit the region for the purpose of religious pilgrimage. The region is a significant tourist and pilgrim destination and is one of the vertices of the Orchha–Khajuraho–Bandhavgarh tourism circuit. The influx of tourists within the city has also transformed its population from a few hundred to more than 8000 residents within 20 years. According to the influx, one can see this reflected in the investment pattern of the residents in the hotel and leisure business (Kalinowski, 2016; Shrivastava et al., 2020).

S. Ramani (✉)
Institute of Architecture and Planning, Nirma University, Ahmedabad, India
e-mail: ramani.sneha@gmail.com; ; sneha.ramani@nirmauni.ac.in

A. Mandić et al. (eds.), *Cultural Sustainable Tourism*, Advances in Science, Technology & Innovation,
https://doi.org/10.1007/978-3-031-10800-6_1

The study region aligns with the watershed basin of the rivers Betwa and Jamni, within which the city of Orchha is situated. Further, it has led to the increased attention of locals and authorities to heritage and its preservation; however, it has also led to the degradation of natural resources within the watershed region of the city. The study intends to comprehend the tourism pressures of the region and propose a holistic conservation framework, using Ian McHarg's framework, as presented in Design with Nature book that is based on managing and protecting the water-dependent cultural landscape. To achieve this, the objectives of the study are divided into three parts—(1) to understand the religious, cultural, and environmental heritage of the region, (2) to delineate and overlay the natural layers in order to identify the ecologically sensitive areas with the delineated region, and (3) to extract suitable sites for landscape restoration for proposing appropriate conservation measures within the setting.

Methodology of the study—The vegetation character mappings were carried out on the field by collecting samples and identifying visually similar sightings (both color and textures were considered). The maps and drawings were prepared using QGIS to delineate and overlay the region's natural features. The village and city survey was conducted to understand people's livelihood dependency and culture. In a similar manner, tourists interviews were conducted to study their profile and their perceptions of the city. Further, to collect data on natural, administrative, and cultural layers and the nature of the region's development plans, Orchha Administrative Offices were visited.

2 Design with Nature—Ian Mc Harg's Perspective on Landscape Management

Ian Mc Harg laid down his theories of ecological planning and contributed to the methodology of overlay analysis of maps. This, eventually, led to the development of the ArcGIS software, which is the basis of land use planning widely across the world. Mc Harg addresses his theory of working and designing with nature not only for social welfare but also for ensuring public safety and endurance against nature. A hydrologist, on the other hand, from his discipline, may incorporate this theory of designing with nature, analyze a terrain where groundwater is prone to pollution, and advocate necessary measures to the welfare of the community. Although here we see that the theory is scientific in approach, the intention and purpose of conducting their practice are backed up by a strong sense of compassion toward social health and ecological health (McHarg, I. L., 1969).

3 Context

3.1 Regional Context

The intended study region, i.e., Orchha, a historic city nestled on river Betwa, is a tourist destination in Madhya Pradesh.

The Orchha upland is situated in the Tikamgarh district of Madhya Pradesh state, India.

Orchha in central India, once the capital of the Bundela Rajput from 1531 to 1783, and now in Madhya Pradesh, is a case study of increasing tourism and pilgrimage at a heritage site. (Sinha & Valderrama, 2014).

The study region comprises the town of Orchha and the villages around Orchha, which extends into two administrative zones, which are the Tikkamgarh district, Madhya Pradesh, and Jhansi and Lalitpur districts, Uttar Pradesh. (as seen in Fig. 1).

3.2 Climatic Context

Orchha has a pleasant climate as its location is in the flat terrains. It is a historic place; most tourists visit from October to April, which are the comfortable months. The region receives rainfall of about 900 mm every year (average). The monsoon winds come from the southwest winds from the Arabian Sea. It lasts for about 3–4 months, during which the river Betwa remains flooded. The region received most of its precipitation in August, about 355 mm yearly.

Orchha has a warm and temperate climate. The region's average temperature is about 34.6 °C, with May being the warmest month and April being the driest month with 10 mm average rainfall. The region receives its lowest temperature in January, which is 8 °C.

The region and its environs have a dry deciduous forest with higher variation in diurnal temperature.

3.3 Hydrological Context

The Betwa is a river in Northern India and a tributary of the Yamuna. It originates in the Vindhya Range just north of Hoshangabad in Madhya Pradesh. R. Betwa flows northeast through Madhya Pradesh and Orchha to Uttar Pradesh and joins the Yamuna River. The river Betwa becomes the lifeline of the city of Orchha and the Surrounding villages and settlements.

The James River is a tributary, which meets the Betwa River at Panchmaria. The two rivers and their conjunction and confluence form the boundary of the Sanctuary.

Fig. 1 Orchha city and its surroundings

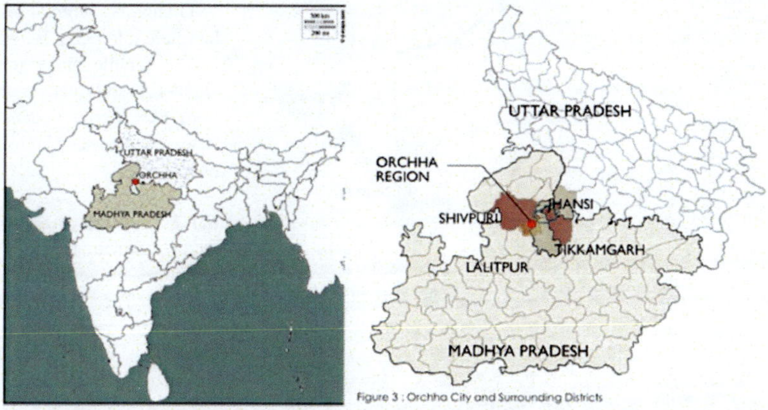

The extent of the basin of the watershed (as seen in Fig. 2), which creates the region of our study, reaches Jhansi in the North West, Babina cantonment in the South West, Bharua Sagar village, fort and settlement in the North East, Talbehat for and lake in the South East, to the reservoir in the South.

4 Orchha—City

Orchha came into existence in the sixteenth century, when accounting for the remote and challenging terrain of the region, Rudra Pratap Singh shifted his capital from Garh Kundar (48 km southwest of Orchha) to present-day Orchha. The Bundela kings began building the town of Orchha, also known as the heart of Bundelkhand, in 1531. The citadel island of the medieval Bundela rulers is situated on the Betwa River. Until 1738, Orchha remained the capital of Bundelkhand.

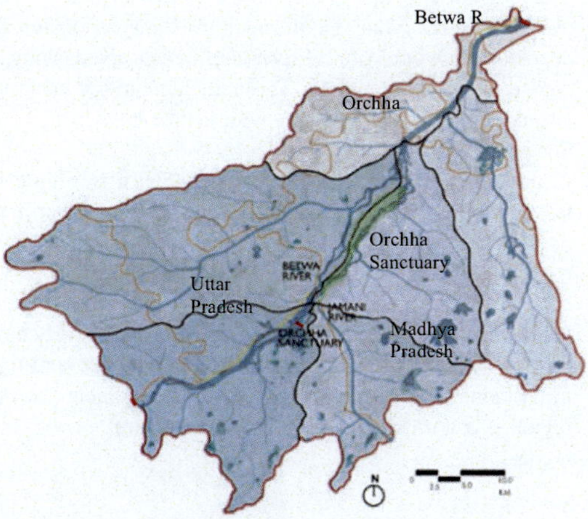

Fig. 2 Delineation of study area. *Source* author

5 Infrastructure

Orchha has Indian Railways connectivity (as seen in Fig. 3) with stations at Orchha, Arjar, and Teharka and sections like Jhansi-Bhopal Jhansi-Manikpur. Its population is 11,511 approximately, with about 2049 households (Primary Census Data Tables—Madhya Pradesh, 2011). The region has rocky terrain with rocky outcrops and stony soil. The soil is not fertile. Furthermore, as seen in Fig. 4, the region attracts a similar proportion of domestic and foreign tourists.

The occupation pattern within the region comprises about 33% of people engaged in agriculture, 20–30% engaged in small business and tourism-related activities, and about 20% engaged in contractual labor. The city's growth, Orchha, has been observed along State Highways-37 toward Jhansi, which is also the route taken by most pilgrims and tourists.

In the older part of the town of Orchha, untreated sewage (0.80 MLD) and sullage are discharged into stormwater drains. The town has 90% septic tank toilets, while 7% uses the field and the remaining 3% uses the public toilets.

5.1 Heritage Significance

5.1.1 Cultural Heritage

The ruins and remnants in the form of the architectural heritage of the Bundelas within the region picture and depict their presence and connection with Mughals and their influence. The significant palaces and temples are located in

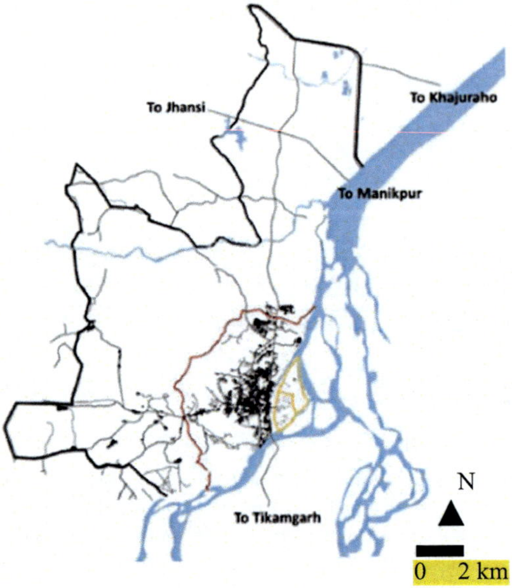

Fig. 3 Connectivity of Orchha city

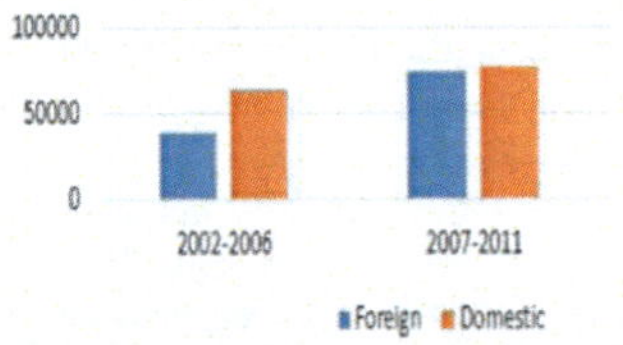

Fig. 4 Tourist population of Orchha

higher areas—like a promontory—(greater than 225 m) within the region like Jahangir Mahal, Chaturuhuj Temple, Ram Raja temple, Raja Mahal, and Lakshmi Narayan Temple. Other places such as mansions, smaller temples, and plazas are situated on low heightened hillocks like Raghuvamsani, Kanhaiyya mandirs, and Radhika Raman (about 220 m). The reflections of the cenotaphs and ghats can be seen in the river Betwa, accounting for its closer proximity to the river. The visual relationship among the buildings, less the street layout, is crucial in understanding the visual order discerned in triangular yantras and circular mandalas. (Sinha & Valderrama, 2014).

The city of Orchha has open spaces ranging from house courtyards to the Betwa River ghats, with a different character.

Orchha is a crucial pit stop in the Kavad Yatra. Pilgrims halt here and carry Betwa water to their destination.

The temple plaza acts as a platform for the rituals in Orchha, a spillover space for pilgrims. Significant celebrations like Navratri occur in the ground behind the temple; however, the space is insufficient when the users are large in number.

5.1.2 Environmental Significance

Orchha Sanctuary is situated between Betwa and Jamni rivers in the erstwhile Bundelkhand region of Madhya Pradesh. The two rivers and their conjunction and confluence form the boundary of the Sanctuary. The Orchha Sanctuary forms a habitat for many species of reptiles, mammals, and birds, making it a must-visit destination for nature lovers.

The total area of the Nature Reserve is 45.86 km^2, which includes both land and water bodies (rivers). It is just 19 km from Jhansi in Uttar Pradesh. The altitude of Sanctuary varies from 207 to 357 m above mean sea level.

Orchha Wildlife Sanctuary has about 200 bird species, including in-house and migrant birds, which can be found accounting for its location near the river ecosystem. Jungle Bush Quail, minivet, stork, kingfisher, owls, woodpecker, peafowls, peacocks, swans, black swan, geese, collared scops owls, and many more can be observed within the Sanctuary environs. Further, faunal species such as Jackal, Nilgai, peacock, blue bull, wild pig, slot bear, and other such species can be spotted within the Sanctuary (as seen in Fig. 5, which highlights spotting of some fauna within the sanctuary environs).

The river's course (river Betwa and river Jamani) within the region has many islands, and therefore, the river is braided. These islands show diverse vegetation types and densities such as tall grasses, open to dense scrubs, dense teak forests, etc. Many landforms formed due to erosion and depositional works of the river can be observed along its course within the region. These include rocky outcrops, islands, rapids, shallow banks with pebble beds, banks with fine sand, etc.

As most of the Betwa is not navigable, it is among the least polluted rivers of the country. The depth of water in the river varies from 1 to 13 m. Shallow depths are observed near Orchha city, whereas deeper water depths are observed near the reservoirs for dams. The diversity in vegetation, landforms, and water depths observed along the river course forms habitats for much aquatic fauna, avifauna, and territorial fauna. Due to the diverse resources available here, the region is a host to 97 bird species rare turtles, among other fauna.

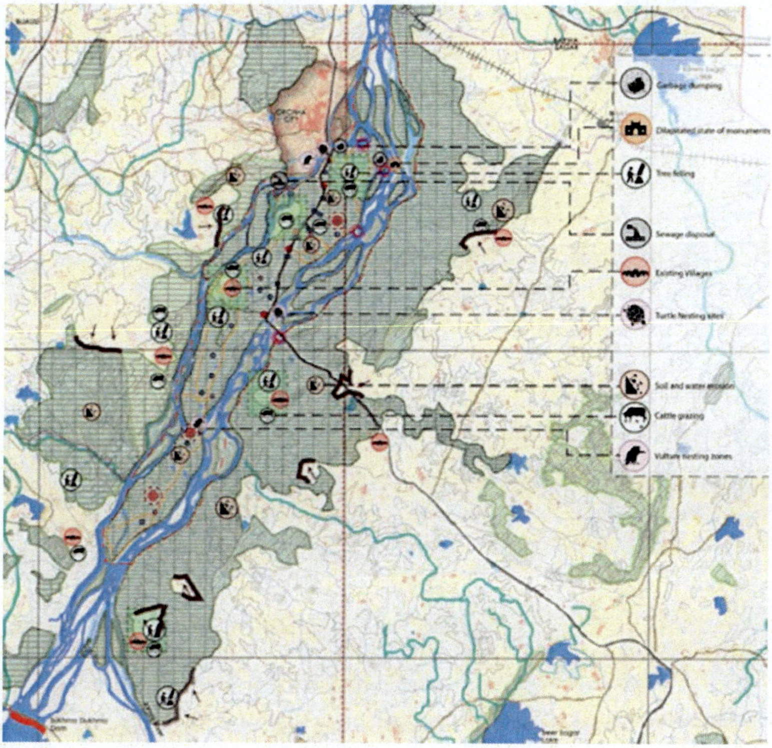

Fig. 5 Orchha Sanctuary—environment heritage

6 Orchha—Regional Study—Natural Features of the Delineated Region

6.1 Terrain—Topography, and Slope

The elevation varies from 120 to 420 m, with an average elevation of 220 m—the land slopes from the southwest to the northeast corresponding to the flow of the rivers. The highest ridges are found along the eastern side of the delineated region. The highest elevation is near the reservoir, while the lowest is near the Parichha thermal power plant (as seen in Fig. 6).

6.2 Hydrology

The river's total length from its origin to its confluence with the Yamuna is 590 km, of which 232 km lie in Madhya Pradesh and 358 km lie in Uttar Pradesh. For the regional study, the length of Betwa River spanning between Matatila Dam (elevation of 310 m above mean sea level) and Parichha Dam (elevation of 200 m above mean sea level) has been considered. (Government of India, 2014).

The region is divided into catchments of Betwa River and its tributaries—Jamini and Gorai. The Betwa River originates in the Bhopal district of Madhya Pradesh at an elevation of about 475 m above the mean sea level (as seen in Fig. 8). After flowing in a generally northeastern direction through Madhya Pradesh, it enters the Jhansi district of Uttar Pradesh. The river further pours down into the Yamuna catchment downstream, which forms a part of the Ganga Basin. Majorly, the stream flows in Monsoon and is generally dry in the summer months. The streamlines follow a dendritic pattern in the region. Based on the SCS-CN method (Soil Conservation Services (SCS) of US)) for run-off calculation.

The total area of watershed = 2215.64 km^2.

Total volume of water coming inside the watershed = 2,306,481,240.00 = 2306.48 mcm.

Yearly discharge, Q = 2,141,114,160 cu. m = 2141.11 mcm.

The volume of surface run-off in watersheds of Orchha region (based on SCS-CN method) (ref Fig. 7 for watershed boundary demarcation):

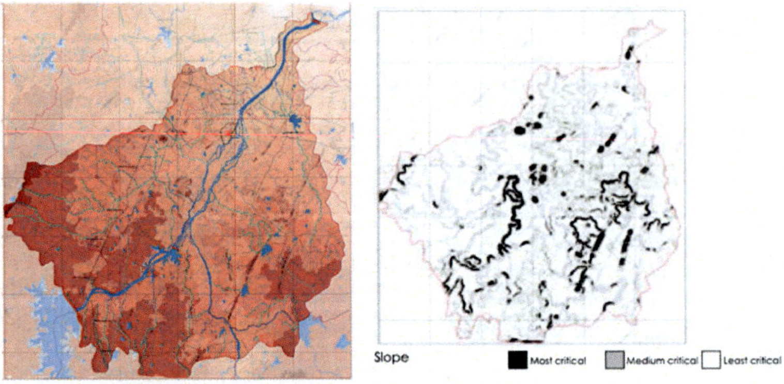

Fig. 6 Topography

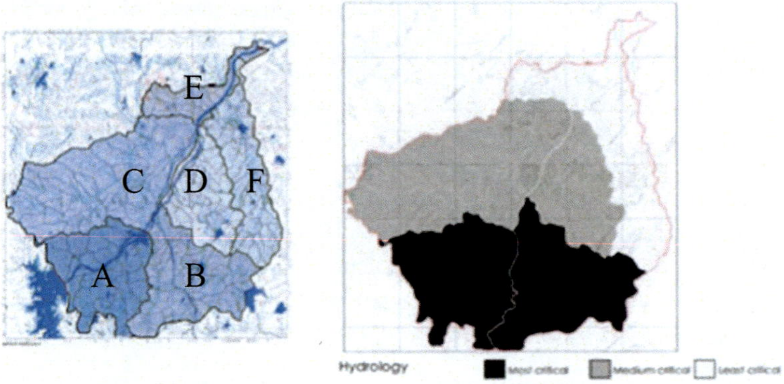

Fig. 7 Hydrology

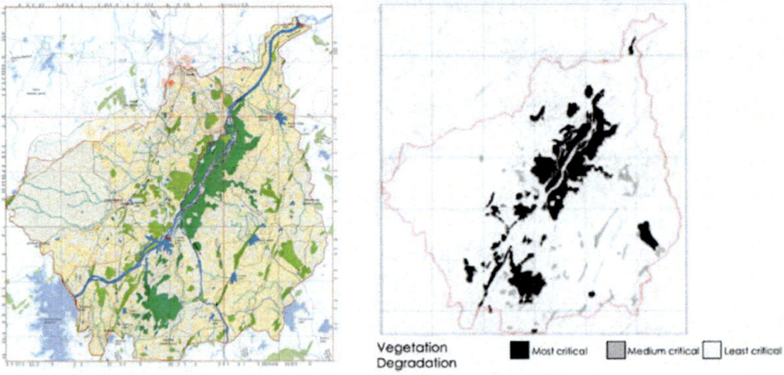

Fig. 8 Vegetation degradation

- Watershed-A—386.87 MCM
- Watershed-B—381.46 MCM
- Watershed-C—681.46 MCM
- Watershed-D—246.77 MCM
- Watershed-E—298.33 MCM
- Watershed-F—146.22 MCM.

The surface run-off differs according to the topography and the land use or cover of the region, hence contributing to understanding the areas that need to be conserved (ones with more surface run-off). Further, regions with agriculture and forest area have good groundwater recharge potential, as seen in Watershed-C.

Flash floods occur during the months of July–August. The average depth of rivers in the watershed area is 1–13 m.

Percentage percolation loss—4.32% (for the region with an average annual rainfall of 1041 mm, by taking the specific yield of weathered granite to be 2% and average change in groundwater level 2.25 m) (Ministry of Water Resources, 2013).

Hence the percolation in the total watershed—99.7038 mcm.

The weirs and dams found within the study region are (Government of India, 2014):

- Matatilla Dam: dam depth—17.3 m; area of dam—65.6 km^2; maximum capacity of the dam—1132.7 McM; average discharge—23,390 m^3/s
- Sukhwan-Dukhwan Dam: dam depth—3.9 m; area of dam—15 km^2; maximum capacity of the dam—57.8 McM; average discharge—15,177 m^3/s
- Parichha Dam: dam depth—5.3 m; area of dam—14.8 km^2; maximum capacity of the dam—78.8 McM; average discharge—21,547 m^3/s.

6.3 Vegetation Mapping

With five main species within the region, Tectona grandis constitutes the highest percentage of the forest vegetation, followed by Anogeissus Pendula.

The flora species Tectona grandis is found abundantly within the region, with a mixed forest of a deciduous type with species such as Mitragyna parvifolia, Boswellia Serrata, Bombax Ceiba, Sterculia urens, Madhuca indica, Soymida febrifuga, Ougenia oojeinensis, Buchanania lanzan, Terminalia bellerica, Albizia procera, etc. The Tectona grandis forest comes under the dry deciduous type of forest can also be categorized as a suitable forest concerning the minor degradation along the fringes in comparison to the rest of the domain (Fig. 8 shows the vegetation degradation mapping with the white part being highly degraded and black being least degraded). (Shrivastava et al., 2017).

Scrubland/wasteland constitutes most of the regional area, either barren or used for purposes other than agriculture. Land near Orchha city is not used for agriculture.

Linear hillocks form a crucial geological formation and help agriculture because of sediment deposits.

6.4 Geology

The geological context is of quaternary type. Rock formations such as Gneiss and Bundelakhand Granite can be found within the region, older than 2500 million years from the earliest period. (Saxena, Jitendra prasad, 2015).

The region is mainly a granite country with rocky outcrops of more extensive diversity in its mineral compositions and sizes (ranging from small patches to large hillocks). These are primarily found in Lalitpur, Chhatarpur, Sagar, Datia, Panna, Jhansi, and Tikamgarh. (Saxena, Jitendra prasad, 2015) (Mineral Resource Department & Ministry of Environment, 2002).

The area is the part of Bundelkhand craton with Bundelkhand Gneissic Complex of Archaen–Paleleo–Proterozoic age (>2500 to 2000 million years old) forming the oldest rock group. (Central Groundwater Board, 2002).

Foundation characteristics are excellent; hence, they provide suitable structural conditions. Steep ridges need to be avoided for construction work.

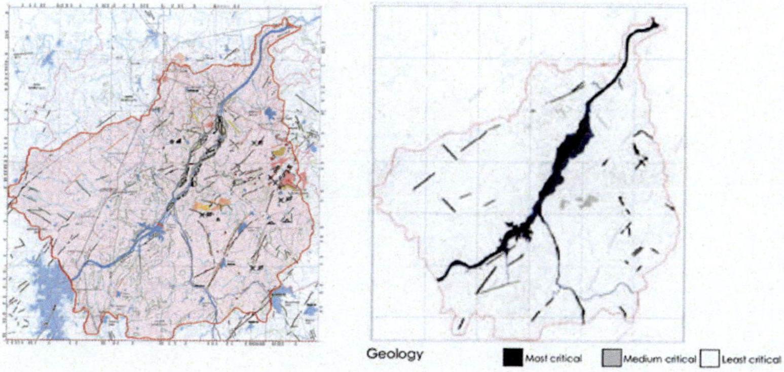

Fig. 9 Geology

As the native material, granite provides easy construction materials like aggregates, sand, and chiseled stone (Ramizr & Mondal, 2015).

For earthquake, region is liable to zone 1 and is classified as low damage risk zone.

Significant areas have shallow soil cover and rocky outcrops with less permeability for groundwater recharge. Water could be collected near dykes and lineaments; as they are good water recharge areas, accounting for the granitic strata, these points are the only respite for groundwater recharge. (The black sections marked in Fig. 9 are good to recharge areas—as they depict dykes and lineaments) (Central Groundwater Board, 2002).

Rocky outcrop can be seen evidently within the Betwa River. The soil formation is a combination of coarse loamy and fine soil. Steep slopes vary from shallow to steep in fine soil. These tributaries flow from the Sukhma Dukma weir, where deep soil depth. (Refer to Fig. 10) (Ramizr & Mondal, 2015).

6.5 Ground Water Recharge Potential

Riverbeds and streams are considered good recharge potential due to the complex rock foundation, i.e., where drainage lineament is considered good potential areas. The area under stable forests is considered good potential, whereas the area under degrading stage forest is considered a potential fair zone. Region aquifer is confined; first, fracture depth is 12 m to 18 m; second fracture depth 24 to 36 m; after that starts from 36 to 42 m. The shape of the aquifer is irregular. Groundwater is mainly tapped from the weathered zone aquifer, whose thickness varies with weathered zone thickness. (as seen in Fig. 11) (Central Groundwater Board, 2002).

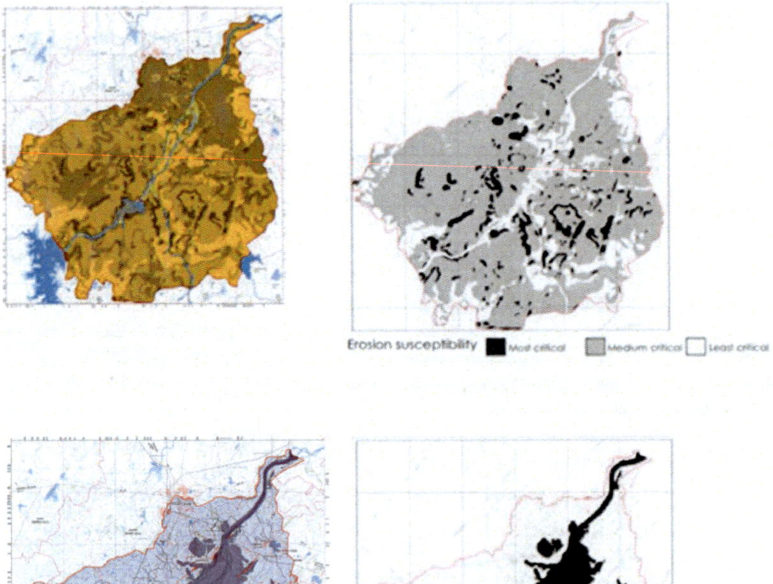

Fig. 10 Erosion susceptibility

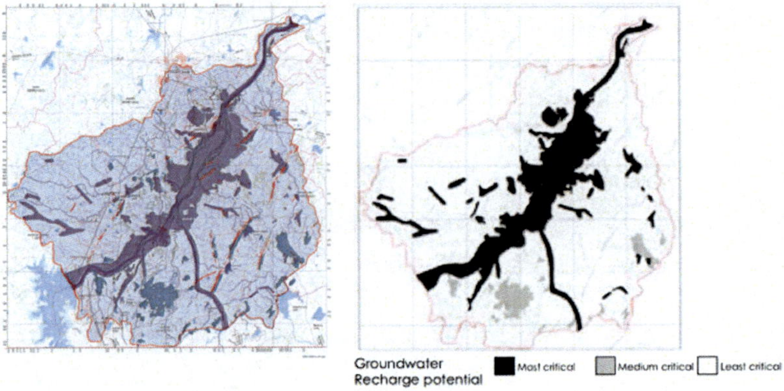

Fig. 11 Groundwater recharge Potential

7 Results and Discussion

7.1 Issues

There is a shortage of water supply in the city and the environs of the delineated study region. Also, there is no sewerage in Orchha, and there is one privately owned STP in the city-owned by a resort, which releases its treated water in Betwa River.

Further, garbage dumping along the riverbed causes much damage to the otherwise clean waters of the Betwa River, which in turn affects the turtle nesting sites on the banks of the river. Soil erosion on steeper slopes and rocky terrains further degrades the forests' status. Faster rainwater run-off due to less ground cover is also noticeable in the sanctuary area. The existing depressions and ponds do not hold enough water to sustain summer. The lack of surface water bodies in the Sanctuary and the drying up of the perennial river is also an issue that affects flora and fauna.

7.2 Watershed Management

Pollution can be observed along the river's course, where the edges have been deforested due to bathing, defecating, and washing. Pollution of water, due to the formation of a thick fly ash layer (fly ash is released from the Parichha Thermal Power Plant), especially during summer months, is observed in the vicinity of the Parichha Dam.

Here, Watershed-A and Watershed-B are chosen as a point of reference for the study, as the quality of water reaching Orchha will be determined by the surface run-off of the Watersheds A and B. Further, within the watersheds, the measures for water recharge are proposed based on six zones identified by overlaying four layers of surface hydrology, groundwater recharge potential, degradation stage of the forest, and land use.

- Further, the first layer will determine the streamlines and water bodies, naturally allowing water to percolate.
- The second layer will determine the spots or regions that give good groundwater yield and the poor ones; hence, the ones with poor recharge could be used to capture surface run-off that will subsequently allow for percolation.
- Further, the third and the fourth zones will again determine to give the regions of good surface run-off, and degraded forests will be the regions with good run-off, hence laying the grounds to propose a measure for capturing the same.

Based on the ground water recharge potential mapping (as seen in Fig. 12) for capturing potential surface and ground water recharge areas, the distribution of surface run of within watershed is (refer to Fig. 13 for percentage distribution of surface run-off in watersheds):

1. Recharge pits = 0.5% (1.93 MCM)
2. Percolation ponds = 1% (9.671 MCM)
3. Farm ponds = 5% (19.3 MCM)
4. Existing water bodies = 6.5% (3.474 MCM)
5. Check dams = 16% (61.76 MCM).

7.3 Landscape Management

The landscape management plan results from the overlay of the different eco-layers such as erosion susceptibility,

Land Use	* Ground Water Recharge Potential	Ground Water recharge: Farm Ponds	Recharge Pits	Percolation Pond/ Forest	Injection Wells	Surface water recharge: Farm Ponds with lining	Check Dams	Rainwater harvesting	Deepening water-bodies
Forest	G		●	●			◔		●
	F		●	●	◔				
	P								
Agriculture	G		◔	◔		●	●		●
	F								
	P	●		◔	●				
Scrub land	G		●	●			◔		●
	F								
	P		◔	●	●				

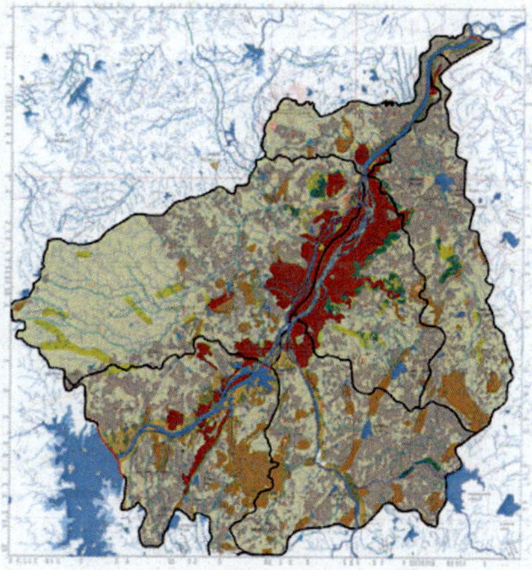

Fig. 12 Groundwater recharge potential mapping

geology, groundwater recharge potential, hydrology, slope, and vegetation. The sub-layers within each of these layers were identified and categorized accordingly based on their eco-sensitivity as most critical, medium critical, and least critical.

The identified zones are then prioritized, where Zone 1 is the most critical eco-sensitive zone within the region, while Zone 5 is the least critical eco-sensitive zone within the region. Further, a proposal module involving rejuvenation techniques for sustainable natural resource management is suggested concerning the natural feature within each zone—forest, agriculture, scrubland, and water bodies.

8 Conclusions

The study takes its theoretical framework from design with nature by Ian McHarg, which provides a method to intervene in urban landscapes—the layering of complex spatial data to discern the site suitability for evolving conservation, restoration, or development plans for the region. The study helped extract suitable sites for landscape restoration with the help of maps of natural and cultural systems in the setting. The overlay analysis also suggests the higher groundwater recharge potential seen in patches with agriculture and forests. Besides this, geological formations like dykes and lineaments are good respites for groundwater recharge. Further, the proposed landscape restoration considers the conservation of heritage and religious structures, accounting for its tourism significance. In addition to this, the restoration proposal considers topography, hydrology, vegetation restoration, geology, erosion susceptibility, and groundwater recharge potential, thus identifying potential areas for groundwater recharge and surface recharge using different techniques (as seen in Figs. 12 and 13).

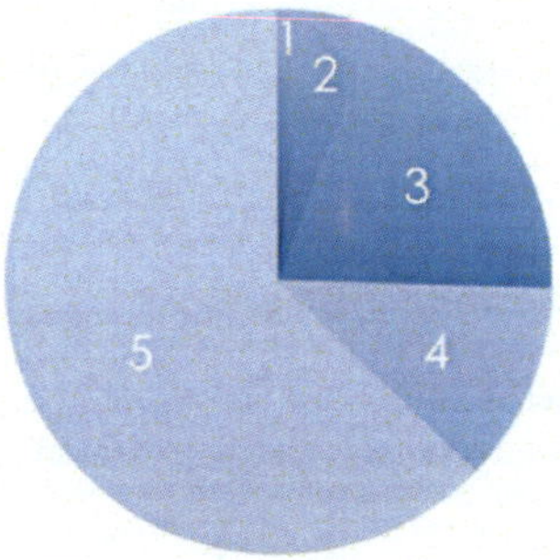

Fig. 13 Percentage distribution of surface run-off in watersheds

Using the approach of overlay analysis, the study illustrates a holistic approach toward landscape restoration by taking into account the natural and cultural significances of the heritage setting. It also takes into account the degradation of forests and agricultural lands and proposes an appropriate module for rejuvenating the natural features (as seen in Fig. 14). It illustrates the analytical framework of McHarg by delineating and overlaying the natural layers in order to identify the ecologically sensitive areas with the region. The proposed site planning and design strategies suggest conservation of native trees and plantations in the Orchha Sanctuary, along with a proposal to plant along with the existing water bodies for soil loss due to erosion. Thus, the proposal limits itself in predicting an 'ideal' scenario, where water and natural resource management is taken up, assuming human participation at all scales of landscape management.

Acknowledgements The data of the essay draws upon the regional studio conducted at Orchha in July 2017, followed by a design studio on campus, and documented in regional landscape study Orchha, Madhya Pradesh; master's program in Landscape 2016–2018, CEPT University, Ahmedabad. The project was done under the guidance of Prof. Deepa Maheshwari, Prof. Sandip Patil, Prof. Tapan Modi, and Prof. Beena Tindwani. We thank all faculties for their incisive critique of the project and helpful suggestions.

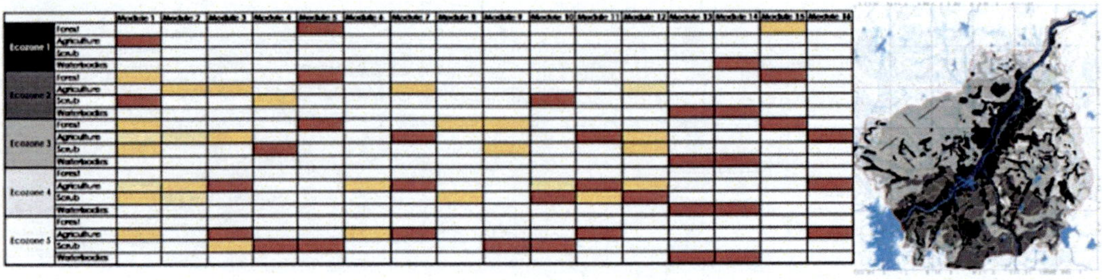

Fig. 14 Landscape management plan based on ecologically sensitive zones identified

References

Central Groundwater Board, C. (2002). India water portal. Retrieved 09, 2018 from Master plan for artificial recharge of groundwater in India: https://www.indiawaterportal.org/articles/master-plan-artificial-recharge-groundwater-india-2002-and-its-critical-assessment-and

Government of India, M. (2014). Ganga basin report. *Central Water Commission, Delhi and National Remote Sensing Centre (NRSC)*

Kalinowski, B. D. (2016). Reawakening in Bundelkhand: Cultural identity in Orchha and the effects of tourism on its creation, preservation, and loss. *Black & Gold, 2*(1), 3.

McHarg, I. L. (1969). *Design with nature* (pp. 7–17). American Museum of Natural History.

Mineral Resource Department District Tikamgarh & Ministry of Environment, Forest and Climate Change. (2002). India Water Portal. Retrieved 01, 2020 from Master plan for artificial recharge of groundwater in India: http://www.environmentclearance.nic.in/DownloadPfdFile.aspx?FileName=PstZW0V2OopXt1nFzHI5lfKp/sYq9DM7jwQS5iUKBlwxXe0Jjlj0lvkHS8ORHxxV&FilePath=93ZZBm8LWEXfg+HAlQix2fE2t8z/pgnoBhDlYdZCxzVPEh4a7F53Cae7tleKGoXIpcOTnXkKLT0lTpjEwmLUjBSrFzGq2Qfht6wPnN2mB/cEj0hCWO+RWWXs53Hp/+iS

Ministry of Water Resources, C. G. (2013). District Groundwater Information Booklet: Tikamgarh District, Madhya Pradesh. Tikamgarh, Madhya Pradesh: Government of India.

Ramiz, M., & Mondal, M. (2015). Orbicular structures near Pichhore, Shivpuri district, Bundelkhand Craton: forerunner for geoheritage site. *Current Science, 109*, 684–687.

Saxena, J. P. (2015, June 9). Agricultural geography of Bundelkhand. Retrieved January 19, 2020 from Shodhganga Inflibnet: https://shodhganga.inflibnet.ac.in/handle/10603/42716

Shrivastava, V. K., Hore, U., Kala, J. C., & Srivastava, A. (2017). Preliminary study on flora of Orchha wildlife sanctuary (Madhya Pradesh) India. *Annals of Plant Sciences, 6*(9), 1681–1685.

Shrivastava, V. K., Hore, U., & Kala, J. (2020). An investigation on eco-tourism potential and possibility of community integration in its development in Orchha, Central India. *Indian Journal of Science and Technology, 13*(23), 2294–2302.

Sinha, A., & Valderrama, A. (2014, August 06). The oracle landscape of Orchha, India: reclaiming the lost heritage. *Journal of Cultural Geography, 21*.

Heritage Tourism: The Case Study of *I Benedettini* (BARI)

Ludovica La Ghezza

Abstract

Heritage tourism is a cultural phenomenon in which examples of historical, monumental, folklore and environmental excellence gather in specific itineraries to enhance destinations and convey different feelings and meanings to different targets. Understanding these dynamics is fundamental to develop an effective and well-integrated marketing strategy for possible heritage tourism destinations. The purpose of this study, therefore, is to understand how to promote a small destination in the province of Bari, San Vito. In particular, we chose to focus the analysis on a social cooperative that runs the restaurant *i Benedettini*: a local enterprise which has been working for years with the aim to introduce disadvantaged people to the job market, pursuing solidarity and social inclusion in a conscious, sustainable and respectful way. Basing our research on qualitative and quantitative survey on tourists' habits, we concluded that San Vito is an important spot to develop a sustainable and community-based tourism. The analysis of the case study relating to *i Benedettini* should be, therefore, a starting point for the development and promotion of sustainability, traditional values and modern lifestyles as an example intangible cultural heritage.

Keywords

Heritage tourism • Landscape • Gastronomy • Intangible cultural heritage • Community-based tourism

1 Introduction

Heritage tourism is one of the most successful tourist phenomena of recent years. A multidimensional activity touches many lives and many different economic activities within a framework where time and location are key variables. In addition, heritage tourism is built upon an intangible set of service and cultural values, so that it can offer travel experiences to satisfy visitor's imagination and expectations. In other words, heritage tourism is a cultural phenomenon in which examples of cultural identity and environmental excellence gather in specific itineraries to enhance destinations and convey different feelings and meanings to different people.

Today, in fact, heritage tourism enables us to understand the world around us, the way in which we travel within it and the relationship between history, regional cultures, traditions and identities and the different places where these interactions occur.

Bound to all of this, there is the need to understand the shape and scope of heritage tourism industry and understanding heritage tourists' own perspectives and motivations. Focusing on concepts such as cultural heritage and authenticity, we will analyse the development of heritage tourism market and its implications concerning sustainability, such as destinations' cultural values, landscape and gastronomy.

The purpose of this study, therefore, is to understand how to promote the cultural and rural heritage of a small site, San Vito, which today is still unknown by locals and tourists. In particular, we will focus on intangible values, such as cultural identity, authenticity, self-responsibility and sustainability. The author's will is, then, to demonstrate that heritage is not just far from what we are and where we live.

San Vito is a small Apulian village in the municipality of Polignano a Mare (Bari), where there is a restaurant—*I Benedettini*—set in the former oil mill of a tenth-century Benedictine Abbey. Here, people have been working for years to help and support disadvantaged workers, and this is

L. La Ghezza (✉)
Alma Mater Studiorum—University of Bologna, Bologna, Italy
e-mail: l.laghezza96@hotmail.com

A. Mandić et al. (eds.), *Cultural Sustainable Tourism*, Advances in Science, Technology & Innovation,
https://doi.org/10.1007/978-3-031-10800-6_2

why the author decided to focus her research on this cease study.

The main purpose of this study, therefore, is to offer a new perspective on heritage tourism destinations, starting from intangible values and merging them with landscape, gastronomy and history of the destination itself.

Our research revolves around historical and geographical sources about San Vito, but above all, we focused on a target-oriented survey to fix the gap between tangible and intangible cultural heritage. We analysed concepts such as authenticity, gastronomy and landscape, as we asked our respondents to give their own ideas about the destination, the restaurant and their travel habits in general. Therefore, we concluded that today it is possible to develop a sustainable tourism destination nearby the world-famous Polignano a Mare and that heritage tourism should be a landmark for those small communities who live there.

We believe that thanks to this study, both locals and tourists will understand the intangible cultural heritage of San Vito, so that it will be possible to increase mutual respect for others' ways of life, as well as to learn about local cultural identity. Above all, this study could also set the basis for further analyses and research about other small communities in Southern Italy, where locals struggle to survive to the strong impact of mass tourists and are afraid of losing their cultural traditions and habits.

Cultural heritage today does not mean only ruins and rural landscape, but people and their customs. There is a real need to save our roots and to create new relationships between locals and tourists, starting from the authentication of lands and of what they cherish from historical and geographical perspectives.

2 Heritage Tourism: New Cultural Perspectives

What is cultural heritage? In 1972, UNESCO gave the first concrete definition of cultural heritage during the convention concerning the protection of the World Cultural and Natural Heritage. By that time, it also introduced the concept of outstanding universal value, explaining that the resources above-mentioned should be protected for their historic, artistic, ethnological or anthropological values (UNESCO, 1972).

However, the first definition of cultural heritage referred exclusively to the monumental remains and collections of objects of cultures. It was only at the beginning of the twenty-first century that cultural heritage gradually started to include new categories, related to traditions or living expressions, for instance, performing arts, social practices, rituals, festive events, knowledge and practices concerning nature and knowledge. The idea of *heritage* currently includes both human, cultural and natural environment and its preservation (Jokileto, 2005).

Because of the redefinition of cultural heritage, tourism industry has seen emerging a new field in which products and services derived from cultural assets can be commodified: that is what we today call *cultural tourism*. The term cultural tourism was coined to define those journeys that include visits to cultural resources, both tangible and intangible. In particular, the World Tourism Organization (UNTWO) has defined cultural tourism as:

> A type of tourism activity in which the visitor's essential motivation is to learn, discover, experience and consume the tangible and intangible cultural attractions/products in a tourism destination. These attractions/products relate to a set of distinctive material, intellectual, spiritual and emotional features of a society that encompasses arts and architecture, historical and cultural heritage, culinary heritage, literature, music, creative industries and the living cultures with their lifestyles, value systems, beliefs and traditions.

Moreover, according to the sociologist Monica Gilli (Gilli, 2003), cultural tourism recently has evolved in niche sub-genres that have the purpose to offer tourists holiday experiences based on cultural identity, nostalgia for the past, revitalization of the rural society and authenticity, including *Heritage Tourism.*

This niche is a different approach that aims at preserving both cultural assets and natural places through mechanisms of education and active participation. What does this actually mean?

Originally, heritage tourism entailed visiting ancient monuments and dwellings, rural and agricultural landscapes or locations where historic events occurred and places of interesting and significant values (Dallen, 2006). Today, these statements go beyond visiting historic sites and monuments; this is why heritage tourism includes experiencing life and habits, local pride, nationalism and mythology of the places visited. This new approach suggests that history is a vital part of the experience, where each place or objects establish a strong connection between local communities and tourists. Nevertheless, when they can relate to their personal, local, regional or national heritage, both visitors and residents are motivated to safeguard historic and cultural resource.

In other words, through involvement and exposure to local historic sites, heritage tourism becomes an educational means and helps visitors understanding the importance of a place and its related culture. These kinds of experiences allow the host community to recognize themselves as living part of cultural heritage and at the same time allow tourists to strengthen their ties with different cultures and traditions (Gilli, 2003).

Heritage tourism bridges the gap between visitors' desire to escape from their daily routines and the authentic tourist experiential consumption, thanks to its relational and close connection with a sense of place and identity. However, these are only theoretical trends towards heritage tourism; for this reason, we believe that it is important to focus also on practical trends and strategies related to the management of heritage tourism products and destinations, analysing the importance of landscape and gastronomy as sustainable products.

3 Landscape and Gastronomy as Sustainable Heritage Tourism Resource

Heritage tourism products and experiences can generate substantial economic and social contribution to local communities. Yet, as all tourism sectors, there are challenges related to target visitors and environmental context that help preserving destinations from deterioration and commodification.

According to a preliminary review of heritage strategies applied on tangible heritage products conducted by Chris Landorf in 2009, it is possible to categorize preservation efforts for heritage tourism sites into three main approaches: physical restoration, adaptation and education. All of these can be employed independently or in combination (Little, 2020).

Creating tourism experiences in heritage sites has significant potential for the destination themselves, as it can contribute to the communication and preservation of history and culture. Today, the symbiosis of tourism and cultural landscape has become a major objective as a new sustainable method in the management and planning of heritage tourist areas.

On one hand, this aims at achieving a balance between economic success, social equity and environmental preservation. On the other hand, it also enhances the role of place-based identity in tourism as a local development tool, stressing that *the active use of landscape in tourism can bring about a positive response to destination competitiveness and development of a tourist site, regarding its positive influence on destination recognition and heritage revitalization.* (INFOTA, 2016).

For these reasons, recently the World Tourism Organization has promoted new approaches about heritage preservation, which draw attention to intangible heritage products. These are a significant resource in the field of heritage tourism for a sustainable development of both existing and emerging sites.

Intangible cultural heritage according to the UNESCO General Conference in 2003 means:

> Practices, representations, expressions, knowledge, skills—as well as the instruments, objects, artefacts and cultural spaces associated therewith—that communities, groups and, in some cases, individuals recognize as part of their cultural heritage.

> Furthermore:
> Intangible cultural heritage is transmitted from generation to generation, and constantly recreated by communities and groups in response to their environment, their interaction with nature and their history, and provides them with a sense of identity and continuity, thus promoting respect for cultural diversity and human creativity.

In other words, intangible cultural heritage represents communities' living culture and communities and has an intimate connection with its creators, as well as the self-identification by communities, groups and individuals. It is the result of the interaction with nature and history and provides communities with a sense of identity and continuity. This means that intangible cultural heritage is a relationship between the community and the context in which they live and a product of this interaction raise in form of cultural heritage (Giudici, 2013).

In addition, the UNWTO clarified that such sustainable development approaches can assist in securing a growth path, where providing for the needs of the present generation does not mean compromising the ability of future generations to meet their own needs. In other words, creating tourism activities and experiences in a sustainable way brings improvement to destinations, satisfies locals' needs and allows future generations to have equal access to heritage, without deteriorating the natural and cultural environment.

Today, the enhancement of cultural heritage, even for tourism purposes, needs to be based on the affirmation of its significance, in particular on the image of the destination in its historical, but also contemporary values.

A prime example of intangible cultural heritage products that constitute the basis for an effective heritage tourism management project are landscape and gastronomy. These two aspects are particularly important for the development of sustainable marketing strategies in the field of heritage tourism.

On one hand, landscapes have a range of cultural and natural values that communities recognize as important and want to conserve, and therefore, it is fundamental that stakeholders must know what values are to be found in their cultural landscapes and consequently reinforce the protection and enhancement of the destination (Esposito, 2006).

However, when speaking about tourist experience, landscape can be subject to different interpretations depending on its evolving appearance over time; it can be designed and created intentionally by man, or a relict or fossil landscape, a cultural landscape which may be valued because of the religious, artistic or cultural associations of the natural element (INFOTA, 2016).

Relating to this, the Italian scholar Paolo Pileri believes that there is a sort of tautological relationship between tourism and landscape. The latter forms a whole, in which the natural and cultural components are complementary, including urban areas, historic towns and cities, rural areas city, as well as taste, art, nature, culture, history, traditions and cultural identity of a certain territory (Pileri, 2014).

Furthermore, Barbara Bender has carried out a multidisciplinary study, concluding that landscape is:

> The world out there as understood, experienced, and engaged with through human consciousness and active involvement. Thus, it is a subjective notion, and being subjective and open to many understandings it is volatile.
> Different people will experience the same place at the same moment differently; the same person will experience the same place, at different moments, differently; the same person may even hold conflicting feelings about a place.
> When, in addition, one considers the variable effects of historical and cultural particularity, the permutations on how people interact with place and landscape are almost unending, and the possibilities for disagreement about, and contest over, landscape are equally so. (Tilley, 2006).

From a heritage tourism perspective, then, the cultural richness of the territory, and in particular that of rural areas that have not undergone phenomena of identity commodity, constitutes an opportunity on which running a tourist business able to reintegrate the heritage in the local economy.

Attracting new visitors means creating a destination where they can feel completely overwhelmed by the heritage landscape they observe, so that their pleasure turns into a desire of being actively part of that landscape and not just tourists (Pericoli, 2015).

Speaking about gastronomy, it has become a single touristic product from being a helping sector to tourism, holding important places in destinations' images. Every cuisine has a unique characteristic, so that some local foods are better known than their origins.

Local food is indeed a powerful and effective tool for an area to become a tourism destination. With a great contribution to the image of the destination, gastronomy is also able to develop rural tourism and contribute to the protection of cultural heritage (Kocaman, 2014).

Culinary heritage, cuisine and food ways are an interesting and vital part of a living culture, and they all play an important role in developing countries. They reflect the geography and place of a region or nation, and other aspects of heritage such as religion and culture. Food invokes and incites emotions and experiences that connect people to the past and the present traditions of a community, creating socio-cultural identities.

Today, there is a growing interest in food and wine tasting as tourist products, especially in those kinds of tourist activities oriented to the rediscovery of nature, rurality, roots and traditions linked to the territory. Therefore, all of the local sourcing and legacy of ingredients, the traditions and meanings associated with preparation and consumption of cuisine and the conservation of particular food often invariably led to the use of food and cuisine in heritage tourism (Richards & Hjalager, 2002).

Gastronomy offers many authentic encounters with different cultures. That is why recently many destinations have chosen to create experiences based on food and culinary heritages and traditions, linking resources in a single or in different regions through gastronomic trails and routes, along which tourists can taste a range of products or compare those products to others from different locations.

Experiencing new local cuisines, therefore, means firstly experiencing a new culture, and this is the prime motivator that leads tourists into searching for authenticity and cultural heritage in societies other than their own. Therefore, regional cuisines are an emblem of particular community's identity, regional values and social connections and can create heritage meanings through the establishment of traditional foods supply chain that includes both human geography, local landscape and environment (Ramshaw, 2015).

In the light of all this, we can conclude that gastronomy can be a driving-force behind the cultural revival of heritage tourism industry and a powerful marketing tool, because seeking authentic local and regional foods is a valid motivation for visiting a particular destination.

Together with a well-structured sustainable effort, gastronomy and landscape allow local communities to communicate their values and traditions without conforming to globalization standards and allow tourist to escape from mass tourism and their frenetic routines and experiencing authentic past values.

4 Research Methodology: How to Develop a Heritage Tourism Plan

To be successful, a heritage tourism plan must be locally driven and should create an incentive for broad local involvement, so that the result accurately reflects your

Fig. 1 The Abbey of San Vito

community's culture, heritage, values and goals. Thus, a community, whether a town, county, region or local business should collectively decide which resources it wants to share with visitors and then develop a strategy.

The small harbour of San Vito, a district of Polignano a Mare (Apulia), has the potential to become a heritage tourism destination to enhance the social identity of the territory and the values pursued by the local community. Here, a small social cooperative has opened a restaurant in the former oil mill of the abbey of San Vito with the aim of bringing back to life the history of that place (Fig. 1).

In a community-based heritage tourism destination, the entire community's activities and attractions have the strength to pull visitors to the destination. In addition, creating a community-based tourism plan with an integrated range of varying experiences will have the advantage of providing a unique selling proposition of the community's *sense of place* and of building residents' pride.

According to the National Trust for Historic Preservation, an effective heritage tourism plan should incorporate the following five principles. The first principle is *Find the Right Fit and Define Your Purpose*. This implies that heritage tourism development should find the fit between community and tourism needs and, at the same time, maintain a community's unique character.

The second principle is *Tell Your Own Story and Focus on Authenticity and Quality*. Visitors are interested in discovering what distinguish a community from the others. Hence, the stories worth telling are the ones that compare the current and previous generations, with a focus on authenticity and cultural values.

The third principle is strictly linked to the second one, because it is *make tourism sites come alive—tell a story*. Today's visitors want the human drama of history, not just names and dates. The stories about a community and its sites should provide accurate, authentic information in an engaging and memorable way that makes the site come alive for the visitor.

Moreover, in order to add appeal to the site, planners can integrate the history and traditions of a community with links across the region or the state. But most of all, to attract visitors and stay competitive, the message should be creative and exciting, identifying how the community fits into overall patterns of the destination history.

The fourth principle focuses on the preservation of community's ethics and resources. Tourism has an impact on the resources and puts stress on infrastructure such as roads, airports, water supplies and public services. When a community's heritage is at the core of the visitors' supply, it is essential to protect the historic landmarks, places and traditions and avoid that possible increasing tourism flows do not destroy the very qualities that attracted visitors in the first place.

Finally, the first principle is *Collaborate with Partners and Form Alliances*. An excellent way to leverage scarce financial and human resources is to form partnerships between the community and other stakeholders. These collaborations are important in planning a community-based heritage tourism destination because they can create opportunities to enhance travel experiences.

In addition, linking historic and cultural resources through the combination of several experiences, stories and themes develops a more complete understanding and appreciation of that area and draws tourists to attractions they might not otherwise visit (Speno & Green, 2010).

The small village of San Vito is settled around an ancient haven on the Adriatic coast, three kilometres from Polignano a Mare, and owes his celebrity to the abbey that stands upon the haven, overlooking the sea. A programme that aims at promoting this territory should focus on sustainable and ethical tourism, preserving the destination from mass tourism and modern attractions.

A possible solution could be recalling the old fame of the Abbey of San Vito through the Grand Tour experience, offering modern tourists the opportunity to spend their time in the majestic venue *I Benedettini*, as the eighteenth-century travellers that stopped there asking monks for hospitality. However, the attempt to revive the history of San Vito requires also an accurate knowledge of modern culture and good creative skills, which are currently fundamental to create an itinerary for modern tourists that will help them to live a memorable experience in the footsteps of the past in Southern Italy.

The classic Grand Tour may be old-fashioned, but it is true that Apulia is still able to connect past and present with its small towns and villages, landscape views and cultural heritage. And yet, these peculiarities can be a springboard for the future of the region and would allow tourists to find the cultural authentic roots in its local realities.

5 Case Study: *I Benedettini*

The tourist offering aimed at the management and promotion of San Vito is not confined to an overview of data and documentary sources, but it also concerns a small business that has been working in the area for years: the restaurant I Benedettini (Fig. 2).

This local business, run by people who focus their mission on helping disadvantaged human beings and on sustainability, represents the starting point for the promotion of a contemporary Grand Tour itinerary and for the preservation of the cultural heritage of San Vito.

Within the framework illustrated so far, the management model applied by the partners of the restaurant is relevant for creating a strong heritage tourism product. According to the scholars Carmichael, Getz and Brown and Urry, in fact, the involvement of both physical and psychological senses can stimulate and determine the tourists' motivations behind the destination choices (Patti & Messina, 2013).

Furthermore, the context in which meals take place is relevant too: venues, settings, décor and staging, as well as the atmosphere and the environment add importance and values to the preservation and promotion of an intangible heritage, such as traditional cuisine. In this perspective, choosing a restaurant would mean choosing a place where it is possible to get involved in an experience that goes beyond a simple meal.

The restaurant is located in the former olive oil mill of the abbey; inside the venue, there are two large rooms, with the original oil presses used by the monks in the past. It opened in 2008 on behalf of Sol&Blu social cooperative, which was founded in 2005 by the chef, Benedetto Ventura and the manager Giuseppe Pagano.

Today, eleven people work in the restaurant, and among them, there are four young men with mental or motor disabilities. So, the ideas behind the social cooperative are that work should be first of all a social activity and that an employer should recognize and valorise the capabilities of every human being, including disabled and disadvantaged people.

These values developed a certain sensibility and nostalgia for the past and history of the venue, where the Benedictine monks worked and welcomed Grand Tour travellers. Therefore, all the crew tried to recreate the same atmosphere described in the old travellers' diaries, making possible for disadvantaged people to become an intangible cultural heritage practitioners and a sustainable tourism resource.

Consequently, this study will provide an integrated project design for the realization of a short trip for those who are mainly interested in cultural heritage, food and wine tourism and rural sites, as well as the locals and casual visitors from the neighbouring cities.

Restricting the claim only to natural, cultural, historical resources of the territory of San Vito does not seem the best promotional and marketing solution, as it would make

Fig. 2 Restaurant layout—I Benedettini

people believe that the tourist product developed has nothing different from the others. Our starting point should be an analysis of the human beings in the time-spatial context of San Vito and the restaurant I Benedettini, so that it would be possible to create a network between the site resources and the actors that exploit them and design a project in which these different types of products blur.

The best way to proceed is to find out what Italian and foreign tourists like the most about San Vito and about the restaurant. The final product must fulfil three standards: (a) promote the preservation of the cultural heritage, (b) create interest of the visitors and (c) make profit for tourism industry and especially for the restaurant.

6 Baseline Survey and Study Findings

Assuming that I Benedettini has a great potential from the point of view of capability and suitability, it is necessary to analyse what is its potential relating to feasibility, that is in other words, gaining insight into how tourists want to experience San Vito as heritage tourism destination.

In order to understand the best solution to promote the destination, we decided to carry out a survey to identify and profile the possible targets and their habits.

The survey is based on a combination of quantitative and qualitative approaches that aim at investigating the habits of visitors and residents who have recently spent their time at the restaurant. It has been launched online from April to August 2020 through the official webpages and the main social networks of the restaurant, such as Instagram and Facebook.

The questionnaire, both in Italian and in English, was individual and anonymous. It included multiple choices questions about visitors' experiences, expectations and about the quality of services provided by the restaurant, that are hospitality, attitude towards visitors, activities and tourist products provided.

Questions were organised in three main sections. In the first section, respondents' sample was firstly asked to indicate their gender, age, level of education and country of origin. Then, they were asked to select from a series of given answers, the one which represented the motivation behind their holiday and travel choices, and which factors influenced their choices.

The second section is about the site of San Vito: the respondents are asked whether they have ever been there, and what they liked most in the destination. These questions are useful to understand tourists' perception of the image of San Vito. Image is becoming an increasingly important element in the competitive struggle to attract cultural tourists, and destinations strive to offer a product that is unique and clearly recognizable.

Moreover, beliefs and perceptions that people have about a destination can influence the growth of a tourist area, so that by exploring the personal meaning of tourists' experiences in San Vito, the author's research sought to strengthen support for preservation and promotion and gain a rich personal insight of the site.

It is also possible to establish cross-relations—if statistically significant—between expectations and experiences of the visitor, on the one hand, and variables like origin, age, education, previous visits, motivations, etc. (Gonzalez Fernandez et al., 2010).

Finally, the third section of the questionnaire examines respondents' opinions about I Benedettini, analysing the key factors for the quality of stay and for possible further visits. In particular, people were asked if they knew the restaurant and, in case of affirmative response, to express their satisfaction, or in case of negative response, to express their curiosity to have an experience there on a scale of one to five.

The analysis of data provided by the survey has showed what are tourists' preferences, motivations and habits when they travel; in particular, data revealed that the average of responses about where tourists want to go on holiday and what they would like to experience depends on their socio-cultural status.

Overall, the sample consisted of 262 people, aged 18 to over 65 (68% women, 32% men), mainly students (55%), office workers (15%) and freelancer professionals (22%). All of them have mostly high-school education or a university degree (see Table 1). 95% of the sample comes from Italy, while only 5% comes from different foreign countries (see Table 1; Fig. 3).

The majority of them (58%) travel once a year, followed by those who travel less than once a year (30%), every month (7%) and every week (0.8%). Only the 3.2% of the sample reported that they have never travelled so far. Besides, most of them usually travel "with a group of friends", "in couple" and "with family», while only a small amount of the sample reported of travelling alone.

Speaking about the motivations behind their travel choices, the answers with most votes were "visit new places" (81%), "make new and/or cultural experiences" (45%) and "escape the routine" (35%), while, among the multiple choices relating to the most important aspects that determine the choice of a holiday destination, the sample chose "landscape beauty", "cultural heritage" and "value for money". Other variables, such as "site reputation", "accessibility", "hospitality services", food and wine tasting", "guided tours" and "festivals" were considered neutral or less important (Table 2; Fig. 4).

In questions 10, when respondents were asked if they have ever been to San Vito, little more than the 50% of the sample gave an affirmative response, but a small 10%

Table 1 Socio-demographic profile of the sample

Gender	Number	Percentage
Male	81	30.9
Female	179	68.3
Not specified	2	0.8
Age		
18–24	128	48.9
25–34	60	22.9
35–44	14	5.3
45–54	42	16
55–64	16	2.1
65+	2	0.8
Education		
Middle school diploma	25	9.7
High school diploma	118	45.6
Bachelor's degree	70	27
Master's degree or higher qualification	46	17.8
Profession		
Student	145	55.3
Housewife	10	3.8
Office worker	40	15.3
Freelance professional	22	8.4
Businessman	9	3.4
Retired	3	1.1
Others	33	12.7

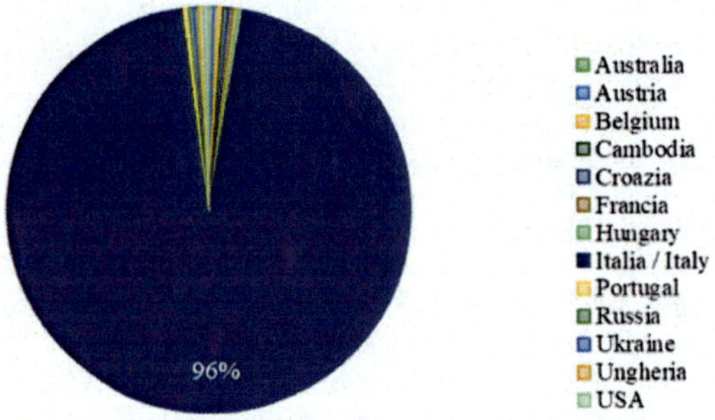

Fig. 3 Sample's home country

asserted of knowing the place. However, it is relevant to notice that, although the 40% has never been there, seeing the picture within the question, 250 respondents have judged the site appealing, with a score from 3 up to 5.

In particular, the survey showed that the cultural heritage and the rural landscape represent the most attractive features of San Vito; only the 15% of the participants consider the site a simple seaside resort.

In the series of answers about I Benedettini, instead, 57% of respondents have never been there, although they knew the restaurant, and yet only the 11% of the participants have never had heard about it. Among those who have been to the restaurant, 127 participants have expressed a positive opinion in terms of satisfaction, with a score that goes from 3 (neutral) to 5 (highly satisfied).

Nevertheless, only the 23% of the sample knew that the restaurant is run by a social cooperative, but, on the other hand, 173 participants agree with the objectives of social cooperatives and 197 would like to live an experience in the restaurant and try its cuisine (Table 3).

The survey included also photograph-based questions, in which respondents were asked to choose the photographs they liked the most about the restaurant. All the pictures were introduced by the question:

Imagine you are making a tour inside the restaurant. Could you tell us what do you like the most about its layout?
Imagine you are having your meal at the restaurant. Could you tell us what do you like the most about the staff?
Try to remember the menu options. Could you tell which course you like the most among those you tasted and why?

Table 2 Travel profile

Frequency of Travels	Number	Percentage
Every week	2	0.8
Every month	20	7.6
Never	9	3.4
Less than once a year	79	30.2
Once or twice a year	152	58
Travel preferences		
In couple	76	29
Alone	20	7.6
With friends	93	35.5
With family	73	27.9
Motivation behind travel choices		
Escape the routine	92	35.1
Have cultural experience	120	45.8
Visit new places	213	81.3
Have new experiences	121	46.2
Enlarge social Relationships	32	12.2
Work	1	0.4
Summer vacation	1	0.4
Know new traditions	1	0.4
Relax	1	0.4
Stay with family	1	0.4
Stress release	1	0.4
Try local food and delicacies	1	0.4

Fig. 4 Destination choices variables

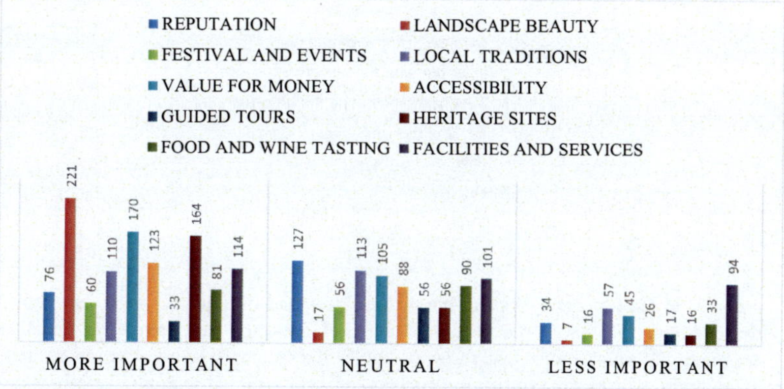

After processing all the answers, the researcher analysed the answers useful and relevant to determine and underline the meaningful relationships between tourist's behaviour and the tourist product presented. From the analysis of the answers emerged that the stone ceiling, the brightness and the atmosphere of the room, the smiles and kindness of the staff and the freshness and flavours of seafood attracted participants.

Moreover, the photograph-based approach confirmed participants' opinions and emotions relating to the dishes

Table 3 Respondents trends relating to San Vito and I Benedettini

About San Vito	Number	Percentage
Have already visited the place	138	52.7
Have never visited the place	98	37.4
Do not know at all	26	9.9
Level of appeal of the site		
High	136	51.9
Good	87	33.2
Normal	28	10.7
Low	11	4.2
Null	0	0
About La Locanda dei Benedettini		
Have already been at the restaurant	150	57.3
Have never been at the restaurant	81	30.9
Have not been at the restaurant yet	31	11.8
Know that it is run by a social cooperative	62	23.7
Do not know that it is run by a social cooperative	200	76.3
Costumers' level of satisfaction		
Very satisfied	40	29
Satisfied	39	28.3
Neutral	48	34.8
Not Satisfied	2	1.4
Totally unsatisfied	9	6.5
About social cooperatives		
Know what their objectives are	126	48.1
Do not know what their objectives are	137	51.9
Totally agree with the objectives	173	70.9
Agree with the objectives	50	20.5
Indifferent to the objectives	20	8.2
Disagree with the objectives	1	0.4
Totally disagree with the objectives	0	0
Level of curiosity about I Benedettini		
High	197	75.2
Good	55	21
Neutral	10	3.8

prepared at the restaurant. Among the pictures provided in the question, 57% chose the picture of the chef cooking spaghetti and another one that shows a bottle of organic extra-virgin oil with focaccia on the background, and 40% chose the picture of a kitchen-girl seasoning some seafood starters (Table 4).

These results are important from the point of view of the analysis of the potential feasibility of I Benedettini. They show that the layout of the restaurant, the staff, the meals and the spirit of the place have a great extent of appeal in relation to what tourists would like to experience in their visit to the destination. In addition, the responses to the Grand Tour questions are important from the point of view of heritage and cultural sustainability, because they show how much visitors care about authenticity in what they experience. Therefore, we can assert that I Benedettini can be site of a community-based heritage tourism destination.

In the end, in order to better identify the potential target and to understand how to create an effective heritage tourism destination, it has been carried out a multiple correspondence analysis of the responses.

Two clusters are relevant for targeting visitors in San Vito: The first one includes about 70% of the sample and is made up of Italian from 18 up to 35 years old, who usually

Table 4 Grand tour questions

Q1: Imagine you are making a tour inside the restaurant. Could you tell us what do you like the most about its layout?	This venue looks rural and I really like it Honestly, I have never been there. But the colour of this pictures make me feel at home. This venue has a story to tell, and I would really know it
Q2: Imagine you are having your meal at the restaurant. Could you tell us what you like the most about the staff?	They look funny and make me feel happy They make you at your ease and convey a sense of peace, which is the overall atmosphere of the venue. Have lunch here is a true culinary experience
Q3: Try to remember the menu options. Could you tell us, which course you like the most among those you tasted and why?	Cavatelli with cheakpeas and clams is a perfect balance of flavours. When you eat them, you feel sea shents in your mouth Tagliolini with jumbo shrimps! A course like this deserves a big spot of tomato sauce on your shirt!

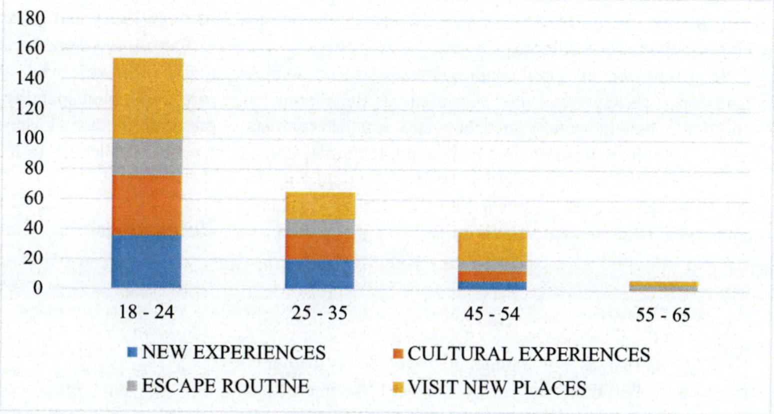

Fig. 5 Targets' travel motivations

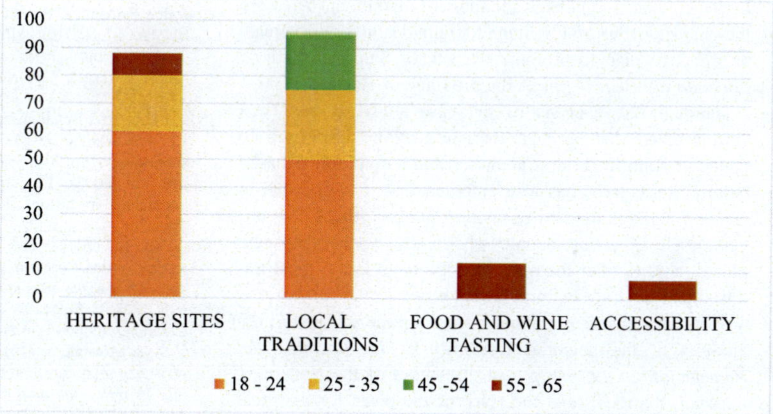

Fig. 6 Key factors for destination choices

travel once or twice a year in couple or with friends. They mostly enjoy travelling to have new experiences and visit new places, but also to escape their routine, and consider heritage sites and local traditions the best options for their holiday.

The second cluster includes about 20% of the sample: Italian participants from 45 to 65 years old, who travel on average once or twice a year with their families. They enjoy travelling to visit new places and escape their routine, but for them food and wine tasting, accessibility and local traditions are relevant aspects when choosing a destination (Figs. 5 and 6).

Thanks to this socio-demographic analysis of tourists we could contextualize the potential target that would visit San Vito, that are people that enjoy travelling with friends or in couple for having cultural experiences or visiting new places, and yet choose their destination to visit heritage sites and discover local traditions.

In conclusion, this study confirmed that San Vito, with its landscape, heritage and the restaurant I Benedettini as additional benefit, totally matches and satisfies tourists' needs. Therefore, it is possible to inscribe the small village and its peculiarities in the field of heritage tourism with the aim to attract an even greater number of tourists, especially foreigners, who are still a minority.

7 Conclusion

Heritage tourism is not defined in terms of its resources themselves, but from the perceptions and the interactions that tourists and stakeholders have with the resources. This research aimed at understanding the concept of cultural heritage, what it implies in the tourism market and what are the characteristics of a heritage tourism destination to create an effective plan to promote the site of San Vito and *La Locanda dei Benedettini* in the province of Bari.

Through the analysis of the case study, it has been demonstrated that heritage tourism may be also used to create or confirm cultural identities and a community-based cultural heritage. It has also emerged that it is possible to create a heritage tourism promotion plan to attract new targets, such as young people who enjoy discovering new places, able to offer them the possibility to know local tradition and discover cultural heritage sites.

These results are important from the point of view of the analysis of the potential feasibility of *La Locanda dei Benedettini*, as they show that the layout of the restaurant, the staff, the meals and the spirit of the place have a great extent of appeal in relation to what tourists would like to experience.

We believe, however, that within the development of a heritage tourism destination the creation of a heritage tourism destination must include an effective marketing, branding and communication plan. To do so, however, a strong connection between I Benedettini and other partners and local administration must be created. Only then, it will be possible to promote the gastronomy and the destination image, enhance the sustainable development of the site and give rise to competition between traditional values and modern lifestyles as an example of the interpretation of the cultural heritage resources.

However, this study offers a first sight on new possible way to promote rural destinations in Apulia without mass tourism attractions. We are aware that there is still much to do, in order to succeed and develop a real sustainable project for both locals and tourists.

Our survey demonstrate how much today is important for tourists this topic and, above all, how much young people care about sustainability and authentic places. In the end, these are the actual aspects to be taken into account to face the new challenges in the contemporary tourism market.

References

Dallen, J. T., & Boyd (2006). Heritage tourism in the 21st century: Valued traditions and new pespectives. *Journal of Heritage Tourism*, 1–16.

Esposito, C. (2006). The World Heritage and Cultural Landscapes. *PASOS. Revista de turismo y patrimonio cultural*, pp. 409–419.

Gilli, M. (2003). *Turismo*. Generi, Generazioni. Bologna: Zanichelli.

Giudici, M. E. (2013). Is intangible cultural heritage able to promote sustainability in tourism? *International Journal of Quality and Service Sciences*, 105.

Gonzalez Fernandez, A. M., Rodriguez Santos, M. C., & Cervantes Blanco, M. (2010). Measuring the image of cultural tourism destination through the collage technique. In G. Richards & W. Munsters (Eds.), *Cultural Tourism Research Methods* (pp. 156–172). CABI.

INFOTA. (2016). Relationship between tourism and cultural landscape: a new sustainable development model. In *Proceedings of TCL2016 Conference*, (pp. 376–384). Budapest.

Jokileto, J. (2005). Definition of cultural heritage. References to documents in history. ICCROM Working Group Heritage and Society.

Kocaman, M. (2014). The importance of cultural and gastronomic tourism in local economic development: Zile sample. *International Journal of Economics and Financial Issues*, 735–744.

Little, C. (2020). Innovative methods for heritage torusim experiences: Creating windows into the past. *Journal of Heritage Toruism*, 3–4.

Patti, S., & Messina, A. (2013). Il viaggiatore del gusto: indagine sulle motivazioni al consumo turistico della gastronomia locale. *Economia Marche—Journal of Applied Economics*.

Pericoli, T. (2015). Il turismo enogastronomico: progettare, gestire, vivere l'integrazione tra cibo, viaggio, territorio. In E. Croce, & G. Perri (Eds.), Milano: Franco Angeli Editore.

Pileri, P. (2014). Turismo e Paesaggio in Il grand tour del XXI Secolo: l'Italia e i suoi territori. In *Italiadecide Rapporto 2014*. Bologna: Società Editrice il Mulino.

Ramshaw, G. (2015). Food, Heritage and Nationalism. In J. T. Dallen (ed.), *Heritage cuisines. Traditions, indentities and tourism* (pp. 55–64). London: Routledge.

Richards, G., & Hjalager, A. M. (2002). *Tourism and gastronomy*. Routledge.

Speno, L., & Green, B. (2010). *Heritage Tourism handbook: A how-to-guide for Georgia.* Historic Preservetion Dividion Georgia Department of Natural Resources.

Tilley, C. (2006). Introduction: Identity, place, landscape and heritage. *Journal of Material Culture*, 7–32.

UNESCO. (1972). World heritage convention concerning the protection of the world cultural and natural heritage. *Art, 1.*

From the Paradise of Beit Shean Valley to the Contested Landscape of the Valley of Springs: Water Amenities, Environmental Justice, and Sustainable Development

Tal Alon Mozes

If paradise is found in the Land of Israel, its entrance is through the gates of Beit Shean (Reish Lakish, third century Jewish sage).

Abstract

This chapter discusses the failure to revive paradise in the fertile lands of the Valley of Springs, located in the northeastern part of the Jordan River Valley in Israel. It focuses on the area's abundance of water and sunshine, which could have served as the essence of life and enjoyment for the various communities of the Valley and also for domestic and international tourists. However, during the 2020s, it provoked a symbolic as well as a physical struggle among the locals and the Israeli public at large. While currently proposed solutions focus on limited legal and spatial resolutions, the chapter suggests examining the Valley of Springs environmental history and current predicament through the perspective of water as an essential substance for living and tourism and an agent of social relationships, which also holds symbolic meaning. As such, and in light of the three pillars of environmental justice: distribution, procedure, and interaction—it proposes an alternative approach for creating a more sustainable future for the Valley's tourism industry and for the region as a whole. This chapter is based on a comprehensive review of local and regional schemes that address water as a prominent, local resource, and the actual water-based recreation sites which were established throughout the Valley since the 1950s. The results indicate the need to adopt a holistic vision for the region, based on water.

Keywords

Water • Recreation • Environmental justice • Sustainable development • Israel

1 Introduction

The Valley of Springs (Figs. 1 and 2), located in the northeastern part of the Jordan River Valley, is one of the hottest areas in Israel. In the summers of 2020 and 2021, the dozens of springs which are dispersed throughout the Valley barely succeeded to cool the air and alleviate the tension between the various Israeli communities that share this mythological gate of paradise (Kershener, 2021). Residents of the town of Beit Shean, belonging to a low socioeconomic status, and thousands of domestic tourists who could not vacation abroad due to the pandemic, fought over the right of access to the Assi Stream and the lavish lawns on its banks, which have been inhabited since the mid-1930s by the affluent agricultural settlement-Kibbutz Nir David. The heated encounter, which repeated itself each weekend, and at times even deteriorated into physical altercations, was loudly echoed in the social networks, in the national media and even in the Israeli parliament. Although it focused solely on several hectares of pastoral European-like landscape, it expressed nearly a century of economic, social, and institutional disparity.

2 Problem Statement and Methods of Investigation

The chapter discusses the failure to revive the mythological paradise in contemporary Israel, in a region with an abundance of water, and a sun which shines almost year-round. It

T. A. Mozes (✉)
Technion, Israel Institute of Technology, Haifa, Israel
e-mail: artal@technion.ac.il

A. Mandić et al. (eds.), *Cultural Sustainable Tourism*, Advances in Science, Technology & Innovation,
https://doi.org/10.1007/978-3-031-10800-6_3

Fig. 1 The Valley of Springs: in the foreground—Kibbutz Mesilot, in the middle ground—the town of Beit Shean, and in the background the—Jordan Valley (Beit Shean Museum, http://www.betshean-museum.co.il/)

presents Spring's Valley environmental history, focusing on the role of recreation planning near water amenities, and suggests discussing the current crisis and future challenges through the perspective of water as a substance for living and an agent of social relationships, which also holds symbolic meaning. As such, it proposes an alternative approach for creating more sustainable tourism for the Valley and for the region as a whole.

The chapter is based on a comprehensive analysis of Israel's Valley of Springs from diverse perspectives: physical, social, cultural, and political—emphasizing the role of planning in making the region a heaven for domestic tourism. It is based on rich secondary sources of information on the Valley and on an in-depth investigation of the role of water in regional and local schemes. The chapter is divided into three parts: The first part lays out the theoretical framework of the research, emphasizing the connection between sustainable tourism/recreation, environmental justice, and water; the second part discusses the Valley of Springs as a case study; and the third and last part proposes a framework for a more sustainable future for the Valley.

3 Literature Review: Sustainable Tourism and Recreation, Environmental Justice, and Water

Since the 1990s, and in parallel to the emerging discourse on sustainability in all fields of development, sustainable tourism became the topic of numerous texts which explored the unique dimensions and challenges of the new paradigm (Buckley, 2012; Hall et al., 2015; Hardy et al., 2002; Waligo et al., 2013). Tourism as a growing industry (despite the Covid-19 pandemic) contributes to environmental changes more than ever (Hall et al., 2015: 4). Consequently, maintaining and developing tourism while aiming to balance its economic, environmental, and social impacts are a great challenge for present and future generations. Sustainability, in general, and sustainable tourism, in particular, are "wicked problems" (Hall et al., 2015; Blahna et. al. 2020). The latter is examined using objective and subjective tools, with the intent to establish a set of defined indicators (Rasoolimanesh et al., 2020). While in theory, the concept of sustainable tourism is a success in academic circles and in policy discourse, on the ground, tourism has also negative impacts as it changes landscapes, natural and man-made systems, and economic and social relationships (Hall et al., 2015: 4).

Among the various aspects of sustainable tourism, environmental justice, or the fair distribution of environmental amenities and hazards, is an issue which has received relatively little attention (Hales & Jamal, 2015). Rigolon et al. (2019) examined environmental justice as part of an ecological model for recreation, emphasizing distributional, procedural, and interactional justice as the three main pillars of environmental justice.

Distributional justice relates to people's spatial access to environmental amenities and their exposure to environmental hazards regardless of their race, ethnicity, social class, gender, sexual orientation, disability status, and other characteristics.

Procedural justice refers to the inclusion and the lack of representation of marginalized groups in decision-making processes which address the location and qualities of environmental amenities and hazards.

Interactional justice describes the experience of marginalized groups in public spaces and recreation settings. It examines whether "these groups are being treated in a discriminatory way, or whether they are targets of harassment, insults, or other rude behaviour" (Low, 2013, p. 298). In addition, interactional justice calls for creating recreational spaces that "suit" the preferences of marginalized groups (Gibson et al., 2018).

This chapter highlights the role of water in the context of sustainable tourism and recreation. It argues that water is

Fig. 2 The Valley of Springs (https://www.masa.co.il/article)

currently considered as one of the most pressing environmental concerns globally (Tempelhoff et al., 2009). From scarcity to excess, responsible for the death of thousands and the migration of millions, water is first and foremost the essence of life, and after that a source for enjoyment, happiness, and recreation activities. Krause and Strang (2016) suggest studying water not only as an object of social relationships or as a natural resource, "but rather as a generative and agentive *co-constituent* of relationships and meanings in society" (Krause & Strang, 2016: 633). They argue that such an approach extends beyond the conventional in analyzing topics which were previously analyzed separately and includes more critical and ethical dimensions. Therefore, in their opinion, it is deserving of more nuanced policies and practices. This framework discusses water as fashioned by the interrelated relationships between geology, climate, institutional arrangements, property relations, market forces, ideologies, and social and power relations. To this extensive list, this chapter adds and emphasizes the role of planning as an organizing force of the landscape.

Below, I discuss the role of water in the Valley of Springs and the failure to create sustainable tourism in the Valley mainly because of the failure to fairly distribute its prominent natural resource among its various communities. The conclusion suggests an alternative approach for the region.

4 Results: The Valley of Springs as a Case Study

4.1 Geology and Hydrology

The Jordan Valley, of which the Valley of Springs is a part of, is the product of two fault systems: the great Syrian African fault line, which runs north to south, and several west to east faults, which uplifted the Gilboa Mountains. These faults are responsible for the abundance of springs in an area with an annual rainfall of only some 100 in. There are 30–35 springs in the Valley, which differ in their annual water flow, temperature and level of salinity, and also a stream—Harod Stream—which flows eastward to the Jordan River. In the past, many of the springs and their tributaries flowed into large swamps, while others flowed into the Jordan River.

4.2 Human Settlement

Well-organized societies and technology to tame water enabled the construction of more than 40 settlements in the Valley, from the Chalcolithic period (4500 BCE) onward. At its peak, during the Hellenistic period (fourth century BCE to first century AD), the city of Scythopolis prospered with great underground reservoirs of water, bathhouses, public fountains, public restrooms with running water, and a municipal drainage system. Nearby, its affluent population even enjoyed a water theater on the banks of Assi Stream. In 749 AD, a great earthquake destroyed the city, and for generations, the region declined. In the mid-nineteenth century, the Ottoman rulers established a small regional center named Beshan next to the ruined old city, but it was the Zionist settlement in the 1930s which changed the landscape. European Jewish settlers brought modern knowledge, great enthusiasm, and firm ideology to this peripheral edge of the Ottoman Empire. They established communal agricultural settlements—kibbutzim—and earned their living from agriculture, growing dates, grains, and vegetables. In 1939, they built the first fishponds, which have since become a main source of income for the Valley residents.

The modern town of Beit Shean was established after Israel's War of Independence (1948–9) and the deportation of the Arab residents of Beshan. During the 1950s-1960s, the town was populated by Jewish refugees from Iraq, Iran, Romania, and Northern Africa. First, they were settled in the deserted Ottoman town, and later, the State housed them in poorly built two-to-three-story housing projects. Since its establishment, the town's residents have belonged to Israel's lower social-economic status (4/10). Currently (2021) 18,500 people reside in Beit Shean, and new schemes aim to increase its population by 50% by the year 2035.

4.3 Water for Living and Enjoyment

The ability to produce water was essential for settling the region. There are numerous irrigation facilities and water-based plants scattered throughout the 180 square kilometers of the Harod Stream drainage basin, some dating back to Roman times (Avisur, 1963). Since the 1930s, Jewish setelers pumped water from Harod Stream and later established a *Water Association Cooperative*, which from 1934 to the 1970s drilled 40 wells. During the 1950s, and soon after its establishment, the State of Israel built a sophisticated regional water plant to supply fresh water for irrigation and living and saltwater for fishponds. The State, which owns and controls all the country's water sources, pooled the various sources and distributed the water in open cement channels throughout the region. In the ensuing years, new water projects were initiated, including the drilling of new wells and the construction of a large reservoir for storing winter floodwater.

In parallel, as early as the 1930s, the Jewish settlers discovered the pleasure of bathing in the Valley's springs during the hot summers and the cold winters (as the springs differ in temperature). Residents of Kibbutz Nir David drained the swamp formed by Assi Stream and built their houses on its banks (Fig. 3). In the 1940s, they proposed establishing a number of recreation sites along the Gilboa fault line, and in 1958, on Israel's tenth anniversary, Israel's president declared the area along the length of Assi Stream as one of the State's first national parks, Gan HaShlosha (Alon-Mozes, 2013). Since the 1960s, due to the improvement in Israel's economy and the establishment of a modern highway system, the area attracted vacationers from all over the country, who arrived to recreate in Harod Spring National Park (Lissovsky & Dolev, 2012) and in other sites, which offered archeologic and scenic amenities. As no swimming pools existed in the nearby Arab villages, the various water amenities attracted this population as well, and after the 1967 war, also people from the occupied territories of the West Bank (Alon-Mozes, 2015). In the following years, the success of the Valley made it popular among Christian pilgrims and tourists from abroad, as reported in 2007 by a Jewish journalist visiting Gan HaShlosha:

> ….It was everything I had imagined I would find in Paradise: As we walked through the site, taking in the majesty all around us, we began to notice something almost as fascinating – the people. Here, lounging in Paradise, was a complete blend of Israelis and Arabs – Jews, Christians and Muslims, side by side, enjoying the park together. Well, maybe not together, but side by side. In fact, with everyone in their swimming suits, splashing in the water, it was sometimes hard to tell them apart. (Ben David, 2007).

Since the 1990s, following the demand for more water-based recreational sites, various proposals emphasized the Valley of Springs' role in benefiting the local community and others, as detailed in Table 1.

Currently, the regional council, in collaboration with State authorities, completed the development of approximately a dozen sites for public use. Table 2 summarizes the amenities of the prominent sites, which differ in their salinity level, amount of available water, on-site recreation facilities, distance, accessibility from the town of Beit Shean, and entry fee.

However, water was not always distributed equally among the inhabitants of the Valley. While the residents of the agricultural settlements were given free access to water amenities—and particularly Kibbutz Nir David, which established its residential area on the banks of Assi Stream—residents of Beit Shean, the only town in the region, barely had access to fresh water for recreation (Fig. 3).

4.4 Beit Shean and Water

Fresh water, in built channels, ran through the Palestinian town of Beshan, and later also through Beit Shean. The water served household uses, and during the 1950s, the channels were buried and disappeared from view and use. Since then, the residents of Beit Shean had limited access to on ground water for various activities. Harod Stream, which used to flow year-round on the outskirts of the town, lost most of its water to irrigation and to the fishponds. Instead, it absorbed the polluted water from the ponds, which turned it into a hazard of poisonous water and a source of foul odors. Furthermore, there were no other water sources within walking distance. For years, Beit Shean residents experienced the water amenities mainly as day laborers for minimum wage, when constructing Gan HaShlosha National Park and other water-based recreation sites. The current urban planning of Beit Shean continues to locate the community with its back to Harod Stream, and future schemes to include the stream as part of the cityscape are very modest in terms of water use, proposing only strolling along the stream and not making it part of the scenery.

4.5 The Symbolic Meaning of Water in the Valley of Springs

Veterans of Beit Shean remember living next to Harod Stream, using its water for drinking, bathing, swimming, and watering the vegetable patches surrounding their houses. In retrospect, not devoid of nostalgia, they recall those days of poverty and hardship as full of joy, confidence, and happiness (Asur, 2021). In parallel, kibbutz members also regarded the fresh water as compensation for their daily difficulties in the harsh climate of the Valley. They cleaned the springs, planted communal gardens around them, and built their houses on the Assi Stream's banks (Fig. 3). They also made extensive use of the local water for establishing their agricultural initiatives, and especially the fishpond industry. Throughout the 1960s and onwards, Gan HaShlosha National Park served as a symbolic icon of hegemonic Israel. It was a preferred site for weekend car trips for middle-upper-class Israelis and functioned as the site for ceremonies marking the end of military trainings in the Valley. These overnight gatherings around the bonfire, in an Eden made by the Zionist early settlers, were very influential in strengthening self- and national identity. The Valley expressed the wish of the third century Jewish sage Reish Lakish, who asserted: "If paradise is found in the Land of Israel, its entrance is through the gates of Beit Shean" (*Reish Lakish, third century*).

But paradise was not equally shared by the various communities of the Valley, and since the 1950s, water and the access to water have symbolized the gap between the two local communities: the relatively small agricultural communities (mainly kibbutzim), which are under the auspices of the regional council, and the relatively large urban community of

Fig. 3 Assi Stream and Kibbutz Nir David (left), Harod Stream and Beit Shean (right)

Table 1 The role of water in the various schemes for the Valley of Springs and Beit Shean

Year	The Scheme	Initiated by	The role of water
1992	Comprehensive local scheme for Beit Shean C 6995 (1992)	Beit Shean local council, and State authorities (Rachamimov)	– Addition of channels, small waterfalls, pools, and fountains to cool the temperature of town – Development of the north–south ancient aqueduct axis expressing the importance of water for the town – Rehabilitation of Harod Stream in collaboration with the regional council and State authorities. Rehabilitation of the stream banks and development of hiking routes and other activities along the stream
2017	Comprehensive local scheme for Beit Shean 203–0,293,993 (2017)	Beit Shean Municipality and the State	Not a part of the town plan. A proposal to establish a hotel next to Harod Stream as part of the national park – Restoration of the stream to revive its fauna and flora – Priority for the stream as a drainage system. In addition, hiking trails, management routes. Any schemes must be coordinated with the drainage authority – Any scheme will be coordinated with the landscape plan of Harod Stream
2007	Local Scheme for Harod Stream C. 13,353 (2007)	Regional council, Beit Shean Municipality, JNF	– Conservation and restoration of the stream as a water system – Controlled development of the stream for recreational uses while maintaining the agricultural/rural character of the region and preventing damage to agricultural areas – Designation of agricultural areas for reservoirs and fishponds – Proposing several areas for intensive tourist development – In 2015, the scheme was revised to include instruction for ecological protection along Harod Stream and its outlying areas
2015	Harod Stream East Part C. 21,331 (2015)		
2004	Harod Stream: Instruction for development the Stream route and planting (2004)	Ministry of Agriculture, Harod Stream Administration, KKl, and others. (Tibi landscape architecture)	– Restoration and preservation of the stream and the open space for the benefit of the public – Improving the qualities and quantities of water for the rehabilitation of the stream system – Development of Harod Stream as a lever for realizing the economic potential of tourism in the area – Landscape development of the stream area – The stream as an educational and tourist theme – Maintaining a continuum of pedestrian accessibility throughout the axis of the stream – Creation a "body" for the stream – Locating existing hotspots and sites along the stream and Harod Valley and the realization their touristic potential – "The Stream Route" as a road designated for pedestrians and cyclists and limited to vehicle maintenance and service only
2009	Master Plan for Spring Valley (D. M. R., 2009)	Regional council, Ministry for Environmental Protection, Ministry of Agriculture, and others	Water is one of the core elements of the region. The plan calls for: – A comprehensive master plan in cooperation with neighboring authorities including the Palestinian Authority and the Jordan Kingdom – Examination of various technologies for improving water quantity and quality – Making water a central theme of research and dialog – Exposing water as a leading motif in branding the region – Exposing the hidden network of water by user-friendly access routes and observation points – Unveiling the local water system and its agricultural achievements – Preserving water sources, mainly by stream restoration – Upgrading existing infrastructure for sustainable use – Examining the feasibility of establishing and R&D center with Jordan

Table 2 Water recreation amenities in the Valley of Springs

	Distance from Beit Shean	Amenities	Accessibility	Price of entrance
Ein Moda	7.6 km 16 min. by car. No public transportation	Fresh water, Eucalyptus trees, picnic tables, no lawn	10–15-min. walk, rent bicycle, or golf carts	Free of charge
Ein Shokek	5.4 km 13 min. by car. No public transportation	Fresh water, Eucalyptus and other trees, natural vegetation, remains of water mill picnic tables, no lawn	15–20-min. walk, rent bicycle, or golf carts	Free of charge
Kibbutzim Stream	4.1 km 7 min. by car. No public transportation	Upper part nature reserve 1.5 km of fresh stream, no lawn	By car	No parking payment
Gan HaShlosha	6.2 km 8 min. by car. No public transportation	Old national park, big swimming pools, lawns, coffee house lifeguard	By car	Almost $13 for adult and $7 for kids. The most expensive national park
Ein Huga	6.3 km 11 min. by car. No public transportation	Large swimming pools, lawns, café lifeguard	By car	$14 for adult

Table 3 Characteristics of the communities in the valley of Springs

	Size (ha)	Population	Density	Socioeconomic status
Beit Shean (1 town)	744	18,463	248.2	4/10
The valley of Springs (24 communities)	25,043	13,415	5.36	6/10

Beit Shean. While the population of the town is 1.5 times larger than the population of the regional council, its area of jurisdiction is almost 35 times smaller, and consequently, the town's population density is almost 50 times greater. Table 3 summarizes the differences between the two communities. Missing from the table are the facts that the jurisdiction area of the regional council includes most of the local springs and that the regional council is wealthy enough to develop these water amenities or to collaborate with State agencies in these development projects. In contrast, the town of Beit Shean, poor in land and financial resources, is unable to take part in these initiatives. The only water amenity included in its jurisdiction area is the polluted Harod Stream.

This uneven distribution of the natural resources reflects the State's historical prioritization of the rural sector over the urban one, and especially the inferiority of new post-independence towns, which suffer from limited resources in terms of land, water, budget, etc.

The symbolic meaning of water is reflected also in the language used in the Valley and the names of its various sites. In the 2000s, the regional council changed its name from Beit Shean Valley to the Valley of Springs, distancing itself from the poor town, and emphasizing an image of a pastoral landscape detached from its actual geographical location. The symbolic meaning of water and water-related amenities is also reflected in the names of the springs, which have changed along the years. Arabic names were replaced with Hebrew ones, as in the case of Assi Stream—which was renamed Amal Stream. Since 2020, in line with the struggle to bring the stream back into the hands of the Valley's inhabitants, its name was switched back to Arabic—Assi. Another example is Jamain Stream, which became Kibbutzim Stream. The latter, although a Hebrew translation of the Arabic word, chose a dual term that also mentions the agricultural settlements (kibbutzim), which are blamed for appropriating the area.

4.6 2020 Events

In the summer of 2020, between the Covid-19 pandemic lockdowns, the pressures on Israel's relatively few water recreation sites became extraordinarily intense (Kershner, 2021). Backed by the law, which defines streams as a public domain, residents of Beit Shean, as well as many other domestic tourists from throughout the country, demanded that Kibbutz Nir David opens its gates and lets them picnic on the grassy riverbank of Assi Stream (Fig. 4). In response, the kibbutz closed its gates and called the police and the struggle quickly reached the press, spread through social media, and was even discussed in the Israeli parliament. It was manipulated by politicians aiming to fuel the generations-old conflict between rich and poor, agricultural settlements and peripheral towns, persons of Ashkenazi and Sephardi (Oriental) origin, and other rifts characterizing contemporary Israeli society. In the ensuing months, the case was brought before the court, which tried to mediate between the adversaries. It forced the kibbutz to allocate a small piece of land along the stream for public use during the summer. However, these small steps taken on site failed to satisfy the involved communities.

Currently, the media is the main battleground between the opposing parties. Kibbutz Nir David established a highly professional Web site, rich with historical data and more recent information, called *Shattering Myths*, (https://www.nir-david.co.il/%D7%A0%D7%99%D7%A4%D7%95%D7%A5-%D7%9E%D7%99%D7%AA%D7%95%D7%A1%D7%99%D7%9D); the opponents, who recruited supporters from all over the country, manage their activity via an aggressive Facebook page of 20,000 followers, named *Free the Assi* (https://www.facebook.com/groups/487207302079116/).

The current debate focuses on legal rights, or the generations-old struggle between hegemonic, privileged Israel as represented by kibbutz members, and "Second Israel," of less affluent people, as represented by those calling to free the Assi. However, in order to escape the numerous pitfalls of conventional planning or problem solving by politicians and the court, I suggest examining the current struggle from the perspective of water, as both the most important natural resource of the area and the carrier of symbolic meaning, power relationships and social interactions. As such, water will become the object of planning, design, and management.

Fig. 4 HaAssi is a public domain. *Photo* Gil Eliyahu, HaAretz, 24.8.2020, https://www.haaretz.co.il/news/education/.premium-1.9073011

5 In Conclusion: Planning Through the Perspective of Water

The following sets out the framework for sustainable planning for Valley of the Springs.

5.1 The Role of the Drainage Basin

As a water-related issue, the drainage basin and the regional water system will serve as the basic unit of investigation, planning, and management. Currently, on a regional scale, two different planning systems relate to the drainage basin of Harod Stream. The first partially covers the jurisdiction area of the regional council, and the second relates to the town of Beit Shean. Consequently, the stream falls between two entities and is not considered as an integral whole. While it serves as an amenity within the regional council, it reaches the town as a hazard. Two decades ago, a comprehensive scheme for the stream was prepared by State authorities; however, due to various reasons it failed and has not been reconsidered since. As similar comprehensive schemes presently work well for other streams (Yarkon River Authority, 2020), it is time to revise the comprehensive plan for the entire region. If this is achieved, it will require any future initiatives to be examined and evaluated within the framework of the entire drainage basin.

5.2 Mapping Water

As a first step, it is crucial to map the current and past water amenities of the region. Relying on old maps, historic accounts and remaining materials (archeological and historical), discussing the environmental history of the Valley of Springs from the perspective of water will serve as a basis for future planning. Such mapping will include the routes of old streams, the flow rate of springs and the quality of their water, the locations of ancient swamps, and evidence of the past flora and fauna of the wet environments. In addition, the investigation will focus on the presence of humans and their various uses of water, as well as on the symbolic meaning that people have attributed to water. Such accounts will be compared with the contemporary role of water for the region, including an in-depth analysis of the water component in the various proposed and implemented regional and local schemes.

5.3 Sustainable Planning

"To sustain the benefits of outdoor recreation for present and future generations, the recreation program must address and work toward a sustainable balance among the three spheres of environmental, social, and economic conditions" (USDA FS, 2010: 4).

These three pillars of sustainability should direct any future development of the Valley of Springs and its water amenities. During an era of climate change, and especially in an area suffering from a scarcity of precipitation, it is extremely important to reconsider the balance between the various consumers of this limited natural resource: nature, human inhabitants, agricultural development, industry, and recreation. It is important to preserve the natural water systems, to avoid over-pumping and the allocation of water for hazardous uses, causing irreversible pollution.

In parallel, it is important to reconsider the various economic aspects of this "water world": to carefully calculate the direct and indirect investments in and revenues from the various water-based initiatives and to question generation-old axioms which are rooted in State ideology and past needs. For example: Is it currently justifiable to maintain fishponds? To continue preserving generation-old agricultural practices and choice of crops? How many recreation sites, based on water, are necessary to serve the country's growing population in general, and the needs of the locals in particular?

In the Valley of Springs, this is extremely important, as previous planning and decision-making processes have ignored the local population, and especially the residents of the town of Beit Shean. Therefore, in order to create a sustainable future for the Valley, it is necessary to discuss its water through the three pillars of environmental justice: distribution, procedure, and interaction, as presented above.

Sustainability and Environmental Justice in the Valley of Springs.

- *Distributional justice* generally relates to the equal spatial allocation of resources and hazards. In the Valley of Springs, fresh water is the precious resource, while the polluted water is considered a hazard. At present, they are not spatially distributed equally, either due to environmental characteristics (fault lines, rock formations, etc.) or to human activities (location of fishponds as a dominant polluting land use, etc.). Technology can distribute water-based amenities fairly throughout the region, as already accomplished by the network of the artificial channels. However, it is necessary to also look at the hazardous components and their spatial distribution, more specifically, the fact that the living stream of the Assi is mixed with the polluted water of the fishponds by the time it reaches the town of Beit Shean, contradicting any distributional justice—an issue that needs to be resolved.

Furthermore, fair distribution relates not only to the resource, but also to the various elements accompanying it. People find recreation on lawns near water more appealing than in grassless shaded groves. They also relate great importance to accessibility by public transportation or by car, to whether or not they have to pay entry fees, and to whether the site is open to visitors throughout the year. As each site is unique—and there is no reason to make them all the same—it is important to evaluate the various components of each site in order to ensure fair distribution of the various amenities.

- *Procedural justice* calls for balanced participation of all communities in the decision-making processes. In the past, kibbutz members voted in a general assembly on all aspects of kibbutz life. Currently, although public hearings are a common procedure in many planning processes, it is not as commonly practiced in low-income communities. The lack of time, knowledge, and experience in dealing with these over-complicated processes distance the community from taking part in shaping its future. In addition, planning procedures in Israel are long and complex; they are most often managed by professional planners and town bureaucrats and at times do not represent the interests of the common people.
- *Interactional justice* refers to the experience of individuals and of the various communities in relation to each other and, in this case, in relation to water. Discussing these individuals' and communities' memories, expectations, and the practical and symbolic meaning they attribute to water might create a space of listening, tolerance, and acceptance as a first step toward more sustainable development. While the stories of kibbutzim members—the first Jewish settlers of the valley—are well documented, very little is known about the experience of the residents of Beit Shean. The first documents in albums, books, and currently—a Web site describe the heroic adventure of drying the swamps, digging pools, and planting Eden next to Assi Stream. The others—the residents of Beit Shean—barely shared their stories of enjoying the water within the town's boundaries or of enjoying the fresh water of Harod Stream. Moreover, their hardship as day laborers building Gan HaShlosha National Park and other recreation sites was never acknowledged. Providing such opportunities for sharing stories might create the first step toward reconciliation and relief.

In conclusion, sustainable tourism is discussed in this chapter as part of a comprehensive framework which perceives tourism as one component among many others in creating a sustainable region for the present and the future. The chapter identified the core resource of the valley—water—which since antiquity has been the source of its prosperity. From the water theater of the Hellenistic period to the contemporary Assi Stream and its cooling water, water has been the essence of the Valley, and the events of 2020 only highlighted its importance for the region as an agent of the Valley's genius loci, economic prosperity, and human interactions. Broadening the perspective of investigation, and consequently of planning and management, challenges past modes of investigation, planning, and decision-making in favor of a more sustainable, refreshing future. The idea must be examined by various local stakeholders and other experts in order to determine its applicability for the Valley of Springs and other regions.

References

Alon-Mozes, T. (2013). If paradise is in the Land of Israel, its entrance is at the gates of Beit Shea (*Reish Lakish*), Israeli National Parks and the Search for National Identity. In E. Carr, E. Shaunn, & R. G. Wilson (Eds.), *Public Nature* (pp. 185–193). Virginia Press.

Alon-Mozes, T. (2015). National Parks for a multicultural society; planning Israel's past and present national parks. In D. Bruns, O. Kühne, A. Schönwald, & S. Theile (Eds.), *Landscape Culture—Culturing Landscapes* (pp. 173–184). Springer.

Asur, D. (2021, January 14) Memories from the Transit Camp. *Saturday Story*. https://pages.facebook.com/merkazmo49/photos/a.1500142763581946/2746304655632411/?type=3&source=48

Avisur, S. (1963). Harod Stream and Beit Shean Tributaries. In S. Avisur, *Survey of Waterpower Facilities in the Land of Israel* (159–176). The Institute for the Knowledge of the People and the Land. (in Hebrew).

Ben David L. (2007, October 7). *Paradise: A Peace of the Middle East, Ynet*. https://www.ynetnews.com/articles/0,7340,L-3421607,00.html

Beit Shean Local council & State Authorities. (1992). *Comprehensive local scheme for Beit Shean, C 6995*. https://www.tabanow.co.il/%D7%AA%D7%91%D7%A2/%D7%91%D7%99%D7%AA-%D7%A9%D7%90%D7%9F/%D7%92/6995

Beit Shean Municipality & State Authorities. (2017). *Comprehensive local scheme for Beit Shean* 203-0293993 https://mavat.iplan.gov.il/SV4/1/2000347543/310. See table 1 2017

Blahna, D. J., Kline, J. D., Williams, D. R., Rogers, K., Miller, A. B., McCool, S. F., & Valenzuela, F. (2020). Integrating social, ecological, and economic factors in sustainable recreation planning and decision making. In S. Selin, L. K. Cerveny, D. J. Blahna, & A. B. Miller (Eds.), *Igniting research for outdoor recreation: Linking science, policy, and action* (pp. 173–188). Department of Agriculture.

Buckley, R. (2012). Sustainable tourism: Research and reality. *Annals of Tourism Research, 39*(2), 528–546.

D. M. R. (2009). Regional Council, ministry for environmental protection, ministry of agriculture & others. *Master plan for the valley of Springs*

Gibson, S., Loukaitou-Sideris, A., & Mukhija, V. (2018). Ensuring park equity: A California case study. *Journal of Urban Design, 24* (2), 385–405. https://doi.org/10.1080/13574809.2018.1497927

Hall, C. M., Gossling, S., & Scott D. (Eds.). (2015). *The Routledge handbook of tourism and sustainability*. Routledge.

Hales, R., & Jamal, T. (2015). Environmental Justice and tourism. In C. M. Hall, S. Gossling, & D. Scott (Eds.), *The Routledge handbook of tourism and sustainability* (pp. 151–164). Routledge.

Hardy, A., Beeton, R. J. S., & Pearson, L. (2002). Sustainable tourism: An overview of the concept and its position in relation to conceptualizations of tourism. *Journal of Sustainable Tourism, 10* (6), 475–496. https://doi.org/10.1080/09669580208667183

Kershner, I. (2021, June 21). Fight over a gentle stream Distills Israel's political divide. *New York Times*. https://www.nytimes.com/2021/06/21/world/middleeast/israel-asi-kibbutz.html

Krause, F., & Strang, V. (2016). Thinking relationships through water. *Society and Natural Resources, 29*(6), 633–638.

Low, S. (2013). Public space and diversity: Distributive, procedural and interactional justice for parks. In Y. Greg & S. Deborah (Eds.), *The Ashgate research companion to planning and culture* (pp. 295–310). Ashgate Publishing.

Lissovsky, N., & Dolev, D.)2012(. *Arcadia: The gardens of Lipa Yahalom and Dan Zur*. Bavel

Ministry of Agriculture Harod Stream Administration, JNF & others. (2004). *Harod stream, instruction for development the stream route and planting*. http://yardend.org.il/wp-content/uploads/2019/12/%D7%98%D7%99%D7%91%D7%99-%D7%AA%D7%9B%D7%A0%D7%99%D7%AA-%D7%90%D7%91-%D7%A0%D7%97%D7%9C-%D7%97%D7%A8%D7%95%D7%93.pdf

Regional Council, Beit Shean Municipality & JNF. (2007). *Local scheme for harod stream* C. 13353. https://mavat.iplan.gov.il/SV4/1/2096927/310

Regional Council, Beit Shean Municipality & JNF. (2015). *Harod stream east part* 21331/c. https://mavat.iplan.gov.il/SV4/1/2000294725/310

Rasoolimanesh, M. S., Ramakrishna, S., Hall, M., Esfandiar, K., & Seyfi, S. (2020). A systematic scoping review of sustainable tourism indicators in relation to the sustainable development goals. *Journal of Sustainable Tourism* (published online): https://doi.org/10.1080/09669582.2020.1775621

Rigolon, A., Fernandez, M., Harris, B., & Stewart, W. (2019) An ecological model of environmental justice for recreation. *Leisure Sciences, 0*: 1–22.

Tempelhoff, J., Hoag, H., Ertsen, M., Arnold, E., Bender, M., Berry, K., Fort, C., Pietz, D., Musemwa, M., Nakawo, M., van Ur, J., Dam, P., Melosi, M., Winebarger, V., & Wilkinson, T. (2009). Where has the water come from? *Water History, 1*, 1–8.

USDA Recreation, Heritage and Volunteer Resources. (2010). *Connecting people with America's great outdoors: A framework for sustainable recreation*. https://www.fs.usda.gov/Internet/FSE_DOCUMENTS/stelprdb5346549.pdf, p. 4.

Waligo, V. M., Clarke, J., & Hawkins, R. (2013). Implementing sustainable tourism: A multi-stakeholder involvement management framework. *Tourism Management, 36*, 342–353. https://doi.org/10.1016/j.tourman.2012.10.008

Yarkon River Authority. (2020). *A Master scheme for the Yarkon River basin*. https://www.yarkon-river.org.il/wp-content/uploads/2021/10/%D7%AA%D7%9B%D7%A0%D7%99%D7%AA-%D7%90%D7%91-%D7%9C%D7%90%D7%92%D7%9F-%D7%A0%D7%97%D7%9C-%D7%99%D7%A8%D7%A7%D7%95%D7%9F_compressed.pdf

Religious Tourism During the COVID-19 Period: The Case of Our Lady of Agony Festival, Viana do Castelo, Portugal

Hugo Martins

Abstract

Religious tourism in Portugal has played a role of great relevance for the economic development of the regions. This type of tourism is more than a visit to sacred places. The festival in honor of Our Lady of Agony is called queen of the festivals of Portugal, inserted in the cycle of summer festivities. This festival always takes place around the 20th of August in Viana do Castelo, Portugal. Given the growing number of visitors that it attracts and the volume of attractions that it has in its program, the Our Lady of Agony festival, over the years, has gained notoriety rivaling other festivals. This stands out, of course, by the religious aspect, but also by the ethnocultural and artistic attributes. The aim of this research is to understand the impact that the pandemic caused in terms of occupancy rate and to what extent conditioned the program of activities for the realization of this festival. Methodologically, we used a content analysis by analyzing statistical data from 2019 and 2020, characterizing the festival with its various moments/events both in the pre-pandemic period and during the pandemic period. The information was analyzed, and the conclusions were drawn. It was possible to verify that Our Lady of Agony festival did not cease to be held because of the pandemic, but underwent many changes in its programming, resorting mainly to digital. The religious part was one of the activities that took place physically. In terms of occupancy rate, although the number of overnight stays in August 2020 had a negative variation of -24%, it was better compared to the national results; the domestic market helped to offset the sharp decrease in the number of overnight stays by non-residents; and the "main stage" of the festival stopped being the physical territory and became Instagram, Facebook and YouTube.

Keywords

Religious Tourism • COVID-19 • Our lady of agony festival • Viana do Castelo

1 Introduction

Tourism as an economic activity brings together a number of elements that influence the territory and, simultaneously and consequently, a wide range of economic and social activities that are part of local and/or regional dynamics. Thus, economic agents increasingly see tourism as an alternative for the development of a particular area or region. The civil and religious heritage and traditions have been in some way appropriated and commodified by the tourism economy becoming part of the segment called religious tourism.

Religious tourism has functioned as a promoter of social and cultural interaction, enhancer of social and personal development and a stimulator of the territory in economic terms, preservation and enhancement of religious heritage (Silva, 2012).

With the pandemic, tourism and pilgrimage to major European shrines fell by 90–95% during the first months of the pandemic (Mróz, 2021). Shrines such as Santiago de Compostela and Lourdes were closed for several weeks during the first wave of the spread of the coronavirus. Europe's biggest Catholic shrines encouraged pilgrims to deepen their ties with the shrine by attending services and prayers online and to make spiritual and virtual pilgrimages (Mróz, 2021). In fact, in-person religious events during the early stages of the COVID-19 pandemic ultimately contributed to the spread of the virus (Olsen & Timothy, 2020): (a) In South Korea, about 5000 cases of the virus were traced to a person known as "Patient 31" who had worshipped at the Shincheonji Church of Jesus in Daegu; (b) in the middle east region, it was attributed to pilgrims visiting religious shrines in Iran and other countries; (c) in California, 71 positive

H. Martins (✉)
University of Maia and CEGOT (Portugal), Porto, Portugal
e-mail: hugomartins@ismai.pt; hugomartins@umaia.pt

A. Mandić et al. (eds.), *Cultural Sustainable Tourism*, Advances in Science, Technology & Innovation,
https://doi.org/10.1007/978-3-031-10800-6_4

cases of the COVID-19 virus were attributed to a Slavic megachurch; and (d) in Malaysia, 513 people tested positive for the virus after attending a meeting outside Kuala Lumpur linked to the Tabligh Muslim missionary movement, attended by 16,000 people (Mróz, 2021). These, and other mass religious events with high rates of viral transmission, have led to several countries banning large assemblies, including religious gatherings, as part of the broader effort to "flatten the curve" by implementing non-pharmacological interventions, policies and practices. Many projects "associated with religious and cultural events at shrines have been reduced to zero. This is an unprecedented event after World War II - nothing else has affected tourism, including religious tourism and pilgrimages, as much as the pandemic of Covid-19" (Mróz, 2021, p. 627). Portugal was no exception.

Due to the existence of a strong set of religious structures and traditions, Northwestern Portugal is a region where religious tourism contributes greatly to achieving this development. This type of tourism has been, in recent years, "the target of attention of several economic, political and scientific agents" (Antunes et al., 2016, p. 276).

In this context, this article aims to understand the impact that religious tourism has in the region, specifically the Our Lady of Agony festival and what the consequences derived from the pandemic caused by COVID-19.

Methodologically, our research question was elaborated to which we intend to obtain answers: "What is the impact that the SARS-CoV-2 pandemic caused in the Our Lady of Agony festival and to what extent conditioned its realization?". To answer our research question, a set of objectives were identified that allowed us to draw some conclusions. We used a content analysis, in which we sought to make a statistical analysis of the data concerning 2019 and 2020. In addition, we sought to characterize the festival with its various moments/events both in the pre-pandemic period and during the pandemic period to understand the changes that the pandemic caused regarding the readjustment of the program and the occupancy rate.

2 Literature Review

Tourism is an extremely important economic activity "and can play a decisive role in terms of development of certain regions" (Martins & Morais, 2016, p. 84), as it can generate wealth and employment (Martins et al., 2021a, 2021b) and also contribute to a relevant role in the preservation and maintenance of heritage, as well as traditions (Martins, 2022).

The practice of tourism can be interpreted as a modern form of pilgrimage (Burns, 2003), and religious tourism is characterized by an intimate connection between the tourist phenomenon and the religious phenomenon. Religious tourism is clearly on the rise since tourists are increasingly looking for authentic products, i.e., that combine cultural and spiritual experience. What attracts visitors with regard to religious heritage is certainly the fact that the properties are sacred places, i.e., a place associated with the divine (Silva, 2012), but also the historical, cultural and artistic character of these same places.

In addition, much of the heritage built by man over time and visited by him is undoubtedly of a religious nature. The most visible relationship between "tourism and religion consists in the huge amount and diversity of sacred buildings of real touristic interest that are visited by tourists" (Antunes et al., 2016, p. 275).

According to (Rinschede, 1992), religious tourism can be understood as a confluence of opposite polarities, in which, on the one hand, there is the sacred aspect and, on the other hand, the profane aspect, in which visitors travel partially or solely for religious reasons.

During the first six months of the COVID-19 pandemic, "the number of pilgrimages and tourists to Catholic shrines in Europe decreased by 90–95%" (Mróz, 2021, p. 625). Many shrines were closed during the first weeks of the pandemic (mid-March to May), and some remained closed until June.

The space for pilgrimages, the sacred space of thousands of shrines around the world, changed from mid-March. The following weeks turned out to be disastrous for pilgrimages. COVID-19 brought isolation, uncertainty, pain, suffering and death to thousands of people. The scenes of the closed Church of the Nativity in Bethlehem (closed from 5 March to 24 May 2020), the Church of the Holy Sepulchre in Jerusalem (closed from 25 March to 24 May 2020), the Major Basilicas in Rome, Pope Francis praying in solitude in St. Peter's Square during the Easter Triduum. Peter's Square during the Easter Triduum, the deserted pilgrimage route to the Shrine of Santiago de Compostela, the empty Massabielle Grotto in Lourdes, hundreds of thousands of closed temples and shrines of all religions around the world have really moved religious people. Passion services, Passions and Easter Triduum celebrations have been canceled. Justin Welby, the Archbishop of Canterbury, celebrated the Easter service broadcast online from his flat in London. Many parishes followed suit.

In statistical terms, the number of pilgrims to the various holy sites has fallen dramatically: The number of pilgrims on the Camino de Santiago registered in July 2020 was five times less than in July 2019 (in July 2020, 9752 people and in 2019, 53,319); in August 2020, it was more than three times less than in August 2019; and in September 2020, the Pilgrims' Office in Santiago de Compostela registered 10,441 people, which is four times less than in September 2018 and 2019 (Mróz, 2021). The number of pilgrims wandering on cross-border routes (Camino de Santiago, Via

Francigena) was reduced several times during the COVID-19 pandemic compared to previous years. The number of "pilgrims on the French section of the Via Francigena during the 2020 pilgrimage season was ten times lower" (Mróz, 2021, p. 635). The Shrine Our Lady of Lourdes, the largest Marian shrine in Europe, until the outbreak of the COVID-19 pandemic, had been visited by about 5 million pilgrims every year in recent years. For two months, until 16 May 2020, the Lourdes shrine remained closed.

Pilgrimage and religious tourism are too economically important to limit religious travel for long. Many "places around the world rely heavily on religious travel for their economic viability; many religious sites serve as anchor attractions that bring considerable fiscal dividends to their communities through tourism" (Olsen & Timothy, 2020, p. 176). The pilgrimage space of shrines has changed greatly, not only in terms of the number of visitors, but also the age range of pilgrims (a clear decrease in the number of pilgrims over 60 years old) and the reasons for going on pilgrimage to shrines (Mróz, 2021).

Therefore, the motivations can be of various kinds, both for pilgrims and for visitors who go to sacred places. While pilgrims travel mainly for reasons of religious practice such as praising the saints, pray or fulfill promises, the religious tourist can be driven by more profane or mixed motivations, wanting to know the traditions, history, nature, gastronomy and traditional music or participate in pilgrimages and religious festivals (Sousa & Pinheiro, 2014). According to Turismo do Porto e Norte de Portugal (2015), the festivals and pilgrimages of the North of Portugal are an authentic invitation to healthy conviviality amidst joyful people and to the visitation of a multifaceted territory, in a perfect symbiosis between the sacred and the profane, one of the anchor products of the Minho region (Martins et al., 2021a, 2021b).

Jackowski (2002) established classifications for festivities' religious sites, calling them international centers, supraregional centers or regional centers, according to their attractiveness. Some of the sanctuaries in Northwest Portugal can be considered supraregional centers, such as the sanctuary of Bom Jesus do Monte in Braga, or the sanctuary of São Bento da Porta Aberta, in Terras de Bouro (Silva, 2012), as they have the power to attract visitors in regional terms, although not assuming an international dimension.

Nolan & Nolan (1992) divided the religious places into three different groups, according to the visitors' motivations: (a) places of pilgrimage, where devotion is the main reason for the visit, distinguishing them according to the greater or lesser power of tourist attraction, the occurrence of seasonal festive pilgrimages; the combination between the artistic/patrimonial value, festive pilgrimages and worship; (b) tourist-religious attractions, whose motivation for the historical or artistic heritage prevails over the religious motivation, not being object of the festivities; and (c) places where religious festivities occur, such as celebrations or processions. According to the authors, there are 321 Christian shrines in Portugal, of which 168 consecrate their veneration to Mary, 124 whose devotion is due to the saints and 29 consecrated to Jesus Christ.

In our territory under analysis, the city of Viana do Castelo, it has long been the stage of festivities, the most important of all being the Our Lady of Agony festival (Martins, 2022), already considered by the media the Queen of festivals. This event has stood out for its diversified program that attracts more and more visitors.

2.1 The Example of Our Lady of Agony Festival: Historical Background

Viana do Castelo is a city on the northern coast of Portugal, subdivided into 27 parishes, integrated in the sub-region NUT III of Alto Minho. Until 1848, the town was called only Viana. After that date, the city was elevated to the name of Viana do Castelo, indicated by the queen D. Maria II. In 1977, it is also elevated to the seat of a Catholic diocese. It currently belongs to the Portuguese Healthy Cities Network. The cycle of festivities in the municipality begins in March, with the festivities, in the parish of Mazarefes. However, its high point is the traditional Our Lady of Agony festival, which takes place in the middle of summer, around 20 August.

The Our Lady of Agony festival is associated with the history of the Chapel of Our Lady of Agony, patron saint of fishermen. Initially, in 1674, a chapel was built in primitive invocation to the Jesus of the Holy Sepulchre of Calvary and, a little above, a small chapel devoted to Our Lady of the Conception. The devotion to Our Lady of Agony occurs from 1751 with the entrance of her image in the Chapel of Jesus of the Holy Sepulchre of Calvary. The name of the saint is currently associated with the Queen of festivals and the multiple traditions of the largest popular festival in Portugal: Our Lady of Agony festival, born from the devotion of the fishermen who came from Galicia and the entire Portuguese coast for the religious and pagan celebrations, which are still repeated annually in the week of August 20, a municipal holiday.

From the second half of the nineteenth century, in 1861, the festival goes beyond the religious aspects: It becomes a festival with all its recreational motivations. In 1871, a bullfight is added to the program of the festival. In 1906, the Costume Festival is born and in 1908 the 1st Agricultural Parade (current historical-ethnographical parade) (VianaFestas, 2021a).

Since the early twenty-first century, in 2002, it is created a promoter association of the city festivities, VianaFestas, in charge of all the logistics and programming of the city festivities. This association has in its origin the institutionalization of the secular Commissions of Festival of Our Lady of Agony, being considered necessary and essential for the continuation of the organization of these festivities, known as the festival of the festivals of Portugal, in view of the involvement not only of financial volume, but, essentially, in its legal and fiscal aspect. This non-profit association has as main objectives: (a) the promotion and realization of festivals, exhibitions and similar acts, especially the promotion of the Our Lady of Agony festival; (b) the promotion of local identity, through the dissemination of its Ethno-folkloric wealth, its heritage in the range of costumes, crafts, gastronomy and traditions; (c) exercise a mobilizing role of experiences between local cultural groups and bearers of collective memories; (d) promote and restore to the community customs and traditions; and (e) contribute for the cultural and social development of the community, in a local perspective (VianaFestas, 2021a).

2.2 The Program of the Our Lady of Agony Festival

The festival in honor of Our Lady of Agony is the greatest exponent of the city's festivities, having received, in 2013, the Declaration of Interest for Tourism. The media calls it the Queen of festivals or even the festival of festivals in Portugal, as it stands out for its program, the number of visitors it has attracted over the years, the culture it carries in the strength of the "Vianesa" costume and the weight of the gold that the "mordomas" display on their chests.

One of the highlights of the festival is the historical-ethnographical parade that has been part of the program since 1908. Initially called the "Agricultural Parade" (in which the floats were pulled by oxen), this is the moment when the parishes of the municipality show their best to the city. The parade is a true living museum, always involving more than 3000 participants, which through a hundred historical performances and dozens of floats form a show that combines history, ethnography and customs. The procession parades in a course of 2300 m along the streets of the city and lasts more than two hours.

Another moment of this festival is the Mordomia Parade. It is an event that brings together more than 500 costumed women who walk through the streets of Viana do Castelo. The Mordomia Parade currently symbolizes a "greeting" of the organization of the Our Lady of Agony to the official entities and marks the beginning of the festival (Fig. 1). The mordomas, women in costumes from the different parishes of the city, parade through the main streets of the city, which gain more shine for the gold they carry. It is also on this day that the mordomas wear different costumes and secular gold pieces. They even carry dozens of kilos of gold, gathering the pieces of family and friends in a single chest. The value of the hundreds of kilos of gold pieces worn by the mordomas in this event is estimated to be around 14 million euros (VianaFestas, 2021a). This moment is seen, in the community, as a great pride in being from Viana do Castelo. It is this very own pride that is called "chieira," because it is with vanity and pride that these women carry the gold and costumes being usual to see in these parades several generations of women.

Another great moment of the festival's program is the "Costume Festival." It is an event characterized by dances and songs, where the beauty and richness of the details of the typical costume of Viana do Castelo are joined, there being moments in which the presenters explain the details of the costumes. It represents a tribute to the ethnography and folklore of Alto Minho, in which it is sought to make known to visitors the origin and history of the uses and customs associated with the costume and the art of dressing well by the genuine young woman of Viana. This is because the linen costume, with various characteristic colors and shapes, is a symbol of the region that the women of Viana do Castelo wore until the end of the nineteenth century depending on the occasion, time of life and their status. The use of gold manifested the wealth of the family, but above all the pride of the woman.

Fireworks are usually characteristic of several festivals and pilgrimages. In the Our Lady of Agony festival, it is an attraction that has always been part of the program, but it was with the "waterfall" on the Eiffel bridge, over the Lima river, that the fireworks of Viana do Castelo became famous, the known "Serenata." In the program, there are three great

Fig. 1 Mordomia Parade (VianaFestas, 2021a)

moments of fireworks: the "Fogo do Jardim," the "Fogo da Santa" and the "Serenata de fogo." The first great firework show of the feast is the "Fogo do Jardim," formerly called "caught fire," a technique that is almost no longer practiced. On August 20, the "Fogo da Santa" is launched near the entrance of the Seaport, at midnight, lighting up the sky of the Ribeira de Viana do Castelo. On the last day, visitors can watch a 30-min "Serenata de Fogo" that involves the launching of fire from the river and several other points along its banks culminating with the "waterfall" at the Eiffel bridge and the traditional saying "Viana is Love."

Another moment of great impact in the festival is the "Gigantones" and "Cabeçudos" to the sound of drums and concertinas that always perform, during the days of the festival, in different spaces of the city.

Finally, the highlight of the religious part of the festival is the solemn procession: The procession of Our Lady of Agony departs from its sanctuary, meeting up in the enclosure in front of the temple with other religions imagens (Our Lady of the Seas, Our Lady of the Assumption and Our Lord of the Afflicted), all coming from St. Dominic's church. Together, amidst the effort of the men who carry them, they walk through the streets of the historic center. The devotion to the patron saint of fishermen brings together more than a hundred participants that, accompanying the saints, dramatize various biblical scenes. It is also traditional for the residents of Ribeira to cover their balconies with bedspreads and throw flowers from their windows as the procession follows, which is received there in complicity and respect, making this the most religious moment of the festival. It should be noted that the litter travels through the streets that are richly ornamented beforehand.

Finally, the so-called Tappet Night, which always takes place from 19 to 20 August, preceding the feast's patron saint's day, transforms the streets of the riverside of Viana do Castelo into a display of popular art. Residents, family members and even volunteers work over 30 tons of colored salt, all night long, to decorate their streets with salt carpets with motifs alluding to the city and fishing. They decorate the six streets of the riverside of Viana do Castelo that receive the procession of Our Lady of Agony, after the sea procession (Fig. 2).

At the end of the procession, the image of Our Lady of Agony returns to the sanctuary, facing the Ribeira de Viana do Castelo, a moment marked by the emotion of thousands of people who gather in the surrounding area applauding and throwing flowers.

3 Methodology

In the context of research methods, our research question was elaborated to which we intend to obtain answers: "What is the impact that the SARS-CoV-2 pandemic caused in the Our Lady of Agony festival and to what extent conditioned its realization?". Once the research question was defined, we were able to turn our attention to the specialized literature in the area, thus initiating the second stage of the research phases suggested by Pizam (1987) and Tuckman (2000).

We used a content analysis in which we aimed at performing a statistical analysis of the data concerning 2019 and

Fig. 2 Procession of Our Lady of Agony through the streets with carpets of salt (VianaFestas, 2021a)

2020. In addition, we sought to characterize the festival with its various moments/events both in the pre-pandemic period and during the pandemic and to understand the changes caused regarding the readjustment of the program and the occupancy rate.

With regard to the literature review, we sought, critically and carefully, to analyze the state of the art of the main theme of the research subject, religious and cultural tourism, focusing specifically on information relating to the chosen territory and its festival. This review included the reading of several articles published in major international journals of scientific nature, books of authors of national and international reference, institutional Web sites and publications of official entities.

To answer our research question, a set of specific objectives were identified: (a) identify the tourist demand of the city of Viana do Castelo in the pre-pandemic period and during the pandemic (2019 and 2020); (b) identify the main emitting markets of tourist demand (residents and non-residents); (c) characterize the events of Our Lady of Agony festival that traditionally take place (pre-pandemic period); and (d) characterize the ways in which Our Lady of Agony festival took place in the period during the pandemic. With these objectives, it was possible to make an analysis of the collected information and draw conclusions.

4 Analysis of the Results

In order to identify tourist demand in the pre-pandemic and pandemic periods, we tried to understand the impact caused by SARS-CoV-2 in terms of overnight stays, in national, regional and local terms, focusing on the month of August, the period of Our Lady Agony festival.

Thus, according to data from the National Institute of Statistics (2021), regarding overnight stays in accommodation in 2019, the pre-pandemic period, 70%, more than 49 million overnight stays (49,051,832 overnight stays) throughout Portugal were by non-residents, while 30% (21,107,132 overnight stays) were by residents (Table 1), in a total of 70,158,964 overnight stays. In Table 1, we verify that the number of non-residents had a great impact in the Portuguese economy, since in every 10 overnight stays, 7 are from non-residents and 3 are from the domestic market. Overall, the number of overnight stays was higher in the Algarve region representing 30% of the overnight stays of the whole country, followed by the Lisbon Metropolitan Area (27%). The northern region recorded the third highest number of overnight stays with 15% (Table 1). In what concerns the resident nights spent by residents, the region with the highest number of nights spent was the Algarve region with around 5 million (4,985,984), followed by the northern region (4,314,067). As regards nights spent by non-residents, the region with the highest number of nights spent was the Algarve region with almost 16 million (15,914,511), followed by the Lisbon Metropolitan Area (14,725,111) (Table 1).

Given the pandemic caused by SARS-CoV-2, the numbers have changed dramatically in 2020. It is important to note that most of the lodgings were closed. According to the information in Table 2, 47% of overnight stays are by non-residents and 53% are by residents. Compared to Table 1, in percentage terms, residents contributed a lot to the hospitality sector, helping to offset the number of non-residents (about 49 million in 2019 vs. about 12 million in 2020). The change was −63% in the total number of residents and non-residents, as there was a sharp decrease from 70,158,964 overnight stays in 2019 to 25,798,299 overnight stays in 2020 (Table 2).

Regarding resident overnight stays, the highest number of resident overnight stays remained in the same regions compared to 2019: the Algarve region, losing about 1 million overnight stays (3,813,998), followed by the northern area, losing more than 1.5 million overnight stays (2,749,967) (Table 2). In what concerns non-resident overnights, the highest number of overnights stayed in the same

Table 1 Overnight stays at tourist accommodation establishments by geographic location in 2019

	Residents		Non-residents		Total	
	f	%	f	%	f	%
Portugal	21,107,132	30	49,051,832	70	70,158,964	100
North	4,314,067		6,496,645		10,810,712	15
Center	4,016,888		3,117,975		7,134,863	10
Lisbon Metropolitan Area	3,913,951		14,725,111		18,639,062	27
Alentejo	1,937,479		1,001,351		2,938,830	4
Algarve	4,985,984		15,914,511		20,900,495	30
Azores Islands	997,799		1,280,006		2,277,805	3
Madeira Islands	940,964		6,516,233		7,457,197	11

Source Self-elaboration based on National Institute Statistics adapted by the author (2021)

Table 2 Overnight stays at tourist accommodation establishments by geographic location in 2020

	Residents		Non-residents		Total		Variation (%)
	f	%	f	%	f	%	
Portugal	13,598,609	53	12,199,690	47	25,798,299	100	−63
North	2,749,967		1,616,089		4,366,056	17	−60
Center	2,614,902		747,109		3,362,011	13	−53
Lisbon Metropolitan Area	1,940,933		3,313,421		5,254,354	20	−72
Alentejo	1,488,438		340,817		1,829,255	7	−38
Algarve	3,813,998		4,076,713		7,890,711	31	−62
Azores Islands	459,427		194,949		654,376	3	−71
Madeira Islands	530,944		1,910,592		2,441,536	9	−67

Source Self-elaboration based on National Institute Statistics adapted by the author (2021)

regions: Algarve and Lisbon Metropolitan Area, both losing about 11 million overnights each (Table 2). The northern region also registered a high loss of overnight stays, having registered a change of −60%, in relation to 2019, losing about 1.6 million overnight stays from residents and 4.9 million overnight stays from non-residents (Table 2).

Making an analysis to August 2020, it was possible to verify that in that month, throughout the Portuguese territory, the tourist accommodation sector recorded 1.9 million visitors and 5.1 million overnight stays, corresponding to variations of −47.2%, respectively, compared to the homologous period of 2019 (Table 3). The overnight stays of residents will have decreased by 2.4%, reaching 3.4 million, representing 65.9% of the total, while those of non-residents will have decreased 72.0%, standing at 1.7 million. The Alentejo continued to show the smallest decrease in the number of overnight stays, compared to the same month last year, with a decrease of 15.1%. The growth of nights spent by residents in Algarve (+9.0%), Alentejo (+4.0%) and Center (+1.1%) (Table 3) should be highlighted.

Table 3 Overnight stays at tourist accommodation establishments, by region, in August (millions)

	Total		Residents		Non-residents	
	August 2020	Variation with 2019 (%)	August 2020	Variation with 2019 (%)	August 2020	Variation with 2019 (%)
Portugal	5088.0	−47.2	3353.8	−2.4	1734.3	−72.0
North	831.8	−41.0	561.5	−2.0	270.2	−67.7
Center	778.8	−27.6	619.0	1.1	159.8	−65.5
AM Lisbon	646.7	−68.6	261.6	−34.2	385.1	−76.8
Alentejo	411.2	−15.1	347.1	4.0	64.2	−57.4
Algarve	2086.2	−39.3	1379.4	9.0	706.8	−67.5
Azores Islands	104.0	−69.3	72.0	−35.5	31.9	−85.9
Madeira Islands	229.3	−72.3	113.1	−21.7	116.2	−83.0

Source Self-elaboration based on National Institute Statistics adapted by the author (2020)

Table 4 Overnight stays at tourist accommodation establishments in Viana do Castelo (Annual)

	2019	2020	Variation (%)
Residents	119,090	90,992	−24
Non-residents	145,268	56,986	−61
Total	264,358	147 978	−44

Source Self-elaboration based on National Institute Statistics adapted by the author (2021)

Regarding the territory under analysis, Viana do Castelo, inserted in the northern region, it can be seen that the overnight stays in tourist accommodation establishments also recorded a large drop in visitors: In 2019, there were 119,090 overnight stays by residents and 145,268 by non-residents, closing the year with a total of 264,358 overnight stays. In 2020, the year of the pandemic, there was a strong negative change in non-resident overnight stays, −61%: 145,268 in 2019 to 56,986 in 2020. Although the domestic market also recorded a negative rate of change to the homologous period, −24% (119,090 overnight stays in 2019 to 90,992 in 2020), we can state that the domestic market ensured the occupancy rate making up for the lack of non-residents. In conclusion, the overall rate of change was −44% in relation to 2019 (Table 4).

However, before the pandemic period, in relation to tourism demand, Viana do Castelo had already been growing in the proportion of non-resident tourists in relation to resident tourists, representing a proportion of 55% (VianaFestas, 2021a). Besides this, Viana do Castelo has also been decreasing seasonality, without decreasing the months of higher demand, in a strategy of sustainability in tourism demand.

Making a monthly analysis, on average, the tourist overnight stays in Viana do Castelo during the month of August, the period when the Our Lady of Agony festival takes place, generate an average between 20 and 22% weight in the annual value, representing these numbers 43,890 overnight stays in 2019, 17.4% more than in August 2018, generating total revenues during August in the order of 3.2 million euros. Although there was a decrease in the number of overnight stays and a negative variation rate of −24% (43,890 overnight stays in 2019 to 33,240 overnight stays in 2020), we consider that the impact was not so huge, comparing in terms of regions (−41.0% in the northern region) or in terms national (−47.2% at national level). In addition, the occupancy rate of residents compensated for the overnight stays of non-residents.

According to VianaFestas (2021a), regarding the 2019 data, Viana do Castelo and the Our Lady of Agony festival are mainly visited by 6 nationalities: the Portuguese, national market, representing 45% of the total visitors, followed by the Spanish, with 12.4% of the total demand; the German market, with 7.43%; the French market, with 6.12%; and the United Kingdom, with 4.4%, being also significant the tourist demand from the American continent, which represents 4.3%, mainly coming from Brazil (Table 5).

Table 5 Main nationalities visiting Viana do Castelo at the time of the Our Lady of Agony festival (2019)

	August 2019 (%)
Portugal	45.0
Spain	12.4
Germany	7.4
France	6.1
United Kingdom	4.4
Brazil	4.3

Source Self-elaboration based on National Institute Statistics adapted by the author (2021)

It should be noted that the average stays of the different issuing markets have a different behavior. Viana do Castelo has an annual average of 1.9 nights per tourist, among the 2.92 nights of the German market, the one that stays longer in average terms, followed by the British market with 2.72 nights and the French, with 2.51 nights, being the national market (1.53 nights) and the Spanish market (1.81 nights) the lowest average (VianaFestas, 2021a).

Making an analysis only to the month of August, the totality of the main issuing markets, comparing the results with August 2019, maintained expressive decreases in that month in 2020, higher than 50%, namely United Kingdom (−79.7%), Germany (−59.60%), Spain (−57.30%) and France (−60.50%) (Table 6).

Despite the data being at national level, we consider that Viana do Castelo and the festival were very penalized in terms of external visitation, considering that the main nationalities visiting the city are in the first five in the Table 6.

Regarding the Our Lady of Agony festival, in the pre-pandemic period, these take place between 4 and 5 days in August (which may vary depending on the year, depending on the day of the week that falls on August 20, municipal holiday), the most significant month in terms of tourist demand, representing a total demand of 1 million visitors. The day of greatest demand is always the day of the historical-ethnographic parade, on Saturday, with an average demand of 40% on that day, followed by the municipal holiday, on August 20, a day not only with the Solemnities

Table 6 Change in overnight stays at tourist accommodation establishments by country of residence (national level)

Main source markets	Year-on-year rate of change August 2020 (%)
1. United Kingdom	−79.70
2. Germany	−59.60
3. Spain	−57.30
4. France	−60.50
5. Brazil	−82.90
6. USA	−93.20
7. Netherlands	−71.20
8. Ireland	−92.00
9. Italy	−74.90
10. Belgium	−67.50

Source Self-elaboration based on National Institute Statistics adapted by the author (2020)

of Our Lady of Agony, but also the day of the sea procession, representing an average presence of 25 to 30% of total visitors to the festivities (VianaFestas, 2021a).

The program aims in tourist terms to maintain the presence and fix in Viana do Castelo visitors for all days of the festival, with several significant moments between the various days of the festival, in one of the most genuine moments of affirmation of the history, culture and ethnography, where the "Vianesa" costume, folklore, brightened by music and fireworks are also a tourist offer of international dimension, already recognized of Interest for Tourism.

In the year 2020, the Our Lady of Agony festival had to adapt to the reality lived. Everything was thought differently, from the poster of the festival to the online publications. This is possible to verify immediately by the choice of the winning poster promoting the Our Lady of Agony festival of 2020.

While in 2019, the poster showed the strong devotion that the costumed woman has for Our Lady of Agony, in 2020 the same devotion is noted, with the costumed woman holding a piece of gold, but with the image of the picture of Our Lady on the wall, with the woman in front of the computer, at home, warning that this year the attractions would be mostly digital, using social networks or short videos. In 2021, a year of some openness, it was possible to try to return to "normal," following the rules imposed due to the pandemic (Fig. 3).

In 2020, with an almost entirely online program evoking the memory of the festival, the celebrations in honor of Our Lady of Agony took place from 19 to 23 August. Due to the COVID-19 pandemic, the festival organization prepared a mainly online program. The main obstacle to the organization of the festival was the question of gatherings since it was necessary to prevent sources of contagion. In this way, the five days of the Our Lady of Agony festival were celebrated through social networks, radio and television. Among the initiatives proposed, there were several videos created for this purpose portraying the most emblematic moments of the festival, as well as the online musical shows, starring national musical ensembles, on 20, 21 and 22 August, respectively (Fig. 4).

Thus, thematic videos were presented at the usual time and day of the program of the festival in previous years with unpublished content and testimonies about the Mordomia Parade, the "cabeçudos" and "Gigantones," the historical-ethnographic parade, the salt carpets of Ribeira, the sea procession, the Solemn procession and the fireworks. An event that the organization created and that was also highlighted in this online content were the faces and stories of 40 people who, year after year, inside and outside, give body and life to the festivals.

Only on the 20th, dedicated to the patron saint, it was celebrated in person, at the sanctuary of Our Lady of Agony,

Year 2019 Year 2020 Year 2021

Fig. 3 Posters of the Our Lady of Agony festival (pre-pandemic period and period during the pandemic. *Source* VianaFestas (2021a)

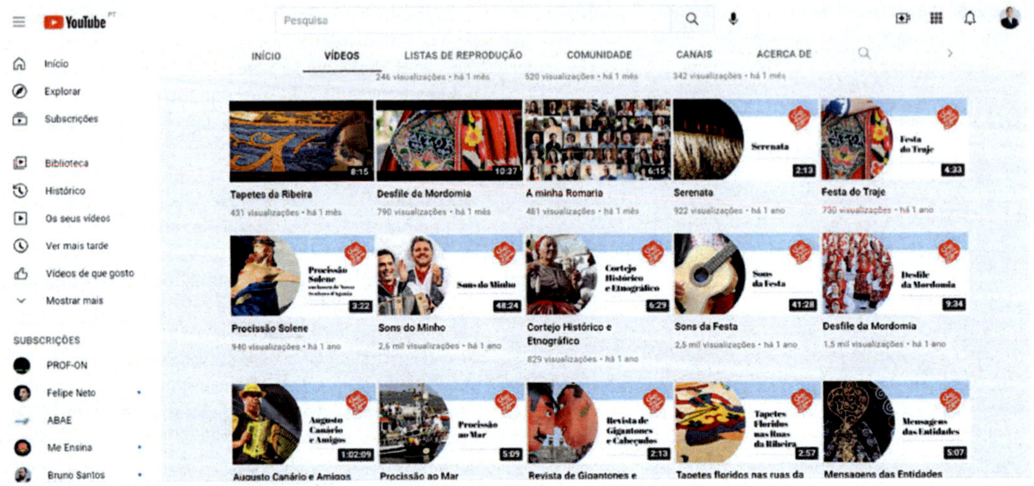

Fig. 4 Videos created for the Our Lady of Agony festival 2020. *Source* Youtube Channel VianaFestas (2021b)

with the limitations determined in relation to liturgical celebrations. The event was broadcast live on the social networks.

In addition, the organization created a street exhibition entitled "Feel the Agony festival," with several images scattered through various streets of the city, depicting the various moments of the festival. In these images, there were Qr codes that allowed viewing multimedia content about the various moments of the festival (Fig. 5).

In 2020, the content available on the social networks of the festival was accessed by 400 thousand people, a number that is added to the success of the new version of the eternal "Havemos de ir a Viana," from the poem by Pedro Homem de Mello and made famous by Amália Rodrigues. This new

Fig. 5 Map of the places where was the street and virtual exhibition. *Source* VianaFestas (2021a)

version, joined by about 50 local artists, from fado to hip-hop, was launched on 11 August 11 2020, having been viewed practically 690 thousand times during the days of the festival, a number that increased over 20% during the 2021 festival—in addition to thousands of shares—thus continuing to assume itself as a new reference. Already in 2021, still strongly conditioned by the pandemic, the "online" contents were accessed by practically 550 thousand people, according to data provided by the social networks Facebook and Instagram, with the reach having increased more than 35% compared to 2020.

According to data provided by the social network Facebook, on 19 August 2020, the first day of the festival, 97,310 users accessed the content of the festival, including daily premieres, a number that grew to 189,500 the following day. On Saturday, there were 129,010 users accessing that content and on Sunday, the closing day of the festival, more than 91,300.

We may refer that the "main stage" of the festival in 2020 ceased to be the physical territory and became Instagram, Facebook and YouTube, centered on the brand "We are all Agony festival." The program was made available on the organization's Web site and through social networks with the main numbers of the festival, information and memories of a festival with about two and a half centuries of history.

5 Conclusions

Religious tourism has functioned as a promoter of social and cultural interaction and enhancer of social and personal development, as well as stimulating the territory in economic terms. In this pandemic period, religious festivities must adapt and try to perform as much as possible within the sanitary conditions allowed. This was the case of the Our Lady of Agony festival.

Traditionally, the festivity takes place between 4 and 5 days in August, the most significant month in terms of tourist demand, representing a total demand of 1 million visitors, among residents, neighboring municipalities of Minho and other Portuguese regions, but also a massive presence of Spanish, German, French, British and Brazilian (VianaFestas, 2021a). The sea procession and the streets of the Ribeira—adorned with flowery carpets of salt—are testimonies of the deep religious devotion that gave rise to the Our Lady of Agony festival. Ethnography has its place in the Mordomia Parade, with hundreds of women parading their costumes with "chieira" (pride, vanity), and also in the historical-ethnographic parade and in the unique Costume Festival.

Considering the initial research question, it was possible to realize that the Our Lady of Agony festival sought to adapt to the new reality. While, in the pre-pandemic period, the festival of the city of Viana do Castelo was visited by about 1 million visitors, in 2020 the main stage of the festivities stopped being the physical territory and started being the social networks.

In terms of activities held in person, it was limited to the part of religiosity, the outdoor mass and the street exhibition entitled "Feel the Agony festival." In terms of occupancy rate, there was a decrease in the number of overnight stays in the city of Viana do Castelo, especially of non-residents. However, the data reveal that despite the number of overnight stays in the month of August 2020, having a negative variation of −24% was better compared to the national results of −47.2%. In addition, the domestic market helped to offset the sharp decrease in the number of overnight stays of non-residents.

While live video streaming of virtual and augmented reality may suffice as a temporary substitute for performing religious rituals or visiting sacred sites during the outbreak of COVID-19, there is ultimately no substitute for "being there" (Mitchell, 2002). Pilgrimage and religious tourism are too economically important to limit religious travel for long. Many places around the world rely heavily on religious travel for their economic viability. Many religious "sites serve as anchor attractions that bring considerable fiscal dividends to their communities through tourism" (Olsen & Timothy, 2020, p. 176).

As a suggestion for future research, it would be interesting to understand with this wealth of information that was created exclusively for 2020, how it will be used and/or monetized in the coming years, considering that in 2021 there have already been some "openings," a sign of a possible return to normality.

References

Antunes, J., Novais, C. B., & Dias, H. (2016). A importância do turismo religioso no desenvolvimento das regiões: O caso do santuário da nossa senhora da lapa. *International Journal of Scientific Management and Tourism, 2*(1), 273–285.

Burns, P. (2003). *An introduction to tourism and anthropology*. Routledge.

Jackowski, A. (2002). Le Système Mondial des Centres de Pelegrinages. *Il Santuário, Spazio per UnAccoglienza Fraterna e Universale. Città Del Vaticano: Pontifocio Consiglio Della Pastorale per i Migranto Egli Itiniranti e Santuário Nostra Signora Di Montserrat*, 115–126.

Martins, C. A., & Morais, A. (2016). O contributo do turismo para o desenvolvimento das regiões. *Gaia Scientia Edição Especial Europa, 10*, 84–91.

Martins, H. (2022). Os impactos económicos da Covid-19 em eventos. *Revista Turismo & Desenvolvimento, 38*, 265–280. https://doi.org/10.34624/rtd.v38i0.25863

Martins, H., Carvalho, P., & Almeida, N. (2021a). Destination brand experience: A study case in touristic context of the peneda-gerês national park. *Sustainability, 13*(21). https://doi.org/10.3390/su132111569

Martins, H., Silva, C., Pinheiro, A., & Gonçalves, E. (2021b). A importância da marca no turismo: o caso da entidade regional Turismo do Porto e Norte de Portugal. *PASOS Revista de Turismo y Patrimonio Cultural*, *19*(4 SE-), 753–762. https://doi.org/10.25145/j.pasos.2021.19.049

Mitchell, H. (2002). 'Being There': British Mormons and the History Trail. In: S. Coleman and J. Eade (eds) Reframing Pilgrimage. London and New York: Routledge, pp. 36–55.

Mróz, F. (2021). The impact of COVID-19 on pilgrimages and religious tourism in Europe during the first six months of the pandemic. *Journal of Religion and Health, 60*(2), 625–645. https://doi.org/10.1007/s10943-021-01201-0

National Institute Statistics. (2020). *Atividade Turística—Agosto de 2020.*

National Institute Statistics. (2021). *Dormidas (N.º) nos estabelecimentos de alojamento turístico por Localização geográfica*.

Nolan, M. L., & Nolan, S. (1992). Religious sites as tourism attractions in Europe. *Annals of Tourism Research, 19*(1), 68–78. https://doi.org/10.1016/0160-7383(92)90107-Z

Olsen, D. H., & Timothy, D. J. (2020). The COVID-19 pandemic and religious travel: Present and future trends. *International Journal of Religious Tourism and Pilgrimage, 8*(7), 170–188.

Pizam, A. (1987). Planning a tourism research investigation. *Planning a Tourism Research Investigation*, 61–75.

Rinschede, G. (1992). Forms of religious tourism. *Annals of Tourism Research, 19*(1), 51–67.

Silva, J. L. F. da. (2012). O Turismo Religioso no Noroeste de Portugal: as atividades económicas dos principais santuários na sua relação com o território envolvente. *Revista Turismo & Desenvolvimento, 3*(17/18). https://doi.org/10.34624/rtd.v3i17/18.13173

Sousa, A., & Pinheiro, C. (2014). O papel do turismo religioso nos territórios-caso de Lamego (Portugal). *Turydes, 7*(17).

Tuckman, B. (2000). Manual de Investigação em Educação Lisboa, Fundação Calouste Gulbenkian. *This Work Is Licensed under a Attribution-NonCommercial-NoDerivatives, 4*.

Turismo Porto e Norte de Portugal. (2015). *Festas e Romarias Norte de Portugal* (1st ed.). http://www.portoenorte.pt/fotos/guias/festas_e_romarias_norte_de_portugal_18697390295cf7e1686240e.pdf

VianaFestas. (2021a). *Our Lady of Agony Festival*. https://www.vianafestas.com/pt/eventos-e-romarias/romaria-sra-da-agonia

VianaFestas. (2021b). *Youtube Channel—VianaFestas*. https://www.youtube.com/channel/UCelPOSJqCv3JBhCSYhqTX7g

Visitors' Perception of Tourist Attractions in a Green Protected Area: The Case Study of the Peneda-Gerês National Park

Hugo Martins and António José Pinheiro

Abstract

Experiences can be subjective and internal responses, characterized by sensations, feelings and cognitions, but also behavioral responses evoked by stimuli. In the tourism context, the destinations promoters aim that visitors enjoy experiences considered pleasant and, ideally, memorable, because, in this way, increase the probability of revisit and recommendation. The protected green areas have gained a greater projection mainly from the beginning of the twentieth century. They are privileged places both in terms of tourist enjoyment and the attention of academic research. This interest is considerably increased taking into account the global health situation that makes this type of areas as more recommendable for tourist practices. The territory chosen in this case study is Peneda-Gerês, the only national park in Portugal that enjoys a certain international reputation. It is the only park in the Iberian Peninsula that integrates the Pan-Parks network, a network of excellence where only the best parks in Europe are included. The research objective was to analyze the experience and perceptions in terms of satisfaction, recommendation and revisit. In methodological terms, a netnographic research was decided considering the comments and evaluations made by visitors of this green area in TripAdvisor, a famous social platform for travelers. Twenty-two tourist attractions were identified, and all their comments were considered ($n = 834$) in a 3-year period. Firstly, it was possible to draw a basic profile of the visitor and to conclude by their high satisfaction resulting from the transmitted evaluations. A content analysis was also performed to categorize visitor perceptions into four major categories: sensory, emotional, behavioral and intellectual, which were further subdivided according to the types of references extracted from the visitors' comments. Satisfaction and recommendation were other perceptions that could be found in the comments. Through the results, it was possible to observe that the visitors' perception focuses mainly on a more sensory and affective component.

Keywords

Perceptions • Tourist experience • Tourism in green areas • Peneda-Gerês National Park

1 Introduction

Any visitor wishing to visit a tourist destination in general or a tourist attraction in particular is faced with the dilemma of only being able to assess the quality of the experience after visited or experienced it at least once. Before Web 2.0, this situation was minimized by the opinion of family or friends, through "word of mouth". As a result of the evolution of the Web, with the search and sharing of information, tourism experiences were being shaped to new realities (Park & Nicolau, 2015). Web 2.0 and its applications have brought a greater interaction among internet users, enabling the dissemination of information quickly, at the distance of a click. The possibility of visitors able to freely comment on their consumer experiences (with those comments being instantly available) was one of the effects of globalization. Some platforms such as Booking, Expedia, Airbnb or even social networks allow people to evaluate through a scoring system, as well as give written feedback about that experience. This is one of the reasons why these sites have become very popular (Filieri, 2015).

H. Martins (✉) · A. J. Pinheiro
University of Maia, Maia, Portugal
e-mail: hugomartins@umaia.pt

A. J. Pinheiro
e-mail: ajpinheiro@umaia.pt

H. Martins · A. J. Pinheiro
Centre of Studies in Geography and Spatial Planning (CEGOT), Coimbra, Portugal

A. Mandić et al. (eds.), *Cultural Sustainable Tourism*, Advances in Science, Technology & Innovation,
https://doi.org/10.1007/978-3-031-10800-6_5

With the use of the internet, platforms such as TripAdvisor have emerged, giving focus to tourist destinations and their attractions. TripAdvisor allows, for example, visitors to post comments and make assessments about the experiences they had in the tourist attractions of a tourist destination. This phenomenon allows internet users to have access to such information, which may become a determining factor in the process of revisiting and recommending to family, friends and others, in general.

Online comments have become very important, from the point of view of service providing companies as well as for customers, they are considered "the best and most reliable sources of information, influencing in a determinant way the consumers' purchase process" (Amaral et al., 2015, p. 48). Currently, companies, whether in the tourism industry or not, also give importance to these ratings and comments, encouraging their customers to make them, because it is a way to monitor the quality of services to improve and build even more customer loyalty.

In terms of literature, there is already some research (Amaral et al., 2015; Arruda et al., 2020; Filieri, 2015; Hu & Chen, 2016; Ye et al., 2009) that tries to understand and study these comments and ratings, focusing on the determinants and motivations of the evaluations made by tourists and how they can be useful to others. However, these studies have been substantially based on quantitative information, such as rankings and number of comments on user-originated content management websites. However, tourists' evaluations go beyond the assignment of a ranking, as they have unstructured comments, whose contents have not been the object of analysis. This work aims to make a more holistic approach, analyzing the comments, understanding the quantitative part, but mainly the aspect of the unstructured content. This work aims to make a more holistic approach, analyzing the comments, understanding the quantitative part, but especially the unstructured content aspect.

Nature-based tourism is a niche with growing popularity in tourism markets (Holden, 2016), which corresponds to an increased research interest (Vespestad & Lindberg, 2011). The experience in nature by tourists is mainly based on the esthetic qualities of the landscape and the wild environment, i.e., places with few human intervention (Holden, 2016), which are very stimulating aspects for the senses. As such, they have a high impact on the emotions felt by tourists and whose perception should be analyzed for a better understanding of the experience in natural areas.

Therefore, we chose to analyze the perceptions that tourists had on the tourist attractions of the Peneda-Gerês National Park (PGNP), present in the TripAdvisor platform. PGNP was chosen for being the only national park in Portugal. Natural areas with protection status are more attractive, with the name "national park" having the greatest effect on attracting tourists (Reinius & Fredman, 2007). On TripAdvisor platform, 22 tourist attractions were identified, 13 of which have a more cultural aspect and 9 a more natural one. We analyzed all the comments of the attractions of the PGNP in the last 3 years, from July 2018 to June 2021. The PGNP, the only national park in Portugal, is a reference associated with nature tourism. It has a vast and rich heritage, not only biotic natural heritage, but also historical, cultural and religious heritage, with the Gerês brand having its own identity that has been acquired over the last decades (Martins, 2020; Martins et al., 2021).

Thus, it was sought to characterize the experiences that tourists had in these tourist attractions. In addition, we sought to analyze in these comments the perceptions in terms of satisfaction, recommendation and revisit. Therefore, our study had as main constructs the experience (comprising 4 dimensions according to some studies), the satisfaction, the recommendation and the revisit. We tried to find these perceptions in the comments that the PGNP visitors left in their TripAdvisor comments.

2 Theoretical Framework

Web 2.0 has led to greater interaction between users and a greater sharing of information and ideas. Additionally, the word-of-mouth has evolved: It has become electronic and is called electronic-word-of-mouth (e-WOMS). The digital word-of-mouth has become so popular that it has become an important factor in the sale of products and services (You & Sikora, 2014), fact by which companies give extreme importance to content in digital format, promoting their products with influencers. Therefore, e-WOM is the new way to influence behavioral intentions. With digital came the possibility of creating communities and groups of belonging on certain tastes or profiles on a global scale (Litvin et al., 2008).

In the business sector, this sharing of information and evaluation makes it possible to raise awareness of the products and services provided by companies, as well as to influence potential customers to purchase these products and services. In this sharing of information published on digital platforms, the comments made through user-generated content (UGC) are in a format of short free texts, where there is the possibility of assigning a rating on a pre-defined scale and sometimes the possibility of sharing photographs and/or videos as a way to validate these comments.

According to the literature, the quality of the information appears as a predictor of purchase intentions (Filieri, 2015; Lee et al., 2008; Park et al., 2007). There is a set of factors that internet users pay attention when read comments to make a decision, namely: (a) updated information (usually the most recent evaluations are the ones that appear first),

(b) understandable information, (c) relevant information, (d) accurate information and (e) valuable information (beneficial for the reader, providing indications on the choice of the product, both positive and negative). In short, the information should be as complete as possible (Cheung et al., 2009; Filieri & McLeay, 2014; Park et al., 2007). As a rule, there are two main ways of analyzing comments: (1) based on the analysis of the ranking seen in the scaled evaluation carried out by internet users (this can be broken down into three types: negative, neutral or positive, the latter being the easiest to analyze and process) (Sayfuddin & Chen, 2021); (2) understand the written part of the commentary which can be analyzed according to relevance, value of information, accuracy of information and its update/the level of trust the information offers (Taecharungroj, 2022). This second way form is much more difficult to analyze suggesting an interpretation of the comments and the assessment of their quality.

The content platforms emerged in the tourism sector. Visitors can share information about a product, service or tourist experience, assigning a quantitative assessment, giving written feedbacks that can be positive, negative or even neutral. On these platforms, the traveler "evaluates, creates and shares content related to various aspects of their vacation, mainly accommodation, restaurants, entertainment, products and tour operators" (Amaral et al., 2015, p. 50). The entities that manage these tourism resources have an updated feedback about the viability of the service and the need to change any shortcomings during the provision of services (Litvin et al., 2008). Usually, the users who make the comments refer the type of experience, allowing those who read them to perceive an image about that tourism offer. This sharing of information "gives potential customers access to updated information about the tourism offer, but also information about the experience, creating a base of initial expectations and consequently allowing them to make a more informed decision" (Amaral et al., 2015, p. 49). These digital platforms have gained a tremendous positive reputation so that the management entities linked to tourism stimulates the creation of comments about the products as also use the same comments in order to promote the products. According to the literature, the comments left by visitors are a way to influence the decision-making process of all stages of planning to a tourist destination, encompassing the before, during and after the visit (Park & Nicolau, 2015).

Some examples of such internet-generated content management platforms in the tourism sector are Airbnb, Booking and TripAdvisor, which have achieved worldwide projection, with a large number of followers and participants who actively contribute. Many potential visitors consult the comments and the votes/valuations of a given tourist destination before making the final decision and preparing the trip. In the specific case of TripAdvisor, the platform allows internet users to make comments and evaluations on various aspects of the tourism offer, including flights, restaurants and attractions of tourist destinations.

There are several studies that investigate TripAdvisor comments, especially in the restaurant/gastronomy field. The potential visitors/customers seek information about the taste of the food and the type of service provided. In the restaurant field, the literature points out that a satisfied customer produces a positive comment to five friends or family members. However, if the customer is dissatisfied by posting a negative comment, it spreads very quickly and can become viral if digital media is used (Gopaul, 2014). Positive comments are considered a predictor of customer loyalty as well as recommendation of restaurants to others. This perception is generated in comparison with the image that the customer had initially projected, according to their expectations. If the performance found in the service and the quality of the food meets expectations, then the customer will project in the comment a neutral evaluation. If it exceeds the expectations, the evaluation will be positive and there will be the propensity to return and to recommend, otherwise, if it frustrates the customer expectations, the comment will have an evaluation in that sense (Chemuturi, 2007) and consequently the reputation of the restaurant will be in question. The same happens in the context of tourist destinations in general and their attractions in specific (Ali et al., 2021).

2.1 The Experience in Tourism

The term experience can have several meanings (Martins, 2018). The most usual definition corresponds to the act or effect of experiencing, being a knowledge acquired by practice, through the sensations of the senses. According to Schmitt (2011), the various definitions can be grouped into two categories: (a) definitions that refer to experience as knowledge accumulated over time; and (b) definitions that consider experience as ongoing perceptions and feelings and direct observation.

The concept of experience can acquire quite different characteristics according to each science, depending on the object of study and the context, as is the case of marketing and tourism: the client experience (Gentile et al., 2007), the consumer experience (Tsai, 2012), the service experience (Hui & Bateson, 1991), the product experience (Hoch, 2002), the consumption experience (Chaney et al., 2018), the purchase experience (Kerin et al., 1992) and the brand experience (Brakus et al., 2009; Zarantonello & Schmitt, 2010). Experiences are not mere evaluative judgments about the product or brand (e.g., "I like this product" and "I like this brand"); experiences include specific sensations, feelings, cognitions and behavioral responses triggered by specific stimuli in the consumer (Schmitt, 2011).

The concept of experience also differs from motivational and affective concepts, such as involvement, brand attachment or customer satisfaction. Involvement is based on the needs, values and interests that motivate a consumer toward an object, e.g., a brand (Jeon & Yoo, 2021; Schmitt, 1999). In the case of experience, it does not presuppose a motivational state. The experience can happen when consumers show no interest, or even if it has a personal connection with the brand (Barnes et al., 2014). Moreover, brands, with which consumers are highly involved, may not necessarily be brands that evoke more intense and memorable experiences (Şahin et al., 2011).

Brakus et al. (2009), within the scope of the study of consumer experiences, after extensive analysis of the scientific literature from various areas, identify the dimensions of the experience associated with the brand. They created a scale whose objective would be to measure in a more holistic way and to what extent a consumer has a sensory, affective, behavioral and intellectual experience. The authors conducted several exploratory qualitative studies to determine whether the conceptualization of brand experience would be in line with the representations of consumers who experienced the product and/or service. The authors created a scale to measure the experience associated with the brand, based on a set of four dimensions of experience: (a) sensory dimension that refers to the way of appealing to the five human senses, through sight, touch, sound, taste and smell; (b) affective dimension that manifests itself through the feelings and/or thoughts of consumers, with the aim of creating affective experiences that vary their intensity; (c) behavioral dimension that seeks, through body experiences, lifestyles and interactions, to enrich the consumers' lives, showing the consumer other ways of getting alternative lifestyles and different interactions; (d) intellectual dimension that appeals to the creativity and innovation of the consumer, possessing the ability to create cognitive experiences. These dimensions can be evoked nominally or in group. In this sense, a company that wants to provide a good experience to its consumers must first formulate a mental module concerning the areas that can affect the consumer's senses. In fact, there are companies whose primary objective is to provide experiences, for example, Odisseias and Cool Gift, currently the two main brands in the market of experiences in voucher form. Another example of companies that sell experiences is Starbucks, which sells not only coffee, but an experience around the consumption of the coffee itself (Marques, 2012). This scale has been tested in tourism by Barnes et al. (2014) and Martins (2018).

Over time, the lasting experiences, stored in the consumer's memory, may affect consumer satisfaction as well as their loyalty (Morais et al., 2004). According to Brakus et al. (2009), the experiences can also occur indirectly, for example, through advertising, marketing communications or websites; they can also occur in an unexpected way, not assuming that there is a motivational state, because it can happen when consumers show no interest.

Within tourism, one of the important studies done on experience was that of Barnes et al. (2014). The authors (2014) sought to determine which dimensions of the experience had the greatest impact on the visitor, using an experience model and a variety of destinations. They conclude that visitors are primarily driven by sensory experiences, which suggest an underlying desire to satisfy hedonic needs. However, despite the prominence of sensory experiences, the study reveals that affective experiences are very important in certain circumstances, and "travel agents and tourism providers should focus more on sensory aspects of visits and to design tourism experiences from a sensory and affective perspective" (Barnes et al., 2014, p. 137). They also considered that behavioral and intellectual experiences seem to be more difficult to achieve in tourism.

The conclusion that the authors draw from this information, like Arnould and Price (1993), is that the entities involved in the tourism sector should give special focus on sensory experiences highlighting the touch, images, sounds, tastes and smells, and entities related to tourism should seek mechanisms that provide hedonistic tourist experiences. This implies that "the design of tourist experiences should provide scope for individuals to learn and to be challenged, and to develop new, social perspectives on life" (Barnes et al., 2014, p. 137). The authors suggest that the profile of the destination experience (brand) is likely to vary according to the specific destination, and the specific experiences that characterize a destination's tourism offer should be carefully selected. In conclusion, they considered that experience is a significant determinant of attracting tourists to that destination and that satisfaction plays a key role in the further processing of tourist experiences.

2.2 Nature-Based Tourism Experiences

Natural areas, with or without protection status, are territories of important and growing tourist demand. The users and the activities in these places are varied and depend both on aspects intrinsic to the visitors and on the biological, morphological, landscape and functional aspects of the natural areas. As a starting point, it can be established that nature-based tourism encompasses human activities carried out within the scope of a visit to a natural area outside the person's ordinary neighborhood (Fredman et al., 2012). Although the definition of nature-based tourism is broad, it is possible to find some consensus around certain constituent elements, such as learning, leisure and adventure in natural settings (Tangeland & Aas, 2011). In operational terms, leisure activities in protected areas can be enjoyed free of charge or as products promoted by operators and entertainment companies (Tangeland & Aas, 2011; Vespestad & Lindberg, 2011).

Generally, it is considered that visitors to natural areas look for special places, with esthetic beauty and possessing characteristics and qualities that are opposite to urban territories, where many tourists come from (Holden, 2016). Indeed, natural areas provide experiences of beauty and serenity, as opposed to an increasingly urban and commercial world (Schroeder, 2002). From a leisure perspective, activities in nature are characterized by considering the natural world as an object and subject of experience (Ried Luci et al., 2018). Vespestad and Lindberg (2011) consider that experiences in nature can be observed from four perspectives: (i) search for authenticity or the "return to nature", (ii) search for fun, (iii) search for sensations to achieve psychological or physical goals and (iv) enjoyment of nature as symbolic meaning of culture.

In a parallel approach, Schroeder (1996) in a reflection on the ways in which people experience natural spaces states that they can be emotional, imaginative and inspirational experiences. This author emphasizes that an experience has the capacity to arouse emotions and motivate actions. In this way, he considers that the value that the tourist attributes to the experience and the emotion is inseparable concepts. Therefore, the stronger the emotions, the greater the value of the experience for the tourist.

Knowing that the esthetic aspects and the authentically natural or little humanized character of the landscapes are crucial for nature-based tourism (Holden, 2016), sustainable management of this type of territories is particularly sensitive. In fact, the tourist use of natural areas occurs in a relationship of forces that can be contrary. If, on the one hand, visitors value a reduced human intervention, the excess of tourist attraction undermines this value requirement. Therefore, it is extremely challenging for managers of natural areas to encourage the appreciation and learning of natural and cultural heritage while seeking to manage and safeguard these resources in the long term, minimizing negative impacts and optimizing positive impacts in social, cultural, ecological and economical levels (Eagles et al., 2001). Understanding the experience gained by visitors will be essential to obtain the best knowledge that supports decision-making in sustainable management of natural areas.

Some studies on the experience and expression of perceptions by visitors of protected areas can be found in the last 25 years. Schroeder (1996) sought to identify the main attributes recognized by visitors in natural areas, having fundamentally detected positive feelings, with beauty being the most mentioned attribute, along with feelings of serenity and peace and the characteristics that nature allow in terms of isolation sensation and contact with wilderness. Air quality, water purity and the local population are also recognized as positive attributes. In another study, Schroeder (2002) compiled the results of 6 surveys based on open-ended surveys on the perception of natural areas, in a time frame of about 15 years. The most valued environmental characteristics were identified, with emphasis on aquatic elements, geological phenomena, viewpoints to the landscape and atmospheric aspects such as light and temperature. Vegetation and aspects of human intervention and occupation are also detected, for example, paths, roads or recreational spaces. Several meanings, values and experiences were also found, the main ones being: the natural character (little or no human intervention), beauty, amazement, serenity, excitement and a sense of refuge and isolation. Conti and Lexhagen (2020), in the analysis of the experience, through publications on the social network Instagram, detected that the appreciation of the experience in natural areas, by the visitors, and mainly based on esthetic aspects coexisting with other values: hedonic (multisensory involvement and fun), relational (experience sharing), emotional and knowledge. A similar study in terms of purposes and methodology is that of Abrahams et al. (2022), considering glaciers and using the analysis of user-generated content on TripAdvisor. Most of the perceptions collected referred to aspects such as satisfaction with the visit, accessibility and weather conditions, concluding that most tourists consider more superficial issues of tourist appeal and leisure rather than environmental and climate concerns. The appreciation of the esthetic component of landscapes is also identified in the study by Zhang and Xu (2020) who found that this element is essential to create the desire to revisit. Loyalty to a natural area is also assessed through a sentiment analysis using TripAdvisor reviews for a park in Canada, in order to understand experiences at various park locations and the effects on revisiting the park (Mirzaalian & Halpenny, 2021).

2.3 PGNP as a Tourist Destination

As a tourist destination, protected areas seek to convey rewarding experiences to visitors. These experiences and impacts also affect local communities at many levels (economic, social and cultural) as well as environmental risks (Martins, 2020). In Portugal, tourism in protected areas has become a national focus, especially since the twenty-first century: first with the National Strategic Tourism Plan (PENT) and currently with the Tourism Strategy 2027. Within the axis valuing the territory of the Strategy 2027, the aim is to economically enhance the natural and rural heritage and ensure its conservation, having as priority projects the "development of nature tourism and rural areas through projects of economic enhancement and active management of natural and rural heritage, which includes the national network of protected areas, biosphere reserves and Geoparks recognized by UNESCO, particularly in the context of promoting the brand Natural.PT" (Turismo de Portugal, 2017, p. 55).

Within the protected areas, the PGNP stands out, both in terms of supply and demand, as a benchmark for nature tourism. By association with an enormous botanical diversity, in PGNP, there are a set of natural habitats that support a rich and varied faunal community, with several endemic species, rare or of limited distribution in Portugal, deserving of highlight at national and international level. Likewise, it has a vast heritage of historical and cultural nature, enhancer of attraction and development of tourism activity (ethnography, gastronomy, historic villages, megalithic, medieval castles and pillories, "Espigueiros" of Soajo and Lindoso, among others).

Currently, the PGNP, compared to other protected areas at the international level, possesses a set of factors that enhance this region as a tourist destination of excellence. In fact, it has a permanent technical staff, a management plan in place, good access conditions (e.g., roads and signposting) and accommodation for tourists, among others (Martins, 2020). In fact, in most developing countries, national parks lack the infrastructures that the PGNP already has (Martins, 2018). In this aspect, it was sought from early on, especially from the 1980s, to know, study and classify the PGNP's heritage, material and immaterial, through the inventorying of the archaeological, architectural and ethnographic heritage of the territory, in addition to the natural heritage, considered tourist attractions.

With regard to tourism demand, this territory, in recent years, has recorded a high growth, as a result of the projection and its identity in national and international terms. In addition to its tourist attractions, over the years, the PGNP has obtained some conservation statuses both nationally and internationally, which gives it greater visibility and projection. This protected area forms a group with the Spanish Baja Limia—Serra do Xurés Natural Park, constituting, since 1997, the Gerês-Xurés Transfrontier Park. In 2009, the cross-border park was considered by UNESCO as a World Biosphere Reserve (Transfrontier Biosphere Reserve "Gerês-Xurés"). At national level, regarding the PGNP, it also has the status of Site of Community Importance (Council of Ministers Resolution No. 142/97 of 28 August). Also at national level, the PGNP, in 2010, the International Year of Biodiversity, was considered one of the 7 Natural Wonders of Portugal, in the category of Protected Areas (Martins et al., 2021).

Therefore, we aimed to analyze visitors' perceptions regarding the tourist attractions of the PGNP that were included in the TripAdvisor platform, namely regarding experience, satisfaction, recommendation and revisit.

3 Methodology

The study of the perceptions and the experiences expressed by visitors to the PGNP was carried out using a qualitative approach. The concrete method used fits into the concept of netnography established by Kozinets (1998) as a qualitative method developed to investigate consumer behavior using the internet. Netnography is being consolidated with regard to tourism research (Tavakoli & Wijesinghe, 2019). For this, we used the comments written by visitors on the TripAdvisor social network. This is a highly reputable platform, with two decades of existence, with ratings and comments on tourist attractions and infrastructures, being the largest online community of travelers (Valdivia et al., 2019). The use of user-generated content as a source of information about the perceptions of visitors and users of tourism services has been gaining expression, as evidenced by the analysis conducted by Prayag et al. (2018). This is a methodological option perfectly justified and used since the beginning of this century (Hall & Valentin, 2005), especially in exploratory studies and case studies. The qualitative approach to collect and analyze visitors' perceptions finds validation in authors such as Schroeder (2002) who considers that the study of the value of experiences lacks qualitative methods, as these allow a better understanding of emotions and feelings than the use of closed questions or mere quantification of cases. The tourist experience in natural areas accentuates the richness of meanings and senses, being fundamentally subjective and hence the qualitative option (Ried Luci et al., 2018).

Having the PGNP territory as background, we identified all tourist attractions of the region that were on the TripAdvisor platform. We recognized, within the territory, 22 attractions: 13 of which are cultural attraction and the remaining 9 are natural attractions. In order to obtain a representative sample of visitors, a 3-year period was established for data collection, to obtain (all comments made from 01-07-2018 till 30-06-2021), at the same time, data on experiences before and during the pandemic of COVID-19. A total of 834 comments was collected. Although fewer natural attractions were identified, it is possible to verify (Table 1) that the number of comments is much higher than the cultural attractions. While the natural attractions are the waterfalls (Tahiti with 112 and Arado with 102) and the Pedra Bela Belvedere with 166, the cultural attractions are the Soajo's "Espigueiros" with 72 comments and the Sanctuary of Nossa Senhora da Peneda with 66 comments.

The reading and extraction of information from the comments were performed using the content analysis technique and the codification technique. It followed the concept and method presented by Corbin and Strauss (1990) called open source. With the reading of the comments, differences

Table 1 Tourist attractions identified in TripAdvisor inside the Peneda-Gerês National Park

Tourist attraction name	Attraction type	f	%
Pedra Bela Belvedere	Nature	166	19.9
Tahiti Waterfall	Nature	112	13.4
Waterfall Plow	Nature	102	12.2
Portela do Homem Waterfall	Nature	89	10.7
Soajo "Espigueiros"	Culture	72	8.6
Nossa Senhora da Peneda Sanctuary	Culture	66	7.9
Lindoso "Espigueiros"	Culture	58	7.0
Castro Laboreiro Castle	Culture	50	6.0
Soajo "Espigueiros"	Nature	27	3.2
Pitões das Júnias Waterfall	Nature	19	2.3
Pincães Waterfall	Nature	18	2.2
Fafião Natural Lagoons	Nature	16	1.9
Santa Maria das Júnias Monastery	Culture	11	1.3
Thermal Park	Culture	7	0.8
Cava da Velha Bridge	Culture	6	0.7
Lindoso Castle	Culture	5	0.6
Ermelo Monastery	Culture	3	0.4
Castro Laboreiro Church	Culture	2	0.2
Vilarinho das Furnas Museum	Culture	2	0.2
Soajo Pillory	Culture	1	0.1
Assureira Bridge and Mill	Culture	1	0.1
Cabril Eco Rural	Nature	1	0.1
Total		834	100.0

Source Self elaboration based on TripAdvisor reviews (TripAdvisor, 2021)

and similarities were found, assigning conceptual labels to the various excerpts. The various similar expressions were grouped forming categories and subcategories. The procedures consisted of three major phases, according to the organization established by Bardin (2016). First, a pre-analysis of all comments was performed to select those that presented written comment component and exclude those that by some lapse did not refer to attractions located in the geographic area of study. In a second phase, the data were coded and aggregated in a scheme of categories and subcategories that were being created as the texts under analysis provided information. This coding had as framework constructs the Experience, Recommendation, Satisfaction and Revisit. In the last phase, data processing was carried out using a specialized content analysis software, namely the Dedoose program version 8.3.35 (SocioCultural Research Consultants, 2015).

When analyzing the visitors' comments, we sought to identify records regarding the experience construct, in the 4 dimensions according to Brakus et al. (2009) and Barnes et al. (2014): behavioral, intellectual, affective and sensory. Therefore, by reading the 834 comments, a set of attributes that were associated with these dimensions was identified.

4 Findings

The sample of visitors of the PGNP (n = 834) collected over 3 years can be characterized in general terms (Table 2). The overwhelming majority are visitors of relative proximity with 72.1% coming from Portugal and 18.9% from the European continent, of which 25.9% are from Spain, which borders the Park. From the total number of national visitors, it can also be seen that 44.6% are from the two metropolitan areas of the country (Lisbon and Porto) which may indicate an urban profile in most visitors. In terms of gender, there is a majority of male visitors (50.7%). Regarding the traveling company, it is clear that these are accompanied visits, especially as a couple (33.1%) and family (22.7%). The satisfaction expressed quantitatively is very high, with an average score of 4.39 in a maximum of 5 points in which 88.3% of the visitors stated their visits as very good or excellent. Finally, in relation to the date of the visit, it is perceived that most experiences, at least 74%, occurred in the pre-pandemic COVID-19 period.

The 834 comments analyzed provided the recognition of 4582 references that were grouped into category and

Table 2 Sample characteristics (n = 834)

Gender (%)		Place of residence (%)		Traveling company (%)		TripAdvisor reviews (%)		Year of visit (%)	
Male	50.7	Portugal	72.1	Couple	33.1	Excellent (5)	54.7	2017	1.6
Female	39.7	Europe	18.9	Family	22.7	Very good (4)	33.6	2018	35.5
Unknown	9.6	America	4.0	Friends	11.5	Average (3)	8.9	2019	36.9
		Asia	1.3	Alone	1.4	Poor (2)	1.7	2020	23.0
		Africa	0.4	Business	0.4	Terrible (1)	1.2	2021	3.0
		Oceania	0.1	Unknown	30.9				
		Unknown	3.2						

Source Self elaboration based on TripAdvisor reviews (TripAdvisor, 2021)

subcategory codes. The first grouping of references was made considering the four constructs under analysis: Experience, Recommendation, Satisfaction and Revisit (Table 3). Clearly, the majority of the references are related to perceptions expressed about the experiences had in the PGNP (73.8%).

According to the literature reviewed, the references related to the construct of Experience (Table 4) were distributed by the dimensions Affective (45.9%), Sensory (33.9%), Behavioral (17.4%) and Intellectual (2.8%). The greater weight of the Affective and Sensory dimensions, which together total 79.8% of the references, is in line with the literature (Barnes et al., 2014; Martins, 2018).

It is important to detail the references for each of the dimensions in order to better understand their scope and relative weight within the respective dimension. Starting with the affective dimension (Table 5), it can be seen that most of the references (66.3%) are expressions of emotions felt in relation to the experiences during the visits to the attractions. The vast majority of the emotions expressed consider the admiration (39.6% of the emotions), for instance "a breathtaking view" or "a wonderful landscape" and the esthetics (37.7% of the emotions), with references like "beautifull place" or "nice houses". With some relevance are the references related to tranquility (7.5% of the emotions), for example "I found peace", curiosity (4.2%), such as "an interesting place" and spiritual (3.9%) like "a magical place". Almost all references to emotions were positive, with a residual number of negative emotions such as fear, sadness or aggressiveness. Another category that had several references in the Emotional dimension is the Valuation of the immaterial with 18.0% of the references. This category considers references that value the natural character (40.9%), the historicity of the places (30.8%) and its authenticity (28.3%). Examples of this references are, respectively: "contact with the purest nature", "a village lost

Table 3 Content references distributed by constructs

Category codes	References	%
Experience	3380	73.8
Recommendation	735	16.0
Satisfaction	432	9.4
Revisit	35	0.8
Total	4582	100.0

Source Self elaboration based on TripAdvisor reviews (TripAdvisor, 2021)

Table 4 Experience construct content references distributed by category codes

Experience category codes	References	%
Affective	1551	45.9
Sensory	1147	33.9
Behavioral	589	17.4
Intellectual	93	2.8
Total	3380	100.0

Source Self elaboration based on TripAdvisor reviews (TripAdvisor, 2021)

Table 5 Affective dimension content references distributed by category codes

Category codes	References	%	Ratio references/total (%)
Emotions	1028	100.0	66.3
Admiration	427	39.6	
Esthetics	388	37.7	
Tranquility	77	7.5	
Curiosity	43	4.2	
Spiritual	40	3.9	
Pleasure	20	1.9	
Affectivity	18	1.8	
Adventure	16	1.6	
Fear	14	1.4	
Sadness	3	0.3	
Aggressiveness	2	0.2	
Valuation of the immaterial	279	100.0	18.0
Natural character	114	40.9	
Historicity	86	30.8	
Authenticiy	79	28.3	
Relationship with others	124	100.0	8.0
Overcrowded	78	62.9	
Hospitality	25	20.2	
Appreciation of solitude	21	16.9	
Safety	94	100.0	6.1
Negative	79	84.0	
Positive	15	16.0	
Cost	26	100.0	1.7
Negative	17	65.4	
Positive	9	34.6	
Total	1551	100.0	100.0

Source Self elaboration based on TripAdvisor reviews (TripAdvisor, 2021)

in time" and "unique constructions, typical of Minho". In third place, with 8.0% of the total references of the affective dimension comes the category relationship with others. Within this category, 62.9% of the references are negative perceptions about the presence of a large number of other visitors at the sites, for example "too many people". The weight of this subcategory may be reinforced by the existence of references (16.9%) that appreciate positive aspects of loneliness, such as "you will be alone in the world". There are also 20.2% of references that value the hospitality, i.e., the positive interaction with the local community. The last two subcategories of the Emotional dimension also deserve to be highlighted: Safety (6.1%) and Cost (1.7%). Although with a small number of references, their importance lies in the fact that most of their references are negative perceptions, especially in the poor safety on some waterfall trails and the need to pay to access parking or circulation in the park.

In terms of Sensory dimension (Table 6), two categories are clearly demarcated: Senses (48.1%) and Functionality (37.8%).

In terms of senses, vision collects the most part of the references (77.4% of the senses), which largely refers to generic descriptions of the landscape and details of watercourses and animals. There is also some expression in the sense of touch, which considers references such as the perception of atmospheric or water temperature. In relation to the Functionality category, most references are made in relation to pedestrian access (41.9%) and road access (35.5%) to attractions. It is important to mention that of the total 182 references to Pedestrian access 63.7% are negative and the remaining positive, i.e., emphasis is given to the difficulty of traveling the paths. As far as Road access is concerned, the situation is inverse as of the 154 references, 68.2% are positive. The remaining two subcategories of the Functionality category were also labeled in positive and

Table 6 Sensory dimension content references distributed by category codes

Category codes	References	%	Ratio references/total (%)
Senses	552	100.0	48.1
Vision	427	77.4	
Tact	80	14.5	
Combination	21	3.8	
Hearing	18	3.3	
Taste	4	0.7	
Smell	2	0.4	
Functionality	434	100.0	37.8
Pedestrian access	182	41.9	
Road access	154	35.5	
Tourist signaling and information	64	14.7	
Infrastructure	34	7.8	
State of conservation and cleanliness	94	100.0	8.2
Heritage	70	74.5	
Environmental	24	25.5	
Spatial Perception	56	100.0	4.9
Negative	42	75.0	
Positive	13	23.2	
Variable	1	1.8	
Insertion with nature	11	100.0	1.0
Total	1147	100.0	100.0

Source Self elaboration based on TripAdvisor reviews (TripAdvisor, 2021)

negative aspects. The Tourist signaling and information obtained 64 references, of which 67.2% refer that there is no signaling or information or that when these exist they are insufficient. In terms of infrastructures, 76.5% of the 34 references are positive (Table 6). Still in the Sensory dimension, the state of conservation and cleanliness category is also registered, where most of the 94 references (74.5%) are related to Heritage and the remaining are Environmental. Within each subcategory, it was noted positive and negative references, being that in heritage terms, most consider a positive state of conservation or cleanliness, while in environmental terms, there is a tie between positive and negative perceptions (Table 6). The remaining two categories show low results and refer to Spatial Perception aspects, i.e., perceptions of the size of the spaces and integration with nature, with considerations on the existence of animals at the visiting sites.

The references in the behavioral dimension are exclusively about the activities reported by visitors during their visit to the park (Table 7). The most referred activity is the mere contemplation of the landscape or the specific existing attractions (45.7%), which can be associated with taking photographs (9.3%). Also with expression is the water leisure (19.0%) practiced in the waterfalls, as well as the sun baths (2.4%) and water sports (0.8%) in the same places of visit and even the realization of walks (11.5%) (Table 7).

Finally, regarding the intellectual dimension, the less expressive in terms of total references, it brings together two aspects: expression of knowledge about the attractions (61.3%), for example, "first century example of Roman engineering" and comparison with other places outside the Park (38.7%), for instance "reminds of Yosemite Park in California" (Table 8).

The recommendation construct was operationalized with references expressed by visitors that seek to influence the behavior of individuals who could potentially become visitors of the park (Table 9). From the total of 735 references that can be included in this construct, it was found that more than half (54.4%) are direct incentives for the readers of the comments to visit the park attractions. Concrete examples are: "strongly recommend" or "come meet". Also directed to the same potential visitors, 21% of the references to recommendations on practical aspects were found. These are the cases of "bring suitable shoes" or "leave the car before the belvedere". These are suggestions and indications for those who will visit the site, and although they differ from the direct incentive recommendations, it is assumed in them an indirect incentive for the view. About 22% of the references within the recommendation construct are critical manifestations of aspects considered negative in the sites visited, some of which work as a disincentive to visit, for example "I don't recommend" and others are warnings for

Table 7 Behavioral dimension content references distributed by category codes

Category codes	References	%
Activities	589	100.0
Contemplation	269	45.7
Water leisure	112	19.0
Walk	68	11.5
Photography	55	9.3
Meals	45	7.6
Rest	19	3.2
Sun	14	2.4
Water sports	5	0.8
Itinerary	5	0.3
Total	589	100.0

Source Self elaboration based on TripAdvisor reviews (TripAdvisor, 2021)

Table 8 Intellectual dimension content references distributed by category codes

Category codes	References	%
Knowledges	57	61.3
Comparison	36	38.7
Total	93	100.0

Source Self elaboration based on TripAdvisor reviews (TripAdvisor, 2021)

Table 9 Recommendation construct content references distributed by category codes

Category codes	References	%
Encourage	400	54.4
Criticism	162	22.0
Practical	161	21.0
Environmental	12	1.6
Total	735	100.0

Source Self elaboration based on TripAdvisor reviews (TripAdvisor, 2021)

Table 10 Satisfaction construct content references distributed by category codes

Category codes	References	%	Ratio references/total (%)
Effect	261	100.0	60.4
Reward	202	77.4	
Disappointment	36	13.8	
Surprise	23	8.8	
Qualitative appraisal	171	100.0	39.6
Positive	163	95.3	
Negative	8	4.7	
Total	432	100.0	100.0

Source Self elaboration based on TripAdvisor reviews (TripAdvisor, 2021)

specific conditions of the view that are not necessarily attempts to discourage, for example "not suitable for children".

Regarding the Satisfaction construct, two categories of references were considered: the one that considers the effect and the one that groups references about the qualitative appraisal (Table 10). The effect category, with 60.4% of the references of this construct, brings together the effects felt after visiting a particular tourist attraction. These can be reward, i.e., expectations before the visit were fulfilled (e.g., "It was worth it"), disappointment, i.e., the perceived reality did not exceed expectations (e.g., "It is not worth the visit") and surprise when the visit to the attraction was not planned and surprised the visitor (e.g., "I was not expecting it"). The categories reward and surprise together represent 86.2% of the references of effect that are perceived as indicators of satisfaction, so that the references of negative satisfaction only come from the 13.8% of the disappointment category. The qualitative appraisal category means a complement to the quantitative evaluation expressed by each TripAdvisor user and maintains the positivity already enunciated, because 95.3% of the references are positive, as for example, "This was excellent".

It is also important to mention that the construct Revisit, with less than 1% of the total of all analyzed references (Table 3) considers only the expressed and direct references of the revisit to the place as a promise (e.g., "to come back" or as something that has already happened (e.g., "I repeated the visit"). As it is an open comment in which no direct question was asked about whether the visitor intends to return, it is perceived that this value is residual.

5 Conclusions

Web 2.0 brought new platforms such as TripAdvisor where it became possible to search and share information about travel, hotels, restaurants and even tourist destinations and their attractions. The electronic-word-of-mouth has become part of people's daily lives, with the sharing of experiences and other perceptions, which can decisively influence the decision-making process (Amaral et al., 2015). In order to create visitor loyalty, it is necessary that stakeholders and other entities responsible for tourism destinations take into account these perceptions and experiences of those who visit it. It is therefore essential that visitors have memorable experiences that lead to a high level of satisfaction, so that they can revisit and/or recommend the tourist destination to friends and family (Park & Nicolau, 2015).

Therefore, we carried out this study using the experience construct and its 4 dimensions identified in studies such as Barnes et al. (2014): Sensory, Affective, Intellectual and Behavioral dimensions. In addition, we seek to find other perceptions, namely Satisfaction, Recommendation and Revisit' constructs.

The visitor experience in the PGNP is essentially the enjoyment of a natural space combined with cultural elements illustrating, above all, rural life. The activities carried out are, firstly, contemplation of the landscape and specific attractions, secondly rest and leisure in a water environment and, to a lesser extent, active tourism with hiking. These findings are in line with the conclusions of Holden (2016). The primacy of contemplation activities indicates a more superficial experience, based on esthetic and visual characteristics present in the territory. In fact, the analysis of the references of the affective and sensory dimensions confirm this perception, corroborating the existing literature, namely Abrahams et al. (2022), Barnes et al (2014) and Martins (2018). In the expressed emotions, esthetics comes in second place with 37.7% of the total references, while visual references represent 77.4% of the total senses references.

This way of using the territory for tourism purposes is clearly satisfactory for the vast majority of visitors to the park. In fact, the average score assigned to the attractions is very high (4.39 out of a maximum of 5 points) and the readings of the content analysis in several parameters point to a positivity pole, particularly at three moments: (i) The satisfaction construct is composed by a strong positive qualitative appraisal (95.3% of the qualitative appraisal references) and by an overwhelming majority of references that recognize that pre-trip expectations were met or exceeded; (ii) in the affective dimension, about 98% of the expressed emotions are positively recognized as well as it is attributed, by visitors, a positive expression to intangible aspects such as natural, historical and authentic character; (iii) the recommendation construct presents robust results in which 75.4% of the references are of encouragement or sharing of practical aspects for future visitors, i.e., positive recommendations that are only made by those who, effectively, feel satisfied with their own experience. Satisfaction and desire to revisit have strong associations with the affective dimension (Mirzaalian & Halpenny, 2021; Zhang & Xu, 2020).

Nevertheless, negative perceptions were detected and can be grouped into two areas. On the one hand, the displeasure with the excess of visitors in certain attractions of the park, particularly in the waterfalls to enjoy the water. This concern, although visible over the 3 years of analyzed comments, seems to gain greater expression and relevance in the experiences carried out during the COVID-19 pandemic period, something that is understandable given the personal health fears and the need for social distancing. However, this is a paradoxical issue, as the critical comments regarding the overcrowded situation are accompanied, in many of the cases analyzed, by references of direct encouragement to new visitors, i.e., the same visitors express a desire for solitude and tranquility in tourist attractions but with their favorable comments encourage an increase in demand (Eagles et al., 2001; Schroeder, 1996, 2002). The second domain concerns the negative perception regarding safety and signage of the pedestrian routes, visible in the sensory dimension in terms of negative functionality of the pedestrian accesses and in the critical recommendations about access difficulties. A predominantly recreational vision of the territory is denoted in a relevant part of these negative perceptions, in which the terrain conditions could hypothetically be modified and artificialized to meet the mobility and enjoyment needs of visitors. This perception is consistent with the relatively superficial, esthetic and visual experience that a significant part of visitors have in this territory (Abrahams et al., 2022).

The analyzed perceptions have practical interest for the management of the PGNP, as they allow a deep understanding of the experience performed, described and characterized by its visitors. It proves the high satisfaction with the experiences provided but also makes clear important aspects that should be considered in the management of this green area. The control of tourist flows is a central issue in the management of this type of territories, something that is confirmed in the perceptions analyzed. The creation of infrastructures and better access conditions enable the experience on the one hand, but at the same time, boost tourist flows, increasing the feeling of overcrowding (Tverijonaite et al., 2018). It is the responsibility of the managers be aware when making decisions about a balance between visitor numbers and the environmental effects allowed (Ghazvini et al., 2020).

There is a need to better manage the expectations of potential visitors, providing better information in the process of pre-travel to the park, in order to raise awareness that an attraction fundamentally of nature has inherent limitations that prevent an optimal transformation for tourist enjoyment. For managers, knowing negative feelings makes it possible to identify aspects of the experience where the destination may have failed and, in the most enthusiastic feelings, to identify aspects that allow to stimulate the revisitation and attraction of new visitors (Mirzaalian & Halpenny, 2021). In the extreme, one can talk about a greater challenge that will be to rethink the tourist experiences toward a greater involvement and immersion in nature and culture of the territory.

Finally, it is considered that the methodology developed and applied in this study has great applicability in other protected green areas, as a form of in-depth evaluation of visitors' experiences. However, it could be expanded to question specific aspects that an open platform like TripAdvisor does not allow. For example, some of the users of this platform could be interviewed directly to understand the motivations for visiting this park, as well as obtain elements to draw a more complete profile of the visitor.

References

Abrahams, Z., Hoogendoorn, G., & Fitchett, J. M. (2022). Glacier tourism and tourist reviews: An experiential engagement with the concept of "Last Chance Tourism." *Scandinavian Journal of Hospitality and Tourism, 22*(1), 1–14. https://doi.org/10.1080/15022250.2021.1974545

Ali, T., Marc, B., Omar, B., Soulaimane, K., & Larbi, S. (2021). Exploring destination's negative e-reputation using aspect based sentiment analysis approach: Case of Marrakech destination on TripAdvisor. *Tourism Management Perspectives, 40*, 100892. https://doi.org/10.1016/j.tmp.2021.100892

Amaral, F., Borges Tiago, M., Tiago, F., & Kavoura, N. (2015). TripAdvisor comments: What are they talking about? *Dos Algarves: A Multidisciplinary e-Journal, 26*, 47–67. https://doi.org/10.18089/DAMeJ.26.2.3

Arnould, E. J., & Price, L. L. (1993). River magic: Extraordinary experience and the extended service encounter. *Journal of Consumer Research, 20*(1), 24–45. https://doi.org/10.1086/209331

Arruda, D., Silva, M., & Mariani, M. (2020). Análise da imagem do destino turístico Rio de Janeiro, com base em comentários publicados no TripAdvisor. *Revista De Administração Unimep, 18*(2), 123–144.

Bardin, L. (2016). *Análise de conteúdo (3rd ed.)* (E. 70 (ed.)). https://ia802902.us.archive.org/8/items/bardin-laurence-analise-de-conteudo/bardin-laurence-analise-de-conteudo.pdf

Barnes, S. J., Mattsson, J., & Sørensen, F. (2014). Destination brand experience and visitor behavior: Testing a scale in the tourism context. *Annals of Tourism Research, 48*, 121–139. https://doi.org/10.1016/j.annals.2014.06.002

Brakus, J., Schmitt, B. H., & Zarantonello, L. (2009). Brand experience: What is it? How is it measured? Does it affect loyalty? *Journal of Marketing, 73*(3), 52–68. https://doi.org/10.1509/jmkg.73.3.52

Chaney, D., Lunardo, R., & Mencarelli, R. (2018). Consumption experience: Past, present and future. *Qualitative Market Research: An International Journal.*

Chemuturi, M. (2007). How to measure customer satisfaction. *Chemuturi Consultants.*

Cheung, M. Y., Luo, C., Sia, C. L., & Chen, H. (2009). Credibility of electronic word-of-mouth: Informational and normative determinants of on-line consumer recommendations. *International Journal of Electronic Commerce, 13*(4), 9–38. https://doi.org/10.2753/JEC1086-4415130402

Conti, E., & Lexhagen, M. (2020). Instagramming nature-based tourism experiences: A netnographic study of online photography and value creation. *Tourism Management Perspectives, 34*, 100650. https://doi.org/10.1016/j.tmp.2020.100650

Corbin, J. M., & Strauss, A. (1990). Grounded theory research: Procedures, canons, and evaluative criteria. *Qualitative Sociology, 13*(1), 3–21. https://doi.org/10.1007/BF00988593

Eagles, P. F. J., Bowman, M. E., & Tao, C.-H. (2001). *Guidelines for tourism in parks and protected areas of East Asia* (UK). IUCN.

Filieri, R. (2015). What makes online reviews helpful? A diagnosticity-adoption framework to explain informational and normative influences in e-WOM. *Journal of Business Research, 68*(6), 1261–1270.

Filieri, R., & McLeay, F. (2014). E-WOM and accommodation: An analysis of the factors that influence travelers' adoption of information from online reviews. *Journal of Travel Research, 53*(1), 44–57. https://doi.org/10.1177/0047287513481274

Fredman, P., Wall-Reinius, S., & Grundén, A. (2012). The nature of nature in nature-based tourism. *Scandinavian Journal of Hospitality and Tourism, 12*(4), 289–309. https://doi.org/10.1080/15022250.2012.752893

Gentile, C., Spiller, N., & Noci, G. (2007). How to sustain the customer experience: An overview of experience components that co-create value with the customer. *European Management Journal, 25*(5), 395–410. https://doi.org/10.1016/j.emj.2007.08.005

Ghazvini, S. A. M., Timothy, D. J., & Sarmento, J. (2020). Environmental concerns and attitudes of tourists towards national park uses and services. *Journal of Outdoor Recreation and Tourism, 31*, 100296. https://doi.org/10.1016/j.jort.2020.100296

Gopaul, M. (2014). *Exploring customer satisfaction with the healthier food options available at fast-food outlets in South Africa.* Citeseer.

Hall, C. M., & Valentin, A. (2005). 15 Content analysis. *Tourism Research Methods,* 191.

Hoch, S. J. (2002). Product experience is seductive. *Journal of Consumer Research, 29*(3), 448–454.

Holden, A. (2016). *Environment and tourism (London)*. Routledge.

Hu, Y.-H., & Chen, K. (2016). Predicting hotel review helpfulness: The impact of review visibility, and interaction between hotel stars and review ratings. *International Journal of Information Management, 36*(6, Part A), 929–944. https://doi.org/10.1016/j.ijinfomgt.2016.06.003

Hui, M. K., & Bateson, J. E. G. (1991). Perceived control and the effects of crowding and consumer choice on the service experience. *Journal of Consumer Research, 18*(2), 174–184. https://doi.org/10.1086/209250

Jeon, H. M., & Yoo, S. R. (2021). The relationship between brand experience and consumer-based brand equity in Grocerants. *Service Business, 15*(2), 369–389. https://doi.org/10.1007/s11628-021-00439-8

Kerin, R. A., Jain, A., & Howard, D. J. (1992). Store shopping experience and consumer price-quality-value perceptions. *Journal of Retailing, 68*(4), 376–397.

Kozinets, R. V. (1998). On netnography: Initial reflections on consumer research investigations of cyberculture. *ACR North American Advances*.

Lee, J., Park, D.-H., & Han, I. (2008). The effect of negative online consumer reviews on product attitude: An information processing view. *Electronic Commerce Research and Applications, 7*(3), 341–352. https://doi.org/10.1016/j.elerap.2007.05.004

Litvin, S., Goldsmith, R., & Pan, B. (2008). Electronic word-of-mouth in hospitality and tourism management. *Tourism Management, 29*, 458–468. https://doi.org/10.1016/j.tourman.2007.05.011

Marques, A. (2012). *Marketing Relacional—como transformar a fidelização de clientes numa vantagem competitiva* (Lisbon). Edições Sílabo.

Martins, H. (2018). *O Turismo No Parque Nacional Da Peneda-Gerês: a Experiência Da Marca Do Destino, O Apego Ao Lugar, a Satisfação, Os Comportamentos Pró-Ambientais E As Intenções Comportamentais* [Universidade de Coimbra]. https://hdl.handle.net/10316/79717

Martins, H. (2020). Turismo em áreas protegidas: o caso do Parque Nacional da Peneda do Gerês. In E. C. Gonçalves, H. Martins, & A. J. Pinheiro (Ed.), *Perspetivas sobre hospitalidade, turismo sustentável e desenvolvimento local* (Edições IS).

Martins, H., Carvalho, P., & Almeida, N. (2021). Destination brand experience: A study case in touristic context of the Peneda-Gerês National Park. *Sustainability, 13*(21). https://doi.org/10.3390/su132111569

Martins, H., Silva, C., Pinheiro, A., & Gonçalves, E. (2021). A importância da marca no turismo: O caso da entidade regional Turismo do Porto e Norte de Portugal. *PASOS Revista de Turismo y Patrimonio Cultural, 19*(4 SE-), 753–762. https://doi.org/10.25145/j.pasos.2021.19.049

Mirzaalian, F., & Halpenny, E. (2021). Exploring destination loyalty: Application of social media analytics in a nature-based tourism setting. *Journal of Destination Marketing & Management, 20*, 100598. https://doi.org/10.1016/j.jdmm.2021.100598

Morais, D. B., Dorsch, M. J., & Backman, S. J. (2004). Can tourism providers buy their customers' loyalty? Examining the influence of customer-provider investments on loyalty. *Journal of Travel Research, 42*(3), 235–243. https://doi.org/10.1177/0047287503258832

Park, D. H., Lee, J., & Han, I. (2007). The effect of on-line consumer reviews on consumer purchasing intention: The moderating role of involvement. *International Journal of Electronic Commerce, 11*(4), 125–148. https://doi.org/10.2753/JEC1086-4415110405

Park, S., & Nicolau, J. (2015). Asymmetric effects of online consumer reviews. *Annals of Tourism Research, 50*, 67–83. https://doi.org/10.1016/j.annals.2014.10.007

Prayag, G., Hall, C. M., & Wood, H. (2018). I feel good! Perceptions and emotional responses of bed & breakfast providers in New Zealand toward trip advisor. *Journal of Hospitality Marketing & Management, 27*(1), 1–20. https://doi.org/10.1080/19368623.2017.1318731

Reinius, S. W., & Fredman, P. (2007). Protected areas as attractions. *Annals of Tourism Research, 34*(4), 839–854. https://doi.org/10.1016/j.annals.2007.03.011

Ried Luci, A., Le Bon, A., Carmody Lobo, S., & Santos Henríquez, R. (2018). *Sentidos del lugar desde la experiencia de ocio y turismo en áreas silvestres protegidas: Una metasíntesis*.

Şahin, A., Zehir, C., & Kitapçi, H. (2011). The effects of brand experiences, trust and satisfaction on building brand loyalty; an empirical research on global brands. *Procedia-Social and Behavioral Sciences, 24*, 1288–1301. https://doi.org/10.1016/j.sbspro.2011.09.143

Sayfuddin, A. T. M., & Chen, Y. (2021). The signaling and reputational effects of customer ratings on hotel revenues: Evidence from TripAdvisor. *International Journal of Hospitality Management, 99*, 103065. https://doi.org/10.1016/j.ijhm.2021.103065

Schmitt, B. (1999). *Experiential marketing: How to get customers to sense, feel, think, act and relate to your company and brands* (Free Press).

Schmitt, B. (2011). Experience marketing: Concepts, frameworks and consumer insights. *Foundations and Trends in Marketing, 5*(2), 55–112. https://doi.org/10.1561/1700000027

Schroeder, H. (1996). Ecology of the heart: Understanding how people experience natural environments. In A. Ewert (Ed.), *Natural Resource Management: The Human Dimension* (pp. 13–27). Westview Press.

Schroeder, H. (2002). Experiencing nature in special places: Surveys in the north-central region. *Journal of Forestry, 100*(5), 8–14.

SocioCultural Research Consultants, L. (2015). *Dedoose Version 6.1.18, web application for managing, analyzing, and presenting qualitative and mixed method research data*.

Taecharungroj, V. (2022). An analysis of tripAdvisor reviews of 127 urban rail transit networks worldwide. *Travel Behaviour and Society, 26*, 193–205. https://doi.org/10.1016/j.tbs.2021.10.007

Tangeland, T., & Aas, Ø. (2011). Household composition and the importance of experience attributes of nature based tourism activity products—A Norwegian case study of outdoor recreationists. *Tourism Management, 32*(4), 822–832. https://doi.org/10.1016/j.tourman.2010.07.005

Tavakoli, R., & Wijesinghe, S. N. R. (2019). The evolution of the web and netnography in tourism: A systematic review. *Tourism Management Perspectives, 29*, 48–55.

TripAdvisor. (2021). *TripAdvisor*. https://www.tripadvisor.pt/

Tsai, S. (2012). Place attachment and tourism marketing: Investigating international tourists in Singapore. *International Journal of Tourism Research, 14*(2), 139–152.

Turismo de Portugal. (2017). Estratégia Turismo 2027. In *Disponível via Turismo de Portugal em*: https://www.turismodeportugal.pt/SiteCollectionDocuments/estrategia/estrategia-turismo-2027.pdf

Tverijonaite, E., Ólafsdóttir, R., & Thorsteinsson, T. (2018). Accessibility of protected areas and visitor behaviour: A case study from Iceland. *Journal of Outdoor Recreation and Tourism, 24*, 1–10. https://doi.org/10.1016/j.jort.2018.09.001

Valdivia, A., Hrabova, E., Chaturvedi, I., Luzón, M. V., Troiano, L., Cambria, E., & Herrera, F. (2019). Inconsistencies on TripAdvisor reviews: A unified index between users and sentiment analysis methods. *Neurocomputing, 353*, 3–16.

Vespestad, M. K., & Lindberg, F. (2011). Understanding nature-based tourist experiences: An ontological analysis. *Current Issues in Tourism, 14*(6), 563–580. https://doi.org/10.1080/13683500.2010.513730

Ye, Q., Law, R., & Gu, B. (2009). The impact of online user reviews on hotel room sales. *International Journal of Hospitality Management, 28*(1), 180–182.

You, L., & Sikora, R. (2014). Performance of online reputation mechanisms under the influence of different types of biases. *Information Systems and e-Business Management, 12*(3), 417–442. https://doi.org/10.1007/s10257-013-0229-9

Zarantonello, L., & Schmitt, B. (2010). Using the brand experience scale to profile consumers and predict consumer behaviour. *Journal of Brand Management, 17*, 532–540. https://doi.org/10.1057/bm.2010.4

Zhang, Q., & Xu, H. (2020). Understanding aesthetic experiences in nature-based tourism: The important role of tourists' literary associations. *Journal of Destination Marketing & Management, 16*, 100429. https://doi.org/10.1016/j.jdmm.2020.100429

Social and Economic Impacts

If we look at tourism activities from a social and economic impacts point of view, it is easy to understand its immense scope and horizontal consequences. In fact, here lay a considerable part of why tourism has become such a relevant sector. For Hall, Williams, and Lew (2014), the circumstance of financial dilemma provides a window where the tourism industry is highly influential on an economic, environmental, cultural, political, and social front; thereby acting as a possibility for the tourism industry to fulfill the social and economic goals of a community and economy.

Moreover, the essence of sustainable tourism development is supported by the balance between the social, economic, and environmental benefits (Rasoolimanesh & Jaafar, 2017; Castanho, Couto, and Pimentel, 2020; Couto et al., 2021) as an effect of regional tourism development (Castanho et al., 2022). Also, the benefits should outweigh the tourism development on a tourism destination's social and economic environment.

Contextually, this part titled *Social and Economic Impacts* is organized into six chapter. Within Chapter "Solidarity Tourism in a Multicultural Society in Southern Italy", Nakabasami highlights the solidarity inherent to developing tourism activities in a multicultural society environment using Southern Italy as a case study. Hoa Nguyen Quynh (Chapter "Tourism Impacts on Heritage Sites in Japan: From Government's View to Local People's Awareness"), investigates the tourism impacts on heritage from the main-actors perspective through case studies in Japan. In the following Chapter "Exploring the Unlimited and Unexplored Rural Tourism in Meghalaya, North East India ", Buam and Naseer provide us with an exciting journey through the unexplored rural tourism in the norteast of India and show us the potential for rural and nature-based tourism potential for development in that specific region. Melo et al. (Chapter "Effects of Market Intelligence Generation, Online Reviews, and Management Response on the Business Performance of Rural Accommodation Establishments in France"), analyze the social side of this issue when the authors address the positive consumer reviews and their relevance in influencing consumers′ purchasing decisions and, therefore, their performance. This research work focuses on the issue of rural tourism accommodation in France. In Chapter "The Fortress of Santa Catarina de Ribamar (Portimão) as a proposal for Good Practices of Military Heritage Preservation", Galamba intends to set a proposal for good practices regarding the preservation and conservation of Military Heritage. The author presents a case study from the Algarve Region in the South of Portugal, where the anthropic pressure due to tourism activities is very high. Finally, in Chapter "Sustainable Tourism in Natural Territories That Have Suffered From Catastrophes: The Perception of Public and Private Stakeholders of the Alva Great Route", Guerra et al. address the problems related to sustainable tourism planning in natural landscapes after suffering from disasters. In this detailed case, the authors use a Central Portuguese territory and analysis the perception of public and private stakeholders.

Rui Alexandre Castanho
Faculty of Applied Sciences
WSB University
Dabrowa Górnicza, Poland
CITUR—Madeira—Centre for Tourism Research
Development and Innovation
Funchal-Madeira, Portugal
e-mail: acastanho@æsb.edu.pl

References

Castanho, R. A., Couto, G., & Pimentel (2020).*Principles of Sustainable Tourism and Cultural Management in Rural and Ultra-peripheral Territories: Extracting Guidelines for its application in the Azores Archipelago*. Cultural Management: Science and Education (CMSE) ISSN: 2512-6962 https://doi.org/10.30819/cmse.4-1.01

Castanho, R. A., Couto G., Pimentel, P., Carvalho, C. B., Sousa, Á., da Graça Batista, M. (2022). *A Preliminary Approach About the Perceptions of Decision Makers versus Entrepreneurs About Rural Tourism Development in the Azores Autonomous Region*. In: Á. Rocha, C. Ferrás, A. Méndez Porras, E. Jimenez Delgado (Eds.), *Information Technology and Systems. ICITS 2022* (Vol. 414, pp. 277–285). Lecture Notes in Networks and Systems. Springer. https://doi.org/10.1007/978-3-030-96293-7_25

Couto, G., Castanho, R. A., Pimentel, P., Carvalho, C. B., Sousa, Á. (2021). *The Potential of adventure tourism in the Azores: Focusing on the regional strategic planning*. In: A. Abreu, D. Liberato, E. A. González, J. C. Garcia Ojeda (Eds.), *Advances in Tourism, Technology and Systems. ICOTTS 2020. Smart Innovation, Systems and Technologies* (Vol. 209). Springer. https://doi.org/10.1007/978-981-33-4260-6_2

Hall, C. M., Williams, A. M. & Lew, A. A. (2014). Tourism: conceptualisations, disciplinarity, institutions and issues'. In A. Lew, C. M. Hall & A. M. Williams (Eds.), *The wiley blackwell companion to tourism*. Chichester: Wiley.

Rasoolimanesh, S. M. & Jaafar, M. (2017). Sustainable tourism development and residents' perceptions in World Heritage Site destinations. *Asia Pacific Journal of Tourism Research, 22*(1):34–48

Solidarity Tourism in a Multicultural Society in Southern Italy

Chieko Nakabasami

Abstract

This article focuses on what makes solidarity tourism sustainable in Camini, a small village within the Calabria region of southern Italy. Since the turn of the century, more immigrants from Africa and Asia have been moving into Camini. With the support of the Eurocoop Camini Jungi Mundu, Camini has been collaborating with volunteers from around the globe and host residents to establish a multicultural society with the immigrants. The affect theory of social exchange complements social exchange theory by incorporating emotion into the exchange process. This affect theory and the balance theory of social network analysis were used to discuss interactivity and mutuality among these players: the Eurocoop, volunteers, host residents, and immigrants. A field survey has been conducted in Camini that explores interpersonal relationships. Results indicate that the key factor in the success of solidarity tourism has been mutual support among all players. Moreover, mixed habitation has been proven to promote mutual support.

Keywords

Community interaction for sustainable tourism • Solidarity tourism • Multicultural society • Affect theory • Balance theory • Southern Italy

1 Introduction

Camini is a small community beside the Ionian Sea in the Calabria region of Southern Italy and has a history extending to the seventh century (Fig. 1). In 2020, it had a population of about 726 people, a number that has not changed much in 20 years (Istat, 2021). With the support of the Eurocoop Camini Jungi Mundu (Cooperativa terzo settore, 2021), Camini has been collaborating with volunteers from around the globe and host residents to establish a multicultural society with the immigrants (Urso, 2021). In fact, between 2013 and 2017, approximately 200 immigrants have settled in Camini. In Camini, we can see young volunteers working with the immigrants as well as the host residents, proving the success of solidarity tourism. Originally, this style of solidarity tourism was imported into the Camini community from the neighboring community of Riace (Fig. 2). Riace is said to be a front-runner of multiculturally inclusive society in Italy. This model of a multiculturally inclusive society is called the "Riace Model," and it has been having a positive effect on the surrounding communities. In the Riace model, solidarity tourism is proposed as one of the driving forces for the strong establishment of multicultural societies (Barillà, 2017; Ricca, 2010; Rinaldis, 2016).

However, in spite of the fact that the Riace model has been advocated from Riace, solidarity tourism does not seem to be working well in Riace. On the other hand, Camini has achieved a multiculturally inclusive society based on the Riace model, and recently, the community has achieved good success with solidarity tourism (Gagliardi, 2017). We will focus on this phenomenon of Camini and discuss what makes solidarity tourism sustainable in Camini using the balance theory in the field of social network analysis. In addition, using the affect theory, we will explain interactivity and mutuality among these solidarity tourism players: the Eurocoop, volunteers, host residents, and immigrants. The affect theory of social exchange complements the social exchange theory by incorporating emotion into the exchange process. Through our discussion, we also hope to give one viewpoint in building a multicultural society in which people can live together with a feeling of strong solidarity that goes beyond cultural and language differences. Since 2016, the author has been conducting field surveys and has been

C. Nakabasami (✉)
Faculty of International Tourism Management, Toyo University, Tokyo, Japan
e-mail: chiekon@toyo.jp

A. Mandić et al. (eds.), *Cultural Sustainable Tourism*, Advances in Science, Technology & Innovation,
https://doi.org/10.1007/978-3-031-10800-6_6

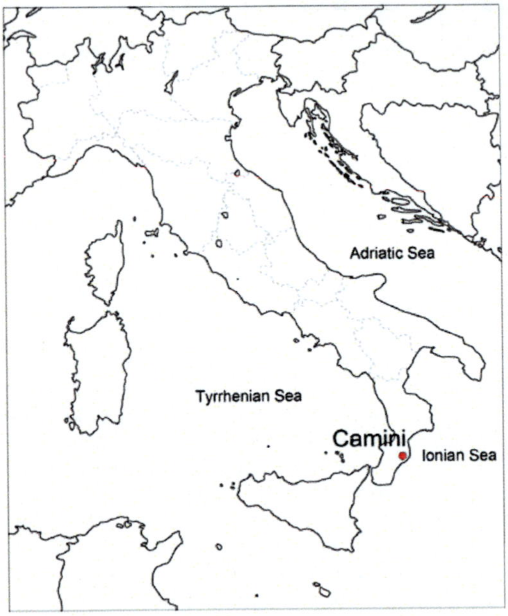

Fig. 1 Map of Camini. *Source* Modified from a free map

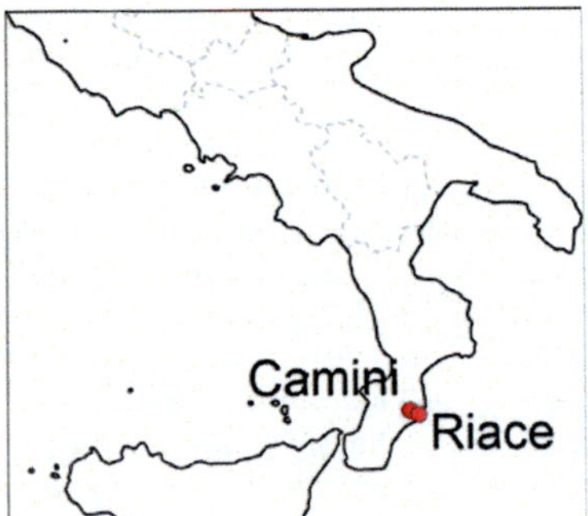

Fig. 2 Map of Camini and Riace. *Source* Modified from a free map

interested in what mechanism can work well for the multicultural societies in Riace and Camini, focusing on the mutuality and interactivity of the society (Miyajima, 2009; Nakabasami, 2019, 2020). *Mutuality* is a kind of material and mental collaboration that is supported by the local people (local Italians) who mutually respect cultural differences. *Interactivity* means that local people share their roles without discrimination because of cultural differences in various situations such as public service and private aid in the community. In this context, *interactivity* means that the immigrants themselves also support other immigrants; for example, the immigrants are in charge of public services at the village office. In Camini, solidarity tourism can play a suitable role in such mutuality and interactivity of the society. The aim of this research is to find answers to what makes solidarity tourism sustainable in Camini. Thus, the author conducted field surveys and interviews to investigate and find answers.

2 Solidarity Tourism

Solidarity tourism is one form of alternative tourism that was proposed at the turn of the century (Smith & Eadington, 1992). Today, diverse tourism styles are emerging as alternative tourism. Alternative tourism is another tourism style different from mass tourism. *Mass tourism* is the phenomenon of sightseeing travel popularized due to rapid economic development after World War II in developed countries. Though mass tourism popularized travel for many people, there are some problems associated with mass tourism. Mass tourism prioritizes the tourist's satisfaction, and this has sometimes caused serious problems for the local people, destroying their traditions and lifestyles and harming the environment by tourist behaviors such as littering. Frankly speaking, mass tourism cannot be said to be sustainable, and, regrettably, mass tourism goes against the current age, in which sustainable development goals (SDGs) were advocated by the United Nations in 2015 to seek for a sustainably better world.

Solidarity tourism can be said to be similar to justice tourism in the field of tourism studies (Higgins-Desbiolles, 2008). Lanfant and Graburn (1992) suggested that alternative tourism could become "tourism in the promotion of a new order." Holden's description of justice tourism is "a process which promotes a just form of travel between members of different communities. It seeks to achieve mutual understanding, solidarity and equality among participants (Holden, 1984; Pearce, 1992)." Many other tourism papers have also discussed the concept of justice tourism.

Among many alternative tourisms, volunteer tourism shares a lot of concepts with solidarity tourism, and almost overlapped. Wearing and McGehee (2013) offer the first comprehensive analysis of volunteer tourism. Wearing (2001) defines volunteer tourism (voluntourism) as "Those tourists who for various reasons, volunteer in an organized way to undertake holidays that might involve the aiding or alleviating of material poverty of some groups in society, the restoration of certain environment, or research into aspects of society or the environment." In this sense, volunteer tourism and solidarity tourism can be regarded almost as the same type of tourism, on the other hand, in solidarity tourism, the author would like to focus on solidarity among potential stakeholders beyond volunteer's behavior.

In this article, the author would like to treat both justice tourism and solidarity tourism equally. In Higgins-Desbiolles (2008), Scheyvens (2002) describes justice tourism as "both ethical and equitable" and says it has the following attributes:

- builds solidarity between visitors and the local people;
- promotes mutual understanding and relationships based on equity, sharing, and respect;
- supports self-sufficiency and self-determination of local communities;
- maximizes local economic, cultural, and social benefits.

Concretely, the following activities are included in justice tourism according to Scheyvens (2002):

> Five forms of justice tourism which include the 'hosts' telling their stories of past oppression, tourists learning about poverty issues, tourists undertaking voluntary conservation work, tourists undertaking voluntary development work and revolutionary tourism.

In addition to the above explanation, childcare and being a teaching assistant are examples of voluntary work mentioned by Scheyvens (2002). In solidarity tourism, activities are basically voluntary, and tourists acting as volunteers are not paid for their activities. Instead, the local people offer hotels and meals for tourists at very reasonable prices. Through solidarity tourism, tourists can make lasting friendships, enjoy the local food every day, and experience beautiful, undamaged nature, as small communities are located in rural areas. Voluntary activities continue for at least two weeks, and sometimes, volunteers stay for more than one year. While on holiday, weekends, or in between activities, volunteers visit the surrounding regions to enjoy the local culture. The fact that young volunteers such as students are the majority of tourists vivifies the local society. Moreover, living closely, using vacant houses in the center of a community can solidify the relationship between the volunteers and local people.

3 Affect Theory and Solidarity Tourism

The affect theory of social exchange was proposed by Lawler (2001) to explain how and when social exchange produces emotions, and that stronger or weaker ties are generated among individuals, groups, or social networks. According to some passages excerpted from Lawler (2001), the theory has various implications for the role of emotions in the production of group solidarity because any social interaction leads to joint activity whether a task is implicit or explicit. The interdependency of the task is greatest if actors find it difficult to separate their individual effects on or contributions to solving the task involved (Lawler said *nonseparability*) and actors perceive a shared responsibility for the task's success or failure. *Nonseparability* means that actors' individual contributions are not as easily identifiable or separable from the contributions of others, and task success or failure heavily relies on to what extent actors are interdependent of each other. The theory also proposed that the greater the inseparability of actors' impact on task success or failure, the stronger will be the sense of shared responsibility.

The theory shows why and how the emotional effects vary across different exchange structures. As shown in Table 1, the theory defined the solidarity of four forms of social exchange in joint tasks based on inseparability, perception of shared responsibility, and global emotions as productive, negotiated, reciprocal, and indirect. Among these forms, the productive exchange structure is the highest in both inseparability and perception of shared responsibility. Productive exchange shows that the inseparability of individual contributions and shared responsibility are remarkable; the corresponding emotions should be felt stronger, both the positive emotion in task success and the negative emotion in task failure (Lawler, 2001).

Solidarity tourism can explain the productive structure of the affect theory. As Lawler (2001) said, the affect theory of social exchange identifies some fundamental ways of exchange processes and reveals implications for the solidarity of exchange. Lawler (2001) said, "The theory assumes a social network with three or more actors who have opportunities to exchange behaviors, goods, or outcomes. Exchanges are dyadic, and dyads are connected; that is,

Table 1 Comparison of the effects of the exchange structures, excerpted from Lawler (2001)

Structure	Nonseparability	Perception of shared responsibility	Global emotions
Productive exchange	High	High	High
Negotiated exchange	Medium	High	Medium to high
Reciprocal exchange	Low	Medium to high	Medium
Indirect exchange	Low	Low	Low

exchange in a given dyad affects or is affected by exchanges in one or more other dyads (Emerson, 1972, 1976)." Individual actors make decisions about whether to exchange, with whom to exchange, and under what terms. Thus, social network analysis can fit very well using the affect theory to analyze the interactivity and mutuality of solidarity tourism, in which players (actors) are working for the same goal as members of certain communities, such as villages.

4 Solidarity Tourism in Camini

4.1 Volunteers

In a multicultural society, solidarity tourism can be adapted for local people, immigrants, volunteers, and the directing organization. According to the field survey conducted by the author, on the occasion of visiting Camini, solidarity tourism was very active. The author's first impression was that Camini had nothing much going for it. Here, only a few local people can be seen walking the steep streets. You cannot tell at first sight whether solidarity tourism has been working well in this village.

The author interviewed an Italian woman who knew a lot about the association in Camini. She provided a lot of useful information regarding the association. In 2011, Eurocoop, an international social contribution organization came to Camini and commenced activities with the mission of regional vivification from various viewpoints, including the economy, community, and culture, while collaborating with local people and immigrants. Since 2016, a solidarity tourism project called Jungi Mundu has been carried out in Camini by the Eurocoop. *Jungi Mundu* means "join the world" in English, and it is also the name of the office in Camini that welcomed everyone regardless of nationality. Jungi Mundu has become a symbolic name representing the association and all the activities conducted by it. Solidarity tourism is one of the Eurocoop's projects. In the solidarity tourism project, international volunteers have been fully accepted and have been collaborating with local people and immigrants living there. In 2018, 102 young volunteers from all over the world worked as volunteers in Camini. Many of them were students.

The staff of Jungi Mundu are composed of professionals and young men born in Camini. The professionals are very diverse—a project manager, art director, event organizer, language teacher, photographer, psychological counselor, etc. Volunteers stayed at vacant village houses scattered in/around Camini (Fig. 3) and were served three meals a day, paying 25 euros per day.

The vacant village houses have been renovated and equipped with Wi-Fi and modern plumbing in the kitchen and showers. A rustic panoramic scenery and old village houses can be seen from the upstairs bedroom of the renovated houses. The houses are shared as apartments adequate for two or three people and equipped similarly to resort apartment hotel standards (Fig. 4).

Fig. 3 Renovated volunteers house (in center). *Source* Photograph by the author in 2018

Fig. 4 Bedroom upstairs in a volunteer house. *Source* Photograph by the author in 2018

The main activities of the volunteers are taking care of immigrants' young children, supporting teachers at school, fabricating and selling artisanal products, and painting pictures promoting multicultural inclusion on the walls scattered around the village. Some volunteers stay in Camini for a short period, while others stay longer. The field survey in Camini was an invaluable experience for the researcher to gain understanding as to how solidarity tourism is being implemented successfully. In the next chapter, the result of the field survey in Camini is reported to acquire a partial understanding about the local people, immigrants, and Jungi Mundu.

4.2 Immigrants, Local People, and Jungi Mundu

We conducted a field survey in Camini by semi-structurally interviewing 18 people, including local people, project staff, and immigrants. The interview mainly focused on interpersonal relationships. Besides demographic information, the following questions were asked:

- From whom do you seek any support/service physically and mentally? (mutuality)
- What activities have you done and with whom for the past five years in Camini? (interactivity)

The two questions above explore mutuality and interactivity among these players: Jungi Mundu, volunteers, local people, and immigrants. As for volunteers, they visit with the intention of working with the rest of the players: Jungi Mundu, local people, and immigrants. In this sense, their support or service activities are very clear. On the other hand, we think that solidarity tourism should not only be about the activities of volunteers but also bidirectional collaboration among all group of players; nonseparability and the perception of shared responsibility should be ensured in solidarity tourism. Thus, in this context, the mutuality and interactivity of the other players in Camini were to be investigated. With these questions, we seek to understand whether mutual support or services are offered and identify interactive activities beyond players' original jobs. In Table 2, activities include both paid and unpaid ones.

For each question, we tried to draw more details from the interviewees with care for their privacy: from whom they receive support or services and the kind of support or services. Table 2 shows the demographic profiles (in part) and the interpersonal data mentioned above. In Table 2, *staff* means the staff of Jungi Mundu, and *group* means players who contribute to solidarity tourism.

From Table 2, although there are few interviewees, some interesting evidence can be seen. The roles of each group visibly overlap. Immigrant p1, p2, and p3 work together with Jungi Mundu to harvest olive oil. Also, p13 is a volunteer who supports other immigrants; at the same time, they are supported by Jungi Mundu. We suppose that they support other immigrants who moved in Camini after them. Mutuality can be seen among the immigrants. P10 works as mason with a local Italian master. P12 is a local Italian who works as a volunteer with his friends. In the same way, p16 is not a staff member but a local Italian who works as a volunteer with their colleagues. Generally, interviewees receive moral support from their friends and families, and, naturally, Jungi Mundu supports everything in life for immigrants.

Besides these three players, volunteers are the fourth player that joins the solidarity activity in Camini. As said in the previous section, volunteers lodge in vacant houses in

Table 2 Demographic profiles and support/service and activity information

ID	Age	Group	Job status	Supports/services	Supports/services from whom	Activity	Activities with whom
p1	18	Immigrants	Temporary worker	Everything in life	Jungi Mundu	Olive oil harvest	Jungi Mundu
p2	19	Immigrants	Temporary worker	Healthcare	Jungi Mundu	Olive oil harvest	Jungi Mundu
p3	19	Immigrants	Temporary worker	Everything in life	Jungi Mundu	Olive oil harvest	Jungi Mundu
p4	20	Immigrants	Volunteer	Everything in life	Adoptive mother	Volunteer	Jungi Mundu
p5	22	Immigrants	Housewife	Moral support	Husband	–	–
p6	22	Local people	Bar tender	Moral support	Parents	Waiter	Bar owner
p7	23	Immigrants	Housewife	Moral support	Husband	Tomato harvest	Friends
p8	24	Immigrants	Housewife	Everything in life	Husband	–	–
p9	29	Staff	Teacher	Moral support	Friends	Volunteer	Friends
p10	32	Immigrants	Mason	Everything in life	Jungi Mundu	Mason	Local Italian master
p11	34	Immigrants	Sanitary worker	Moral support	Brother-in-law	Cleaning	Friends
p12	34	Local people	Shopkeeper	Everything in life	Elder brother	Volunteer	Friends
p13	35	Immigrants	Volunteer	Everything in life	Jungi Mundu	Volunteer	Friends
p14	35	Staff	Researcher	Everything in life	Wife	Volunteer	Friends
p15	35	Staff	Interpreter	Moral support	Friends	Volunteer	Friends
p16	38	Local people	Teacher	Moral support	Friends	Volunteer	Colleagues
p17	38	Staff	Volunteer	Moral support	Friends	Volunteer	Colleagues
p18	40	Local people	Driver	Moral support	Mother	Driver	Colleagues

the village, and their entire stay, including meals, is supported by Jungi Mundu. During their stay, the main activity is settlement support for the immigrants, but they also interact with the local people to help according to their needs as described in the previous section.

5 Balance Theory of Social Network Analysis

5.1 Balance Theory

In Camini, there are four sorts of players in the promotion of solidarity tourism. They each play a vital role in communicating each other. It is appropriate to describe the relationships among the players using an interplaying network. Based on the results of the field survey in Camini, we would like to discuss the situation of solidarity tourism using the balance theory from the perspective of structural balance in networks and to interpret the affect theory's impact on it. The principles underlying structural balance are based on theories of social psychology dating back to the work of Heider (1946) and generalized and extended to the language of graphs beginning with the work of (Cartwright & Harary, 1956; Harary, 1953).

Since then, there have been many applications of the research on the balance theory, the work by Davis (1963) being an example. We explained about the balance theory from (Easley & Kleinberg, 2010).

In balance theory, if we look at two people in a group, we can label the edges between them with + (positive relation) or – (negative relation). In other words, they are either friends or enemies. Considering a set of three people at a time, there are four different ways to label the three edges between three people with + 's and − 's. See Fig. 5.

- Given a set of people *A*, *B*, and *C*, when all edges have a plus sign, the three people are friends.[as in 1 of Fig. 5], and the triangle is said to be stable.
- Having a single plus and two minuses in the relations among the three people is also stable: It means that two of the three are friends, and they have a common enemy in the third. [See 2 of Fig. 5].
- Two plus and one minus triangles [as in 3 in Fig. 5] correspond to person A, who is friends with each of B and C, but B and C are not friends with each other. In this type of situation, A is implicitly forced to make B and C to be friends (trying to change the edge label between B and C to +). Or A tries to be on the side of B or the side of C. This unstable situation makes A swing between B and C. A will want to change one of the labels to minus.
- All three edges are minuses. This indicates that A, B, and C are enemies [as in 4 of Fig. 5]. This situation has the possibility of making two of the three people friends. Thus, the remaining one becomes the common enemy to the two people. As the result, one of the labels becomes plus. However, which edge becomes minus changes depending on the situation, and in that sense, it is an unstable situation.

We refer to triangles with one or three pluses as *balanced*, since they are stable, and we refer to triangles with zero or two pluses as *unbalanced,* marked as unstable. In balance theory, unbalanced triangles are causes of stress. People try to minimize them in their personal relationships, and hence, unbalanced triangles are fewer in real social settings than balanced triangles.

Excerpted from Easley and Kleinberg (2010, pp. 121); explanation added by the author.

5.2 Social Networks in Camini

Based on interviews in Camini, we first considered three stakeholders as the members of a multicultural society and assigned each stakeholder to three different network nodes. It is indispensable for representing interactivity and mutuality among the stakeholders to draw interpersonal networks. Here, we consider a social network before the volunteers' visit because the immigrants arrived at Camini at the beginning, and since that time, a multicultural society has been shaped gradually. Jungi Mundu started asking for volunteers from around the world to establish the system of solidarity tourism. Camini's multicultural society began without volunteers, but solidarity tourism with volunteers

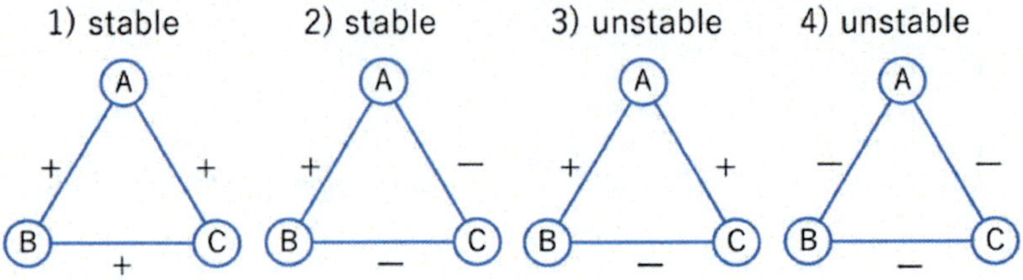

Fig. 5 Four types of signed networks with three nodes

was needed for the multicultural society to thrive. By observing the process of the multicultural society before/after the volunteers arrived, we think that the role of solidarity tourism can be clarified better. Analyzing the transition of social networks in Camini could give some implication for the success of multicultural societies as well as solidarity tourism. First, we consider three-node networks without volunteers, and then four-node networks with volunteers.

The three nodes without volunteers (A, B, and C) are the following:

A: Jungi Mundu

B: Immigrants

C: Local people

These three nodes were connected to draw a signed network for Camini. We considered the transition of the network during the recent situation in Camini as shown in Fig. 6.

Figure 6 is explained as follows: In Camini, in the beginning, Jungi Mundu initiated their activities by applying the Riace Model; Jungi Mundu were complete strangers to the local people, as shown in (i). Since then, Jungi Mundu started a project to renovate old houses throughout the entire village and provided immigrants with jobs by establishing a sound relationship with the local Italian people in (ii). In due course, the immigrants settled and live happily on an economic basis in cooperation with the local people, shown in (iii). It can be said that all three nodes—Jungi Mundu, immigrants, and local host people—are balanced because all three nodes are linked together with a ' + '.

Jungi Mundu's most distinguishing feature is that it is a nonprofit organization and has no relation to the administration of Camini. Although the mayor and Jungi Mundu are always exchanging opinions and information about their activities, Jungi Mundu is not interested in politics but is exclusively concentrated on the activity of settling immigrants. This independence from the administration seems to be a key to sustaining the friendship between Jungi Mundu and the local people. Thus, the balanced network with all three nodes can be kept without any problems in Camini. However, in reality, network (iii) is the ideal, and we supposed that sustainable tourism could play an important role for the transition to (iii). Thus, we consider volunteers to be the fourth player in the network.

In the balance theory, three-node networks can be extended easily to four-node networks. According to the balance theory, structural balance property is described as follows:

> Structural Balance Property: For every set of three nodes, if we consider the three edges connecting them, either all three of these edges are labeled +, or else exactly one of them is labeled +.
>
> Excerpted from Easley and Kleinberg (2010, pp. 122).

Thus, for more than four nodes networks, each set of three nodes should satisfy the structural balance property above. The network is balanced if all set of three nodes are balanced, on the contrary, the network is unbalanced if at least one set of three nodes is unbalanced.

At the time of i) in Fig. 6, Jungi Mundu started to welcome volunteers as players for solidarity tourism. At this point, as shown in Fig. 7, volunteers are included in the interpersonal network. Node 'D' represents volunteers. Now, we add one more node D to the above three-node network.

A: Jungi Mundu

B: Immigrants

C: Local people

D: Volunteers

In Fig. 7, the four-node network on the left side is shown after adding node D (volunteers) to the three-node network of Fig. 6. The positions of each node are irrelevant. At the moment the volunteers arrived, the social network of Camini

Fig. 6 Social network transition in Camini (by the author)

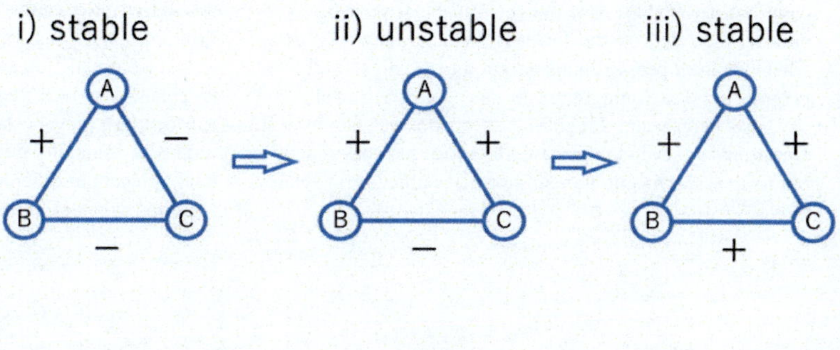

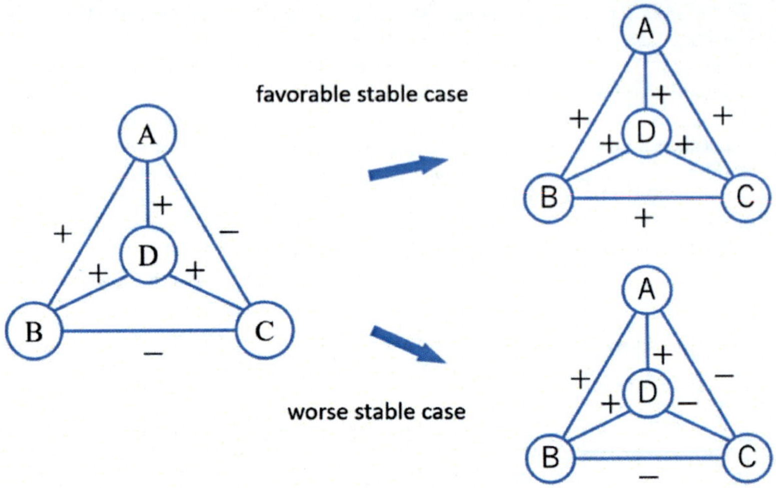

Fig. 7 Flow of solidarity tourism with volunteers (by the author)

was considered to be the network farthest left in Fig. 7. Naturally, volunteers want to get along well with every villager; therefore, ' + ' signs are set on every edge to the other nodes. Then, in the upper right network, all edges have + signs. This case means that solidarity tourism can work well with all players, including volunteers. First, to make the network function with all + signs, a good relationship between the support organization and the host people should be established. It can be explained that Jungi Mundu offers volunteers vacant houses in the village with the help of local people. The volunteers and local people have become neighbors, and they communicate often. The volunteers can vivify the atmosphere of the street, and local people sometimes enjoy telling their various story to the volunteers. Moreover, the immigrants and local people have gotten along well with the help of the volunteers. The volunteers collaborate with the immigrants to repair something in the village (e.g., repainting walls, cleaning up streets). The volunteers communicate with the immigrants, encouraging them to participate in activities in the village. Local people appreciate the work contributed by the immigrants and the volunteers, leading to the establishment of good relationships among all three players.

However, another stable case is possible. It is a worse case, but it is stable. Also indicated in Fig. 7, a worse scenario is indicated on the lower right. This network occurs when: (1) local people do not accept the volunteers, (2) local people and the immigrants do not get along well, and (3) local people and Jungi Mundu do not get along well. Unfortunately, even in this case, the social network is stable, but neither sustainable tourism nor the multicultural society works.

It is important for solidarity tourism to avoid such worse cases. Thus, in the next section, we propose some idea by incorporating the affect theory into the social network in Camini.

6 What Makes Solidarity Tourism Sustainable

6.1 The Affect Theory on the Balanced Network in Camini

Immigrants could not establish interpersonal relationship with local people when they first started living in the village. They were only strangers. The volunteers can play an important role between immigrants and local people. Volunteers start some activities with the immigrants to contribute to the village, and then, local people have become reliable to the immigrants as well as the volunteers. The volunteers can be said to be critical mediators for both immigrants and local people to build better relationships. Without the volunteers, the immigrants might hesitate to start activities that contribute to the village.

Moreover, interactivity among all four nodes is necessary for solidarity tourism to be sustainable. Interactivity means role exchange among the players in the same group. Repeated interactivity means sharing each role interchangeably. All players should be inseparable. At the same time, all players share the same responsibility for their joint tasks. Mutuality means the exchange of support/service among players in different groups, and shared responsibility is ensured when joint tasks have mutuality.

Table 3 Role exchange among all players in Camini

Player	A Jungi Mundu	B Immigrants	C Local people	D Volunteers
A	Work together as staff	Support/service for settlement	Introduce the volunteers for helping them with some troubles in daily life	Arrange activities, offer everything needed for village life
B	Help support/service other immigrants as supporters (interactivity)	Support each other (mutuality)	Do something useful in the village with volunteers	Work together
C	Arrange vacant houses for immigrants and volunteers	Tell their stories	Help each other in daily life, especially mental support	Tell their stories
D	Help support/service immigrants	Help them physically and mentally	Do something useful in the village with volunteers	Work together, live together in the village

6.2 Discussion

In the previous section, the affect theory was applied to the social network in Camini. In the affect theory, inseparability and shared responsibility are key concepts for productive exchange among all players. This means interchangeability among all players. Table 3 summarizes the role exchange in solidarity tourism through the field survey in Camini.

In Table 3, the direction is from the rows (A, B, C, D) to the columns (A, B, *C*, D). For example, Jungi Mundu (A) is offering support/service for settlement to immigrants (B); on the other hand, immigrants help support/service other immigrants as supporters of Jungi Mundu (A).

Mutuality and interactivity are realized in the solidarity tourism of Camini. It can be said that inseparability and shared responsibility are assured through solidarity tourism in Camini. Inseparability and shared responsibility entail strong solidarity. Therefore, mutuality and interactivity ensure strong solidarity and can make solidarity tourism sustainable.

Moreover, we would like to emphasize the habitation status of the immigrants and the volunteers in the village. Through field survey, we found that mixed habitation has promoted mutual support among them. As shown in Table 3, various interactions have been made among all players, especially between local people and other players; mixed habitation is an essential factor in creating interpersonal networks that work well. As indicated in Fig. 7, whether good relationships can be established depends on the relation to local people. In this sense, becoming neighbors of local people is vital for sustainable tourism. Local people would like to tell their stories, including village history, daily life and culture, local products they are proud of, etc. Talking with each other whenever they like is the best way to interact, and the volunteers can better understand what local people need. Renovating vacant houses can also provide jobs to immigrants, which realizes mutuality among the immigrants. For example, immigrants work as masons or wall painters, and then other immigrants who come later live in the house renovated by earlier immigrants.

7 Conclusions

This paper discusses what makes solidarity tourism sustainable in a multicultural society in Camini of southern Italy. With the support of the Eurocoop Jungi Mundu organization, Camini has been collaborating with volunteers from around the globe and host residents to establish a multicultural society with immigrants. Through a field survey on the solidarity tourism of Camini, some implications can be considered.

The affect theory of social exchange explains how and when emotions were produced by social exchanges, generating stronger or weaker ties to relations, groups, or networks. The theory shows why and how the emotional effects of exchange vary across different exchange structures, among which productive exchange ensures inseparability and shared responsibility. Solidarity tourism can be explained with the productive structure of the affect theory. The theory assumes a network context with three or more actors who have opportunities to exchange valued behaviors, goods, or outcomes. Thus, social network analysis can fit very well using the affect theory to analyze the interactivity and mutuality of solidarity tourism in which players are working for the same goal as members of certain communities, such as villages. In a multicultural society, the social network structure of solidarity tourism can be adapted for local people, immigrants, volunteers, and the directing organization.

Based on interviews in Camini, we first considered three stakeholders as the members of a multicultural society and assigned each stakeholder to three different network nodes: Jungi Mundu, immigrants, and local people. We observed the process of multicultural society before/after the volunteers arrived, and the role of solidarity tourism can be clarified better. Analyzing the transition of the social network in Camini carries some implications for the success of multicultural society as well as solidarity tourism. First, we considered a three-node network without volunteers, then a four-node network with volunteers. Based on the interviews, we discussed the situation of solidarity tourism using the balance theory from the perspective of structural balance in networks and to interpret solidarity based on the affect theory.

We understood that the role of volunteers was very important. Volunteers communicate with immigrants, encouraging them to participate in activities in the village. Local people appreciate the work contributed by the immigrants and the volunteers, causing all three players to establish good relationships. Without volunteers, the interpersonal network with local people might not advance in a better direction.

From the balance theory, within the transition from an unbalanced network to a balanced network, unhappy situations can be found that are dependent on the relationship between local people and other actors, including volunteers. To avoid such unhappy situations, solidarity tourism in Camini ensures interchangeability of roles among all players, which means that inseparability and shared responsibility are sustained. These two concepts are required for strong solidarity in the affect theory so as to realize mutuality and interactivity. Mutuality and interactivity are realized in the solidarity tourism of Camini. It can also be said that inseparability and shared responsibility are assured through solidarity tourism in Camini. According to the affect theory, inseparability and shared responsibility require strong solidarity. Therefore, mutuality and interactivity ensure strong solidarity in solidarity tourism and are considered to make it sustainable.

We conclude that the key factor in the success of solidarity tourism has been the mutual support among all players. Moreover, mixed habitation has proven to promote mutual support. In the future, more interchangeability could be realized; for example, as seen in Table 3, immigrants, local people, and volunteers will also be able to become members of Jungi Mundu. It is also possible that volunteers' lodging, now managed by local people, could be delegated to immigrants or volunteers in the future. Such unceasing efforts toward mutual support will make solidarity tourism sustainable in multicultural societies.

Acknowledgements This work was supported by Japan Society for the Promotion of Science Grant-in-Aid for Scientific Research(C) Grant Number JP18K11788.

References

Barillà, T. (2017). *Mimì Capatosta. Mimmo Lucano e il modello Riace*, Fandango libri.

Cartwright, D., & Harary, F. (1956). Structure balance: A generalization of Heider's theory. *Psychological Review, 63*(5), 277–293.

Cooperativa terzo settore. (2021). *Eurocoop Servizi "JUNGI MUNDU" cooperativa sociale*. Retrieved 27 Jan 2022 from https://eurocoopcamini.com/

Davis, J. A. (1963). Structural balance, mechanical solidarity, and interpersonal relations. *American Journal of Sociology, 68*(4), 444–462.

Easley, D., & Kleinberg, J. (2010). *Networks, crowds, and markets* (Vol. 8). Cambridge University Press.

Emerson, R. M. (1972). Exchange theory, part II: Exchange relations and networks. *Sociological Theories in Progress, 2*, 58–87.

Emerson, R. M. (1976). Social exchange theory. *Annual Review of Sociology, 2*, 335–362.

Gagliardi, C. (2017). *Un Mondo Nuovo*. Rubbettino print.

Harary, F. (1953). On the notion of balance of a signed graph. *Michigan Mathematical Journal, 2*(2), 143–146.

Heider, F. (1946). Attitudes and cognitive organization. *Journal of Psychology, 21*(1), 107–112.

Higgins-Desbiolles, F. (2008). Justice tourism and alternative globalisation. *Journal of Sustainable Tourism, 16*(3), 345–364.

Holden, P. (Ed.). (1984). *Alternative tourism: Report of the workshop on alternative tourism with a focus on Asia Chiang Mai, April 26 May 8 1984*. Ecumenic. Coalition on Third World Tourism.

Istituto Nazionale di Statistica (Istat). (2021). *Popolazione residente al 1° gennaio 2020 per sesso, età e stato civile Comune: Camini*, Retrieved 27 Jan, 2022, from http://demo.istat.it/popres/index.php?anno=2020

Lanfant, M. F., & Graburn, N. H. (1992). Chapter 5. International tourism reconsidered: The principle of the alternative. In V. Smith & W. Eadington (Ed.), *Tourism alternatives: Potentials and problems in the development of tourism*. University of Pennsylvania Press.

Lawler, E. J. (2001). An affect theory of social exchange. *American Journal of Sociology, 107*(2), 321–352.

Miyajima, T. (2009). Tabunkakyousei no mondai to kadai. *Gakujutu no doko, 14*(12), 12_10–12_19.

Nakabasami, C. (2019). Communication and social capital in a multi-lingual society. *IPSJ SIG Technical Reports, 2019-DC-114* (14).

Nakabasami, C. (2020). Communication network on solidarity tourism in southern Italy. *IPSJ SIG Technical Reports, 2020-DC-118*(1).

Pearce, D. (1992). Chapter 1. Alternative tourism: Concepts, classifications, and questions. In V. Smith & W. Eadington (Ed.), *Tourism alternatives: Potentials and problems in the development of tourism*. University of Pennsylvania Press.

Ricca, M. (2010). *Riace, il futuro è presente. Naturalizzare «il globale» tra immigrazione e sviluppo interculturale* (Vol. 83). EDIZIONI DEDALO.

Rinaldis, A. (2016). *Riace il paese dell'accoglienza: Un modello alternativo di accoglienza*. Imprimatur editore.

Scheyvens, R. (2002). *Tourism for development: Empowering communities*. Prentice Hall.

Smith, V. L., & Eadington, W. R. (Eds.). (1992). *Tourism alternatives: Potentials and problems in the development of tourism*. University of Pennsylvania Press.

Urso, K. (2021). Understanding international migrations in rural areas: New processes of social innovation and rural regeneration in Southern Italy. *Italian Review of Agricultural Economics, 76*(1), 59–75.

Wearing, S. (2001). *Volunteer tourism: Experiences that make a difference*. Cabi.

Wearing, S., & McGehee, N. G. (2013). Volunteer tourism: A review. *Tourism Management, 38*, 120–130.

Tourism Impacts on Heritage Sites in Japan: From Government's View to Local People's Awareness

Hoa Nguyen Quynh

Abstract

Culture and heritage have been a major part of travel as their contribution to a tourism destination's appealing. Recently, heritage tourism is the fastest-growing segment of the tourism industry as the volume of tourists seeking adventure, culture, history, archeology, and interaction with local people is increasing. Heritage tourism gives positive economic and social impacts, establishes and reinforces identity, and helps preserve the cultural heritage. However, as the volume of traveling rises, there are negative impacts on local communities in heritage sites that create tensions and conflicts among different stakeholders. Therefore, understanding the tourism policymakers' view and local people's awareness of tourism impacts is not only useful in the conservation of a heritage site, but it also contributes to promote the image of that destination and balance the stakeholders' benefits. This paper aims to study (1) the tourism impacts on local community, (2) the local people's awareness and consideration between positive impacts and negative impacts, and (3) the difference between government policymakers and local people's points of view about tourism impacts and development. The data were collected from a qualitative survey on Japanese government officers and academic people, and a quantitative survey on 245 local people in some Japanese heritage sites. From the findings, some gaps between government policymakers and local people's point of view were found and analyzed to contribute to the future tourism policy development for Japan heritage conservation.

Keywords

Japanese heritage tourism • Tourism impact • Tourism policy • Local people

1 Introduction

In the late decades, cultural heritage tourism was booming around the world as more people seeking for cultural experiences, heritage exploration, historical destinations, and interaction with local communities (Garrod & Fyall, 2000). Many types of attractions fall into the heritage tourism sector, such as architecture, festivals, hand-made works, art and folk music, and traditional cuisine. Heritage tourism would give positive effects on economic development and social enhancement, establishes, and promotes a destination's identity, and helps preserve the traditions. According to Moli (2011) and Hall and McArthur (1998), heritage tourism facilitates harmony and understanding among local communities, supports culture, helps renew tourism, and enables local people to define who they are. By providing a source of income based around local culture, heritage tourism can encourage communities to value their culture and heritage more highly. However, many researchers have warned about some unwanted ecology footprints left from the tourists that would give a significant pressure on heritage sites and local resident (Girard & Nocca, 2017; Manniche et al., 2017; UNWTO, 2018). Once they were negatively affected by tourism, the local people retaliate by exhibiting hostile behavior toward tourists. How local resident perceive tourism development and impacts informs their involvement in both economic development and conservation supports (Nicholas et al., 2009; Walpole & Goodwin, 2001; Wang & Pfister, 2008).

H. N. Quynh (✉)
School of Economics and Management, Hanoi University of Science and Technology, Hanoi, Vietnam
e-mail: hoa.nguyenquynh@hust.edu.vn

A. Mandić et al. (eds.), *Cultural Sustainable Tourism*, Advances in Science, Technology & Innovation,
https://doi.org/10.1007/978-3-031-10800-6_7

1.1 Economic Impacts of Tourism

Even though recently COVID-19 has taken significant impact on the global economy and travel, it is still true to say that tourism has huge effects on the development of all countries and regions. The role of travel and tourism in the economy are considered in terms of its contribution to the total GDP and employment of a country. Tourism can generate financial sources for public investment in health, education, medical system, and other assets, provide infrastructure, improve social capital, strengthen sustainable management of natural resources, and create a demand for local communities' livelihood improvement.

According to WTTC (2009), in many developing countries, tourism has contributed a relatively higher percentage to the total GDP and employment than the average. It may be not the only source of income and employment to a nation, but clearly serves as a prime factor in the balance of payments of many countries, especially developing ones. Therefore, the development of tourism industry has got attention from government, regional and local authorities, and other economic stakeholders. It also provides opportunities for the poor by changing the access to the assets. Local people may be aware the effects of tourism to their livelihood—whether positively or negatively—such as cultural pride, a sense of control, good health, and reduced poverty.

Tourism can be used as a tool for raising global awareness, introducing local product and achievements. Therefore, countries can use tourism to stimulate its trading, investment, cultural exchange, improve its image worldwide, and enhance the international reputation through media exposure. Beside revenue, governments want to promote tourism as sign of peace and goodwill (Hall, 2003). Ioannides and Debbage (1998) indicated that one benefit of tourism is to help a nation earning more foreign exchange with less difficulty than other products. Therefore, many governments allocate a large proportion of their budgets to tourism development policy, planning, and promotion to encourage international tourists to visit their countries.

However, without sustainable development planning and policy, tourism may create tensions and conflicts among different stakeholders (Porter & Salazar, 2005) by reducing local access to natural resources, drawing heavily upon local infrastructure and disrupting social network. There were examples in developing countries where local residents lost their accesses to local natural resources to outside corporations (Neef, 2019; Shah & Gupta, 2000). Rural heritage and agriculture land were turned into large resorts and golf courses. As a direct effect, farmers and fishermen lost their traditional jobs and livelihood environment. These negative impacts might be not justified in terms of the economic benefits and new jobs from tourism to local people. However, if the local people perceive tourist will bring bad effects and behaviors to their communities, such as drug abuse, sex, alcohol drinking and gambling, beside the raising of living costs and new taxes, they may express their hostile attitude toward tourists and do not welcome tourism activities in their places (Andereck et al., 2005).

1.2 Environmental Impacts of Tourism

Some previous economic studies suggest that tourism would balance the goals of environment conservation and economic development in and around the nature protected area, such as tourism can increase funds for the protection of natural area, heritage sites, and promote awareness of local people and tourists for biodiversity issues (Ashworth & Van der As, 2006; Figgis et al., 2007; Hoa, 2016). In Liu and Var (1986), local resident agreed that tourism was not the reason for ecological destroy, as it can provided "more parks and recreation areas". If there are proper sustainable developing policies, the local habitats may perceive improvement of their communities' appearance and environment awareness (Perdue et al., 1990). Once the local people realize the benefits from tourism, they would preserve the surrounding environment, cultural and natural resources for their long-term development. It would raise the awareness among local people and tourists and stimulate planning—administrative controls in tourism destinations.

On the other hand, there are doubts from previous studies that the more economic increasing, the more environmental problems would incur. Aref et al. (2009) indicate that environment got more concerns than the economic effects, especially where tourism have developed rapidly without sustainable development policy. The local resident in these places were worried about pollution, wildlife destruction, and other natural resources damage the most. Researchers pointed out that there were adverse effects from tourism development on natural environment and ecosystem around tourism sites because of the facilities building process (Holden, 2000; Hunter & Green, 1995; Pearce, 1989; Telfer & Sharpley, 2008). As the result of poorly planned tourism development and policies, many developing countries had become the victims while mass tourism could significantly damage the local people's living environment and over-exploit their natural heritage (Adongo et al., 2017; Sharpley, 2009; Winter et al., 2008). Moreover, uncontrol mass tourism would increase visual pollution, stress, and annoyance to the local community (Sharp, 2008).

Hence, there has been increasing skepticism among researchers toward positive relationship between tourism and environment sustainability. As local people are suffered the most from environmental degradation, their awareness and behaviors would be important to understand, and their voice should be fully listened to in decision-making process of tourism development.

1.3 Socio-cultural Impacts of Tourism

It is no doubt that the growth of tourism leads to the development of rural areas, accelerates the urbanization and modernization process, promotes free trade, and reduces border barriers. These positive impacts have restructured employment portfolio (people from agriculture sector move to service sector) and encouraged the improvement of technology, especially information technology, AI, smart transport, and e-money. Therefore, tourism has brought huge socio-cultural impacts to the human life.

Tourists are fond of observation and experience of different cultures and traditions. Picard and Robinson (2006) implied that culture is the center of international tourism, helps the tourism industry grow and enable diverse societies to get involve in the development process. Therefore, it not only provides local resident chances to meet new people, promotes their place as a cultural destination, but also fosters pride among them, brings them closer, and gives them chances of relaxation and entertainment.

While acknowledge that tourism brings benefits to the economic growth and enable diverse societies to get involve in the development process, tourism researchers and tourism policymakers have been emphasizing the importance of culture retain. On one hand, they hope tourism can help to preserve local culture and traditions. On the other hand, they criticized international tourism for undermining of local culture, traditional ways of life and encourage Western culture influence on less developer communities. In Tourism Concern (2017), it pointed out that international tourists not only leave physical footprints on a landscape, but also make intangible socio cultural impacts on local traditions, value systems and ways of life. This process would lead to loss of local autonomy, authenticity, and cultural degradation, further creating the concept of increased homogeneity among cultures (Andereck et al., 2005; Nash, 1989; Meethan, 2003; Smith, 2009; Sinclair-Maragh & Gursoy, 2015).

Pizam (1978) pointed out that the negative impacts of tourism on the local resident are from introduction of undesirable activities such as prostitution and gambling, excessive concern for material gains, loss of cultural identity. He concluded that the local people's economic dependency on tourism, income, and occupation are the best predictors of their attitude toward tourism. The study showed that the less dependent a resident is economically on tourism, the more negative his attitude is toward it. And the less attachment to a place, the more negative the one's attitudes toward tourism. However, other studies (Harrill, 2004; Jaafar et al., 2015; Um & Crompton, 1987) disagree with Pizam (1978) as they all indicate from their findings that residents who are more attached to their place perceive tourism development more negatively.

Some of the most notable negative socio-cultural impacts of tourism are related to the flow of traffic and crowded conditions. Crowdedness not only can ruin the resource being conserved, but also spoil the visitor experience. Kim (2016) found that the local people complain about noise pollution and littering due to the increasing number of tourists. In addition, the major negative impacts on local people's life in many famous tourism sites is overcrowding which increases invasion of their privacy and traffic congestion. Moreover, Postma and Schmeuker (2017) and Dogan (1989) indicated that the potential conflicts between tourists and residents may occur from crowding to privacy, from lack of adaptivity to feelings of strangeness in one's own place.

Besides, there is a risk of crime rising together with the increase of tourist arrivals to a place. A higher rate of crime in a destination may lead to the reduce of its attractiveness. Child labor, alcoholism, drug additions, prostitution, and terrorist attacks are problems that got high concerns of tourism stakeholders, policymakers, and local communities (King et al., 1993). These are the main factors that lead to the disruption of the local resident's quality of life.

The challenge of managing the balance between minimizing negative tourism impacts on environment, socio-culture, and optimizing the economic development is the core of a country's tourism policy. Porter and Salazar (2005) pointed out that heritage tourism can also create tensions and conflicts among different stakeholders. Therefore, understanding the tourism policymakers' view and local people's awareness of tourism impacts is not only useful in the conservation of a heritage site, but it also contributes to promote the image of that destination and balance the stakeholders' benefits.

Among the OECD countries, Japan was one of the first countries to recognize the value of intangible cultural heritage (Boyd & Timothy, 2003; Estol & Font, 2016; Kakiuchi, 2014; OECD, 2016). The awareness of heritage protection for the sake of the entire nation was started since the beginning of Japan's Meiji government (1868–1912) as a part of its public policy (Kakiuchi, 2014). Through 150 years with a lot of socio-economic change, natural and cultural heritage conservation has always been the core of Japanese law and policies which makes provisions for the support of cultural activities by all the stakeholders and has played a great role in heritage tourism development in this country.

This study aims to (1) identify the tourism impacts on local communities in Japanese heritage sites, (2) understand the local people's awareness and consideration between positive tourism impacts and negative tourism impacts and their responses, (3) investigate the difference between Japanese government policymakers and local people's points of view about tourism impacts and development, and (4) compare the difference of tourism impacts practice in Japan heritage sites and other destinations in previous studies.

2 Methodology

The research framework for this study is generated from previous literature and research works which shows the impacts of tourism on local community and the efforts of government tourism policymakers to control the negative impacts and promote the positive ones. The framework (Fig. 1) highlights the importance of understanding the perception, attitude, and needs of the local communities in the Japanese heritage sites toward tourism impacts.

The study was divided into two phases as it based on both qualitative and quantitative surveys on local people living in Japanese heritage sites, Japanese government officers, academic people, and people working in tourism area.

- In-depth interviews with some governments officers of Japanese Ministry of Justice, Agency of Cultural Affair (Ministry of Education, Culture, Sports, Science and Technology), local government authority in some Japanese heritage sites and academic people about their ideas and views toward the development of heritage tourism and its impacts on local communities. The interviews were conducted from July 2019 to December 2019.
- A questionnaire was designed to explore the local people living in some heritage sites in Japan about their perceptions, attitudes, and awareness on tourism policy and tourism impacts to their places and life. In this study, 31 items measuring the negative impacts and positive impacts of tourism on social-cultural life, economic development, and environment at the sites and local people's attitude toward tourism development and policy were examined. The items chosen are widely used in international travel literature. A 5-point rating Likert scale where 1 = strongly disagree, 3 = neutral, and 5 = strongly agree were applied to quantify the responses to the items.
- From November 2019 to February 2020, the respondents were recruited on-site. After that, due to the risk of COVID-19 and Japanese government's declaration of "Emergency Situation", the questionnaire had to conduct online via some social media forums and travel blogs related to Japanese life, heritage, and tourism. People were asked if they are living in some heritage areas in Japan and were invited to participate voluntarily into the survey. Up to 15 April 2020, a total of 266 answers was collected which 180 of them were from on-site and the rest were from online method. After checking, 245 answers were usable for analysis. It is believed that all respondents answered the questionnaire honestly as it was anonymous and self-administered.

3 Findings and Discussion

3.1 From Government's View

The tourism management framework in Japan has some differences to other countries. While some other developed countries have well developed systems of heritage tourism law and regulation to conserve and manage the tourism impacts in both general and local levels, the Japanese government separates tourism and heritage laws and policies

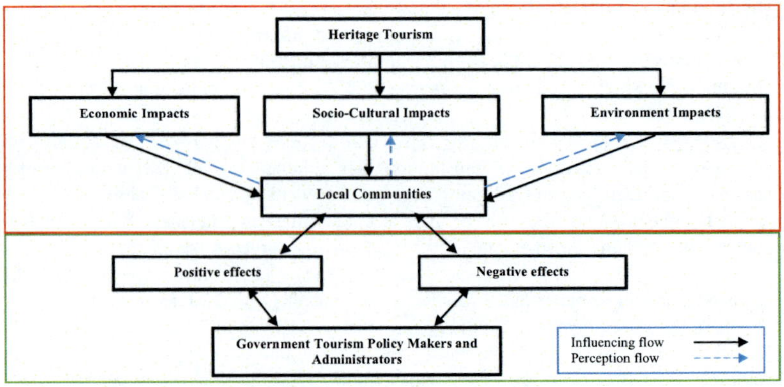

Fig. 1 Research framework

(Fig. 2). Since 1919, the detailed and specialized policies have been given to each type of heritage: cultural heritage, natural heritage: *satoyama* (for mountain) and *satoumi* (for ocean) strategy, agricultural heritage (GIAHS), *onsen* heritage and festivals.

In Japan, the local government in each municipality or prefecture plays an important role in tourism policy and control. Central government and ministries only give out the general directions and support upon the request of local governments. Within the heritage tourism's stakeholder network (Fig. 3), the local government plays as the central node to communicate with other nodes and control the tourism activities and policies within its area. However, NGOs and academic people contribute highly to tourism development policy planning and suggestion to the local government. NGOs and academic people also work closely with local communities to understand their difficulties and needs and help them to solve out the problems or take note for the local government.

However, there are some problems that both the central government officers and local governments find they are not easy to have a solution in the short term:

3.1.1 Aging Population and Cultural Heritage Fading

Japan has been experiencing the issue of aging population to an unprecedented degree. More than 20% of Japan's population is over 65 years old, the highest proportion in the world (Ministry of Internal Affairs and Communications, Statistical Handbook of Japan, 2018). This not only takes effects on the economic growth, but also makes change in family and social structures in the world's third largest economy.

Recently, the aging problem is becoming more and more serious. In rural and remote areas, where retain many Japanese cultural and natural heritage, it is only mostly elderly people who are left behind after the young have moved to urban areas for schools and jobs. According to the government officers, it would be a threat to the cultural heritage retaining in Japan. Cultural heritage includes tangible and intangible heritage which pass from generation to generation. Therefore, the cultural heritage belongs to and live within the local community, through the conservation of its people. As there is a little proportion of young people living in the rural and heritage sites, people have less chances to know or learn their cultural heritage. If the elderly people who possess the cultural heritage passed away, there would be a risk that some cultural heritage will disappear.

Some of the interviewed government officers and academic people share the same concern of many unique cultural traditions and knowledge are being lost as Japan ages and rural towns face depopulation. According to a survey by Kyodo News in January 2017, 60 cultural events in 20

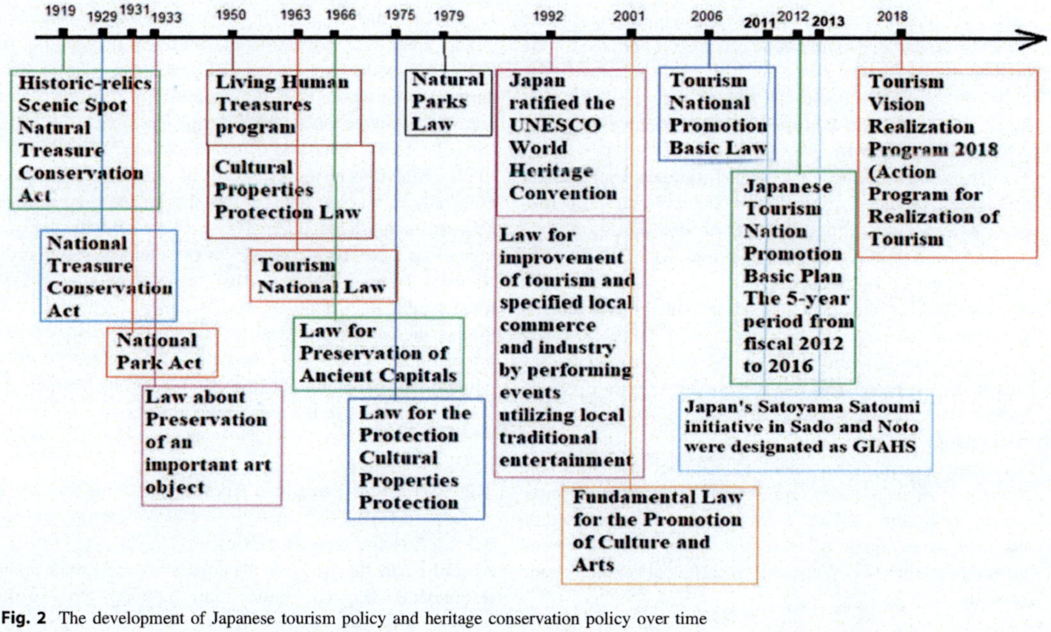

Fig. 2 The development of Japanese tourism policy and heritage conservation policy over time

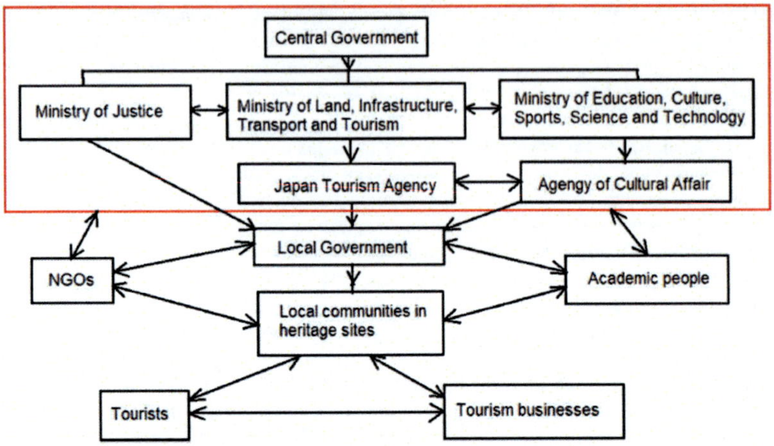

Fig. 3 Management and information flows among Japanese heritage tourism stakeholders

prefectures have been shelved due to falling populations and aging. Mr. Hideo Nigata (Deputy executive director of the Nippon Matsuri Network) indicated that "the disappearance of Japanese festivals means the disappearance of communities". Mr. Hiromichi Kubota (Head of intangible folk cultural properties section at the National Research Institute for Cultural Properties) agrees with this idea as he pointed out that the health of regional traditions reflects the overall health of the community.

However, all the interviewees agree that tourism, especially heritage tourism, would be an ideal solution for heritage and culture conservation. As the people recognize the benefit from it, they would like to retain their heritage for the next generation.

3.1.2 Constrained Budget for Cultural Heritage Conservation

The Japanese government has tried to allocate a support fund for cultural heritage conservation and cultural properties protection. However, this amount is very small. As the national level, the Agency for Cultural Affair (ACA) has prime responsibility for culture, and its budget has remained as only 0.1% of the total general account of the national government for several decades (Kakiuchi, 2018). Looking at a breakdown, roughly 60% of the ACA budget is now allocated to heritage and the rest for arts support. Of the budget for heritage, roughly 40% of the ACA budget is allocated to heritage protection and 15% is allocated to maintenance and management of national museums and theaters of heritage protection (ACA, 2013). Therefore, the budget for cultural heritage conservation is mainly based on the local government in each prefecture, and the central government officers play the role as technical supporters and advisors.

3.1.3 Shortage of Labor Force in Tourism Industry

The aging population also leads to another problem of Japanese tourism, especially in rural areas. That is the shortage of labor force in tourism industry. The working-age population (15–64 years old) percent in Japan is decreasing significantly. It is forecasted that after 2055, the projection of working-age population would fluctuate around 50% of the total Japanese population and the child population (0–14 years old) is only around 10% (Statistical Handbook of Japan, 2018). However, currently, in many rural areas in Japan, there are no children or youth. In some villages in remote areas, all the local people are over 60 years old, and many of them are over 70 years old. In Japanese urban areas, the ratios and numbers of seniors are also increasing (Institute of Population & Social Security Research, 2013; Ministry of Internal Affairs & Communications, 2015).

This situation raises a concern of the shortage of labor, especially in tourism industry. As the Japanese government plans to increase the international tourist arrivals and turns Japan into a "tourism country" as per the declaration of the National Tourism Policy (2016), the lack of tourism labor force would be a problem to the tourism authority. Even though the Ministry of Justice has planned to attract more international labors to work in Japan, the lack of working-age people in Japanese rural areas and heritage sites is still a question to the local government.

3.2 To Local People's Awareness

3.2.1 Descriptive Statistics

According to the frequency analyses on the profile of the respondents (Fig. 4), most came coming from Tokyo

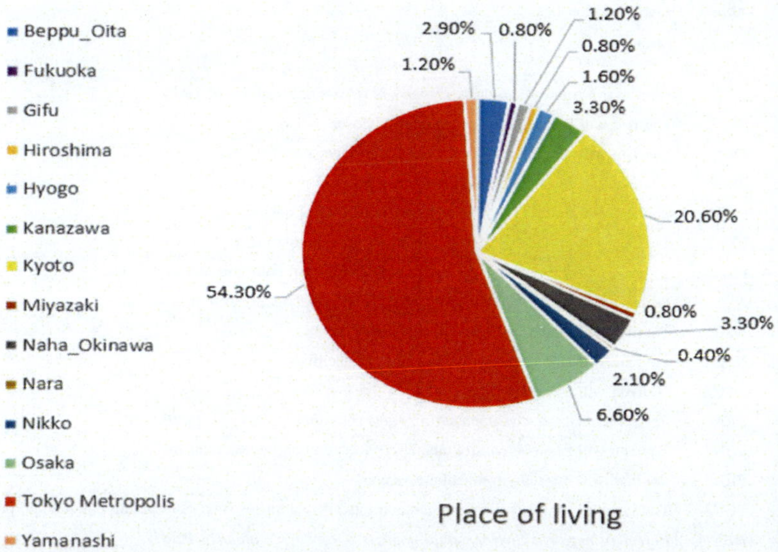

Fig. 4 Provision of "Place of living"

Metropolis area (54.3%) and Kansai area (Kyoto, Osaka, Hyogo, and Nara) (29.2%). These two areas are home to many of Japan's cultural heritage and natural heritage sites. According to the annual statistical reports of the Japan National Tourism Organization (JNTO, 2019), Tokyo and Kansai areas got the largest number of international tourist arrivals. Therefore, there are higher concerns of tourism impacts in these areas.

There are 53.5% of the respondents were local born, while 46.5% of them were coming from another places. Most of the respondents were from 25 to 40 years of age (44.4%). People under 25 years old were 24.3% and those people from 40 to 60 years of age were 28.4%. The rest (2.9%) was over 60 years old. One third of the respondents are working in tourism or tourism-related area (hotels, restaurants, transport, logistics, etc.)

Table 1 indicates the mean and standard deviation values of the 31 items. In general, almost all had a score above the neutral point of 3. The highest mean of 4.70 was found for "Tourism will increase business opportunities". All other items in the "Perceived positive economic impacts" category have high mean from 4.47 to 4.70, which mean that local people realize that tourism bring many benefits to them in economic development: business opportunities, employment, infrastructure, and public service improvement. In fact, the items in this category have the highest means among all perceived positive impacts categories in this study. Local people are also aware that heritage tourism helps to improve their destination image and reputation worldwide, so that they feel proud of their heritage and would like to retain it for their next generation, especially the cultural heritage.

Regarding negative impacts of tourism, local people found the most threatened issues to their life are "disease spreading" (NEN6 = 4.05), "littering increasing" (NEN4 = 3.97), "tax rates and living costs increasing" (NEC1 = 3.57), and "overcrowded of local facilities usage" (NSC2 = 3.54). These problems are not only in Japan, but also incur in many other famous tourism destinations around the world. Many studies recently have indicated that local communities are tired and annoyed of too many tourists coming to their place which leading to negative impacts to their life. Besides, the respondents in this study showed their concerns of "disease spreading" the most as some of them were learning from the COVID-19 pandemic circumstance.

Only two items "Tourism will disrupt residents' quality of life" (NSC1) and "Tourists will bring some bad behaviors to the local people" (NEC3) have the score below neutral point of 3, which are 2.67 and 2.86, respectively. The rest of negative impacts' means are slightly above the neutral point, from 3.1 ("Tourism will increase crime—NSC3") to 3.49 ("Tourism will increase air pollution—NEN5"). This means that Japanese local people in heritage tourism sites are aware of negative tourism impacts, but they do not think these negative impacts could disrupt their quality of life. In other words, the negative impacts of tourism are not high in the perception of local resident.

As mentioned above, the local people found the most positive impacts from tourism on the economy, however, they found tourism have least positive impacts on their

Table 1 Descriptive statistics

	Items	Mean	Std. deviation
PSC	*Perceived positive socio-cultural impacts (Cronbach's alpha = 0.838)*		
PSC1	Tourism will bring the local community closer	3.87	1.106
PSC2	Tourism will provide residents a chance to meet new people	4.52	0.815
PSC3	Tourism will foster pride among local resident	4.49	0.784
PSC4	Tourism will promote this place as a multi-cultural destination	4.45	0.838
PSC5	Tourism will provide residents relaxation and entertainment	3.94	1.035
PSC6	Tourism will strengthen local community bonds and cohesion	4.31	0.863
NSC	*Perceived negative socio-cultural impacts (Cronbach's alpha = 0.810)*		
NSC1	Tourism will disrupt residents' quality of life	2.67	1.208
NSC2	Tourism will lead to overcrowding of local facilities	3.54	1.196
NSC3	Tourism will increase crime	3.10	1.207
PEN	*Perceived positive environmental impacts (Cronbach's alpha = 0.805)*		
PEN1	Tourism will improve environmental conservation and protectionism	3.60	1.065
PEN2	Tourism will raise environmental awareness	3.77	1.023
PEN3	Tourism will stimulate planning and administrative controls such as recycling policies and pollution controls	3.88	0.981
NEN	*Perceived negative environmental impacts (Cronbach's alpha = 0.884)*		
NEN1	Tourism will damage the natural environment	3.18	1.10602
NEN2	Tourism will increase noise pollution	3.47	1.12531
NEN3	Tourism will increase visual pollution	3.15	1.08458
NEN4	Tourism will increase littering	3.97	1.07334
NEN5	Tourism will increase air pollution	3.49	1.15838
NEN6	Tourism will spread disease faster	4.05	0.99234
PEC	*Perceived positive economic impacts (Cronbach's alpha = 0.885)*		
PEC1	Tourism will provide locals employment opportunities	4.68	0.62523
PEC2	Tourism will improve the provision of public services and infrastructures	4.50	0.74064
PEC3	Tourism will increase business opportunities	4.70	0.55696
PEC4	Tourism has led to the regeneration and redevelopment of towns and cities	4.51	0.74062
PEC5	Tourism will enhance this place's international reputation through world media exposure	4.60	0.63075
PEC6	Tourism will improve this place's image worldwide	4.47	0.76757
PEC7	Tourism will foster pride of the local people about their cultural traditions (dance, folk song, history, food, handicraft, etc.), and the local people will retain these heritages for their next generations	4.56	0.67385
NEC	*Perceived negative economic impacts (Cronbach's alpha = 0.627)*		
NEC1	Tourism has led to increased tax rates and living costs for local resident	3.57	1.11986
NEC2	The large investment required to develop tourism cannot be justified in terms of the economic benefits that will be generated for residents	3.29	1.15321
NEC3	Tourists will bring some bad behaviors to the local people	2.86	1.06118
OAT	*Overall local people's attitude toward tourism*		
OAT1	Overall tourism's positive impacts will outweigh its negative ones	4.34	0.76149
OAT2	The local government should hold more tourism events to promote and develop the tourism potentials in this place	4.40	0.78803
OAT3	Overall, I support tourism development and tourism policy in this place as a resident	4.32	0.81048

Note Likert 5-point scale including strongly disagree (1), neutral (3), strongly agree (5) environment. The mean scores for the items in this category are from 3.60 to 3.88.

Despite of those perceived negative tourism impacts, the local people are proud of their heritage value. They believe that "tourism's positive impacts outweigh its negative ones" (OAT1 = 4.34) and they wish "the local government should hold more tourism events to promote and develop the tourism potentials in their places" (OAT2 = 4.40). Therefore, they agree to "support tourism development and tourism policy" in their places (OAT3 = 4.32).

3.2.2 Factor Analysis of Perceived Tourism Impact Items

A principal component factor analysis (Table 2) with Varimax rotation (Hair et al., 1998; Tho, 2012) using 28 dependent variables was undertaken to determine the dimensions underlying the perceived tourism impact items.

The 28 items consist of six factors with eigenvalues higher than 1.0. The factors accounted for 65,576% of the variance and were labeled: "Perceived positive social-cultural impacts", "Perceived negative social-cultural impacts", "Perceived positive environmental impacts", "Perceived negative environmental impacts", "Perceived positive economic impacts", and "Perceived negative economic impacts". All items revealed factor loadings of over 0.5 and communalities values for each variable, which accounts for the variances explained by the factors, ranged from 0.513 to 0.781, indicating that each variable contributes to forming the factor structure.

3.2.3 Differences in Local People' Perception and Attitude Toward Tourism Impacts in Heritage Sites in Japan According to Sociodemographic Variables

The differences in local people's perception and attitude toward tourism impacts in Japanese heritage sites according to their "living places", "place attachment" (whether they were born in their living place or not), "job" (whether their jobs relate to tourism area or not), and "age" were tested using a one-way ANOVA and independent samples *T* test.

According to "Place of living", the respondents were divided into 3 groups: (1) those who are living in Tokyo Metropolis area", (2) those who are living in the Kansai area (Kyoto, Osaka, Hyogo, Nara), (3) people who are living in other heritage sites in Japan. Using one-way ANOVA test to compare the difference of perceived tourism impacts of these groups, the people in Kansai area and other heritage sites believe that heritage tourism would bring them closer, stronger, and more cohesive than those are living in Tokyo Metropolis, as they have higher mean scores of PSC1 (4.14 in Kansai, 3.95 in other areas, compared to 3.70 in Tokyo Metropolis) and PSC6 (4.46 in Kansai, 4.53 in other areas, compared to 4.17 in Tokyo Metropolis). Besides, the NSC1's mean score of Tokyo Metropolis local people is 2.89 which is highest among all other areas, indicates that even though their perception regarding "Tourism will disrupt residents' quality of life" is below the neutral point, there are more people in Tokyo area concerned of this issue than those in Kansai area (NSC1 = 2.39) and other heritage sites (NSC1 = 2.45).

In one other hand, people in other heritage sites in Japan have highest "perceived positive economic impacts" mean score of PEC1 (4.88), PEC2 (4.75), PEC5 (4.75), and PEC7 (4.75) in comparison with those live in Tokyo Metropolis (4.64, 4.39, 4.57, and 4.48, respectively) and Kansai area (4.66, 4.56, 4.58, and 4.59, respectively). This means that people in more rural or remoted areas believe that tourism would bring more benefits to their economy as it increases "employment and business opportunities", improves "infrastructure and public service", enhances the place's image worldwide and local economy, and conserves the "heritage for the next generation" than those in urban areas.

According to the question "is there any difference in perceived tourism impacts between people who were born in that heritage sites and people are coming from other places?", it is interesting to know that people who were born in the places they are living perceive more negative socio-cultural impacts of tourism than people from other places, assuming that local-born people are more attached to where they live. They concern more for disruption of life quality (NSC1 = 2.89), overcrowding of local facilities (NSC2 = 3.66), and crime increase (NSC3 = 3.25) while comparing with people coming from other places (NSC1 = 2.43; NSC2 = 3.40; NSC3 = 2.94). However, both the two groups have similar perception about other tourism impacts. From this result, it agrees partly with previous studies of Jaafar et al. (2015), Harrill (2004), Um and Crompton (1987) as they indicated that residents who are more attached to their place perceive tourism development more negatively. And this result disagrees with argument of Pizam (1978) that the less attachment to a place, the more negative the one's attitudes toward tourism.

However, when analyzing the perceived tourism impacts of people who their jobs relate to tourism area, such as hotels, restaurants, transport, stations, there are cognitive differences in "Tourism will disrupt residents' quality of life" (NSC1), "Tourism will lead to overcrowding of local facilities" (NSC2), and "Tourism has led to increased tax rates and living costs for local residents" (NEC1) between people whose jobs relate to tourism and those whose unrelated jobs. Those people whose jobs relate to tourism have higher negative perception in these items (NSC1 = 2.90, NSC2 = 3.79, NEC1 = 3.88) than people not working in tourism area (NSC1 = 2.56, NSC2 = 3.41, NEC1 = 3.42).

In some heritage sites in Japan, such as Kyoto, Osaka, Kanazawa, Beppu, tourists pay extra tax on their stays at

Table 2 Principal component factor analysis with Varimax rotation

	Factor loading						Communalities
	1	2	3	4	5	6	
PEC5	0.846						0.751
PEC7	0.791						0.713
PEC1	0.779						0.711
PEC3	0.775						0.678
PEC4	0.662						0.553
PEC2	0.616						0.576
PEC6	0.608						0.563
NEN5		0.826					0.743
NEN4		0.820					0.729
NEN2		0.735					0.706
NEN6		0.704					0.580
NEN1		0.687					0.669
NEN3		0.664					0.631
PSC3			0.782				0.714
PSC2			0.725				0.639
PSC1			0.721				0.643
PSC6			0.656				0.595
PSC4			0.635				0.513
PSC5			0.571				0.551
PEN1				0.847			0.735
PEN2				0.812			0.764
PEN3				0.711			0.657
NSC2					0.782		0.781
NSC3					0.781		0.745
NSC1					0.722		0.655
NEC2						0.804	0.671
NEC1						0.625	0.580
NEC3						0.552	0.513
Eigenvalues	4.363	4.153	3.432	2.388	2.217	1.809	
% of variance	15.582	14.831	12.256	8.528	7.917	6.462	

Extraction method: Principal component analysis
Rotation method: Varimax with Kaiser normalization
[a]Rotation converged in 7 iterations

hotel and traditional *ryokan* inns. These accommodation taxes first appeared in large metropolitan areas. Tokyo adopted one in 2002, followed by Osaka Prefecture in 2017 and Kyoto city in 2019. This policy is followed by Nara, Kitakyushu, Fukuoka, and Kanazawa. Roughly, 20 municipalities are considering doing so, as a survey by Nikkei found in 2019. The revenue from this tax is typically used for purposes such as building tourism infrastructure and providing information to visitors.

People working in area related to tourism have better perceived positive environment impacts as their mean score of "Tourism will improve environmental conservation and protectionism" (PEN1 = 3.82) and "Tourism will stimulate planning and administrative controls such as recycling policies and pollution controls" (PEN3 = 4.05) are higher than those of people whose jobs are not related to tourism (PEN1 = 3.48 and PEN3 = 3.80). This result agrees with Pizam (1978) that the less dependent a resident is economically on tourism, the more negative his attitude is toward it.

According to age, there are differences in the perceived negative environment impacts between people over 60 years of age and people in other groups. The people over 60 years old seems less negative about the tourism impacts on environment. The mean scores of "Tourism will damage the natural environment" (NEN1 = 2.57), "Tourism will increase noise pollution" (NEN2 = 2.29), "Tourism will increase visual pollution" (NEN3 = 2.43), and "Tourism will increase air pollution" (NEN5 = 2.71) of this group are all below the neutral point of 3, which mean they are inclined to disagree with these statements. However, people in groups "Under 25", "From 25 to 40" and "From 40 to 60" years old have the mean scores above neutral point, indicate their concern of negative environment impacts from tourism in these issues.

In conclusion, although there are some differences in local people's perception of tourism impacts in some items, people in Japan heritage sites generally have relatively equal attitudes and perceptions about the tourism impact on local economy, environment, and socio-culture. They highly appreciate the positive effects of tourism on local economic development and heritage value and are most concerned about the negative impacts of tourism on the local environment. However, they believe the local government's tourism policy will help to solve these problems. Therefore, people in all heritage sites agree that tourism's positive impacts will outweigh its negative one and they support tourism activities and government's tourism development policy in their places.

4 Conclusion

The results from the in-depth interview with Japanese government officers, local government people and academic people, and the quantitative research on local people's perception and attitude about tourism impacts on their place help to understand and answer the research questions of this study. Although there are some concerns of tourism impacts on environment and socio-culture of the heritage sites, people agree that tourism brings many benefits to the local community development, especially in economy and reputation. The local people are proud of their heritage and would like to introduce it to the outside world. All the government, academic people, and local people believe that tourism's positive impacts will outweigh its negative ones and the local government should hold more tourism events to promote and develop the tourism potentials in the heritage sites.

However, there are gaps between the government's view of tourism development and local people's needs. As Japanese government wants to boost up the number of international tourist arrivals to Japan in the coming years, the government officers want to have more transportations to connect Japanese main cities to rural heritage sites, especially shinkansen. However, from the result of the survey, local people are concerned the most about the overcrowding of their local facilities, littering, noise, and air pollution due to the increase of number of tourists. The Japanese government officers do not think these are big problems if the tourists will be educated and informed clearly how to behave properly during their travels in Japan.

In another hand, Japan has a good welfare and pensions for its elderly, many local people in rural areas and heritage sites indicate that they participate in the tourism activities not for the economic benefits. Living in rural areas, they have the habit of self-cultivating and raising based on nature as their own food source. Many Japanese people follow the *ikigai* philosophy or minimalist living so that they do not have many personal needs for luxury goods or entertainment. Some people expressed their concerns of fading culture and traditions, that is why they want to participate into heritage tourism as they want to educate the young tourists about the traditional knowledge, culture, and working methods. The local people in Japanese rural heritage sites are mostly elderly, so that they would be tired if there are too many tourists come to their place in a short time. They are eager to have guests but within a limited number. Some senior local people are worried that after they became too old or passed away, no one will continue to operate heritage tourism in their places, as all the young people wants to go to urban cities to pursue other careers. In the meantime, the local government wants to have more international tourists coming to its heritage sites to contribute to the economic benefits of the local communities and introduce their culture and beauty nature to the globe.

In conclusion, even though there are some issues need to have improvement for sustainable development of the local community in the Japan heritage sites, the local people are supportive to the tourism activities in their place and government policies toward the heritage tourism development.

References

Adongo, R., Choe, J. Y., & Han, H. (2017). Tourism in Hoi An, Vietnam: Impacts, perceived benefits, community attachment and support for tourism development. *International Journal of Tourism Sciences, 17*(2), 86–106. https://doi.org/10.1080/15980634.2017.1294344

Agency for Cultural Affair. (2013). White papers and Annual Report. https://www.bunka.go.jp/english/report/annual/

Andereck, K. L., Valentine, K. M., Knoft, R. C., & Vogt, C. A. (2005). Residents' perceptions of community tourism impacts. *Annals of Tourism Research, 32*(4), 1056–1076. https://doi.org/10.1016/j.annals.2005.03.001

Aref, F., Redzuan, M., & Gill, S. S. (2009). Community perceptions toward economic and environmental impacts of tourism on local communities. *Asian Social Science, 5*(7), 130–137.

Ashworth, G. J., & Van der Aa, B. J. (2006). Strategy and policy for the world heritage convention: goals, practices and future solutions. In *Managing world heritage sites* (pp. 173–184). Routledge.

Boyd, S. W. & Timothy, D. J. (2003). *Heritage Tourism*. Pearson Education.

Doğan, H. Z. (1989). Forms of adjustment: Sociocultural impacts of tourism. *Annals of tourism research, 16*(2), 216–236. https://doi.org/10.1016/0160-7383(89)90069-8
Estol, J., & Font, X. (2016). European tourism policy: Its evolution and structure. *Tourism Management, 52*, 230–241. https://doi.org/10.1016/j.tourman.2015.06.007
Figgis, P., Bushell, R., & Eagles, P. F. (2007). Tourism as a tool for community-based conservation and development. In *Tourism and Protected Areas: Benefits Beyond Boundaries: The Vth IUCN World Parks Congress* (p. 101). CABI.
Garrod, B., & Fyall, A. (2000). Managing heritage tourism. *Annals of Tourism Research, 27*(3), 682–708. https://doi.org/10.1016/S0160-7383(99)00094-8
Girard, L. F., & Nocca, F. (2017). From linear to circular tourism. *Aestimum, 70*.
Hall, C. M. (2003). Politics and place: An analysis of power in tourism communities. *Tourism in destination communities* (pp. 99–114).
Hall, C. M., & McArthur, S. (1998). *Integrated heritage management: Principle and practice*. Stationery Office.
Hair, J. F., Black, W. C., Babin, B. J., Anderson, R. E., & Tatham, R. L. (1998). *Multivariate data analysis*. Prentice Hall.
Harrill, R. (2004). Residents' attitudes toward tourism development: A literature review with implications for tourism planning. *Journal of Planning Literature, 18*(3), 251–266. https://doi.org/10.1177/0885412203260306
Hoa, N. Q. (2016). Marketing cultural heritage to promote tourism growth in peripheral areas of vietnam for sustainable economic development. In *Proceedings of the Annual Vietnam Academic Research Conference on Global Business, Economics, Finance Social Sciences, Hanoi, Vietnam, August* 7–9.
Holden, A. (2000). Winter tourism and the environment in conflict: The case of Cairngorm, Scotland. *International Journal of Tourism Research, 2*(4), 247–260. https://doi.org/10.1002/1522-1970(200007/08)2:4%3c247::AID-JTR214%3e3.0.CO;2-X
Hunter, C., & Green, H. (1995). *Tourism and the environment: A sustainable relationship?* Routledge.
Institute of Population and Social Security Research. (2013). *Regional population projection for Japan: 2010–2040*.
Ioannides, D., & Debbage, K. G. (Eds.). (1998). The economic geography of the tourist industry: A supply-side analysis. Psychology Press.
Jaafar, M., Noor, S. M., & Rasoolimanesh, S. M. (2015). Perception of young local residents toward sustainable conservation programmes: A case study of the Lenggong World Cultural Heritage Site. *Tourism Management, 48*, 154–163. https://doi.org/10.1016/j.tourman.2014.10.018
JNTO. (2019). *Japan Tourism Statistics*. https://statistics.jnto.go.jp/en/
Kakiuchi, E. (2014). Cultural heritage protection system in Japan: Current issues and prospects for the future. *National Graduate Institute for Policy Studies, 2*, 1–12.
Kakiuchi, E. (2018). *Cultural creative industries from a cultural policy perspective: The case of Japan*. Edward Elgar Publishing.
Kim, S. (2016). World Heritage Site designation impacts on a historic village: A case study on residents' perceptions of Hahoe Village (Korea). *Sustainability (switzerland), 8*(3), 258. https://doi.org/10.3390/su8030258
King, B., Pizam, A., & Milman, A. (1993). Social impacts of tourism—Host perception. *Annals of Tourism Research, 20*(4), 650–665. https://doi.org/10.1016/0160-7383(93)90089-L
Liu, J. C., & Var, T. (1986). Resident attitudes toward tourism impacts in Hawaii. *Annals of Tourism Research, 13*(2), 193–214.
Manniche, J., Topsø Larsen, K., Brandt Broegaard, R., & Holland, E. (2017). *Destination: A circular tourism economy: A handbook for transitioning toward a circular economy within the tourism and hospitality sectors in the South Baltic Region*.
Meethan, K. (2003). Mobile cultures? Hybridity, tourism and cultural change. *Journal of Tourism and Cultural Change, 1*(1), 11–28. https://doi.org/10.1080/14766820308668157
Ministry of Internal Affairs and Communications. (2015). *Population Census Report—Japan*.
Ministry of Internal Affairs and Communications. (2018). *Statistical handbook of Japan, 2018*. Statistics Bureau.
Moli, P. (2011). Community-based eco cultural heritage tourism for sustainable development in the Asian region: A conceptual framework. *International Journal of Social Ecology and Sustainable Development, 2*(2), 66–80. https://doi.org/10.4018/jsesd.2011040106
Nash, D. (1989). Tourism as a form of imperialism. *Hosts and Guests—The Anthropology of Tourism*.
Neef, A. (2019). Tourism, land grabs and displacement—A study with particular focus on the global south. *Report, Tourism Watch* (pp. 27–29).
Nicholas, L. N., Thapa, B., & Ko, Y. J. (2009). Residents' perspectives of a world heritage site: The Pitons Management Area, St. Lucia. *Annals of Tourism Research, 36*(3), 390–412. https://doi.org/10.1016/j.annals.2009.03.005
Tho, N. D. (2012). Scientific research methods in business. *Labor and Social Publishing House*.
OECD. (2016). *OECD tourism trends and policies 2016*. OECD Publishing. https://doi.org/10.1787/tour-2016-en
Pearce, D. (1989). Social impacts of tourism. In *The social, cultural and environmental impacts of tourism*. New South Wales Tourism Commission.
Perdue, R. R., Long, P. T., & Allen, L. (1990). Resident support for tourism development. *Annals of Tourism Research, 17*(4), 586–599.
Pizam, A. (1978). Tourism's impacts: The social costs to the destination community as perceived by its residents. *Journal of Travel Research, 16*(8), 8–12. https://doi.org/10.1177/004728757801600402
Picard, D., & Robinson, M. (2006). Remaking worlds: Festivals, tourism and change. *Festivals, Tourism and Social Change: Remaking Worlds, 8*, 1–31.
Porter, B. W., & Salazar, N. B. (2005). Heritage tourism, conflict, and the public interest: An introduction. *International Journal of Heritage Studies, 11*(5), 361–370. https://doi.org/10.1080/13527250500337397
Postma, A., & Schmuecker, D. (2017). Understanding and overcoming negative impacts of tourism in city destinations: Conceptual model and strategic framework. *Journal of Tourism Futures, 3*(2), 144–156. https://doi.org/10.1108/JTF-04-2017-0022
Shah, K., & Gupta, V. (2000). *Tourism, the poor and other stakeholders: Experience in Asia*. Overseas Development Institute.
Sharp, A. (2008, 13 March). Petrol prices and drought-hit Victorian tourism. *The Age*. Retrieved January 22, 2015, from http://www.theage.com.au/news/management/petrol-prices-and-drought-hit-victorian-tourism/2008/03/13/1205126092295.html
Sharpley, R. (2009). *Tourism development and the environment: Beyond sustainability?* Routledge.
Smith, M. K. (2009). *Issues in cultural tourism studies*. Routledge.
Sinclair-Maragh, G., & Gursoy, D. (2015). Imperialism and tourism: The case of developing island countries. *Annals of Tourism Research, 50*, 143–158. https://doi.org/10.1016/j.annals.2014.12.001

Telfer, D. J., & Sharpley, R. S. (2008). *Tourism and development in the development in the USA and Canada*. Routledge.

Tourism Concern. (2017). *Ethical travel guide*

Um, S., & Crompton, J. L. (1987). Measuring resident's attachment levels in a host community. *Journal of Travel Research, 26*(1), 27–29. https://doi.org/10.1177/004728758702600105

UNWTO. (2018). *Overtourism? Understanding and managing urban tourism growth beyond perceptions*. ISBN: 97892-844-2064-3.

Walpole, M. J., & Goodwin, H. J. (2001). Local attitudes towards conservation and tourism around Komodo National Park, Indonesia. *Environmental Conservation, 28*(2), 160–166. https://doi.org/10.1017/S0376892901000169

Wang, Y., & Pfister, R. E. (2008). Residents' attitudes toward tourism and perceived personal benefits in a rural community. *Journal of Travel Research, 47*(1), 84–93. https://doi.org/10.1177/0047287507312402

Winter, T., Teo, P., & Chang, T. C. (Eds.). (2008). *Asia on tour: Exploring the rise of Asian tourism*. Routledge.

WTTC. (2009). World Travel and Tourism Council.

Exploring the Unlimited and Unexplored Rural Tourism in Meghalaya, North East India

Haziel Mercy Buam and M. A. Naseer

Abstract

Northeast India is a region that boasts of various wonders that are classified under natural, cultural, and heritage which makes them unique from the rest of India. Among the seven states in the northeast region, the State of Meghalaya is also known for similar resources, and hence, tourists venture into these wonders throughout the year. Cherrapunjee in Meghalaya being branded as the wettest place on earth, it became quite natural for the tourists to visit the State, especially during the monsoon season. Consequently, with this development, it started piquing the interests of the researchers and scientists leading to the discovery of the Meghalayan Age in 2018 and the recent discovery in 2019–2021 of the remains of fossils of sauropod dinosaurs dating back to about 100 million years. It is undeniably confirmative that Meghalaya has many more wonders to offer in the tourism sector. At present, the tourism sector of the State has explored adventure and nature tourism only. Meghalaya being a State with villages having unique characteristics, with unique agricultural produce and food preservation techniques, there is a tremendous opportunity for the State to explore rural tourism along with its unparallel cultural heritage. Being landlocked for several years in the past, and having a unique character in terms of demography, socioeconomic conditions, and the environment from the rest of India, tourism is one of the sectors yet to be explored in its entirety. Therefore, this study delves into the nuances of tourism in the rural areas of Meghalaya.

H. M. Buam (✉) · M. A. Naseer
Department of Architecture and Planning, National Institute of Technology Calicut, Calicut, India
e-mail: hazielmercybuam@gmail.com

M. A. Naseer
e-mail: naseer@nitc.ac.in

Keywords

Rural and cultural tourism • Northeast India • Meghalaya • Agriculture

1 Introduction

In contemporary times, nations have propelled themselves forward by advancing into a new era of urbanization. This holds to be true for developing nations as much as it does for developed ones. Urbanization has promoted the development of the tourism sector. The way tourism opens up communities, heritage, and cultural spots to the open world making it more accessible to those who wish to explore the rich heritage and get first-hand experience of the community, ought to adapt to the character of such communities and places, to preserve the same, in their original and most authentic form. This holds to be even more true with the northeastern region of India, where a myriad of tribal communities exists who inhabit a vast landlocked region, sharing borders with Myanmar, Bangladesh, etc. Meghalaya, one of the eight states in the northeastern region of India (Fig. 8.2), has seen a gradual rise in the number of tourist footfalls in the recent past. The State is also seeing the community lands that are being taken care of by the clan or the community, being lost to privatization in this race of urbanization. This has further led to impacting the agricultural sector where more than 80% of the population are primarily involved in. The State is protected by the Sixth Schedule of the Constitution of India, for the sole reason of preserving its ethos and character. Therefore, there exists an immediate requirement of tapping the potential for those still unexplored, especially the Rural Tourism in Meghalaya. This paper is an attempt to look into the potential sectors of tourism in the State of Meghalaya and explore the possibilities of promoting unexplored rural tourism. The rich culture and its preservation, in line with the respective community values, would

A. Mandić et al. (eds.), *Cultural Sustainable Tourism*, Advances in Science, Technology & Innovation,
https://doi.org/10.1007/978-3-031-10800-6_8

remain as the central topic when trying to advance the tourism industry in the State of Meghalaya through Rural Tourism.

2 Methodology

The methodology adopted for the study is depicted in the following figure (Fig. 8.1). The preliminary literature survey conducted led to formulating the aim and objectives of the study. A descriptive type of research design was adopted where a detailed literature review and a primary survey were conducted to understand the present scenario of tourism in the State of Meghalaya as well the potential for future development. In addition to the traditional sectors of tourism, the indigenous practices of various tribes in the State in terms of landholding and land management were studied in detail. The unique style of farming and the agricultural produce that are unique from the State were also studied in detail using an exploratory type of approach through surveys with experts and community members to establish the scope of rural tourism in the State. The above objectives were studied and explored using primary data and secondary data. The primary data was obtained through the primary survey, and secondary data included data obtained from various sources like the Tourism Department of Meghalaya and the Meghalaya Basin Development Authority. The conclusive part of the methodology proposal for promoting rural tourism for the sustainable development of the indigenous communities as well as the State of Meghalaya as a whole is suggested.

3 Meghalaya and Tourism: The Current Scenario

Meghalaya is in the North-Eastern region of India (Fig. 8.2). It has a population of 2,966,889 according to census 2011, and an area of approximately 22,429 km^2. Natural "hot-spots" are present across the region, with diverse topographical rewards. It is an entryway to south-east Asia, and the northeast's folk music, dance, and handicrafts are measured as beautiful works of art. Meghalaya can also be defined by the presence of villages across the State adding up to more than 6000 villages. The Meghalayan Age which followed the identification of the wettest place on Earth has marked Meghalaya as one of the major upcoming tourist hubs in Northeast India. According to the 2011 census, India has about 104 million tribal populations consisting of about 8.6% of the total population of the country (Office of Registrar General & Census Commissioner, 2011). It consists of 365 tribes grouped under 58 tribal communities. These tribes constitute 85.33% of the total population of Meghalaya which is above the national average (8.08%). Meghalaya consists of predominantly three tribes, namely Garo Tribe, Jaintia Tribe, and Khasi Tribe.

Fig. 8.1 Methodology

Fig. 8.2 Map of Northeast and Meghalaya

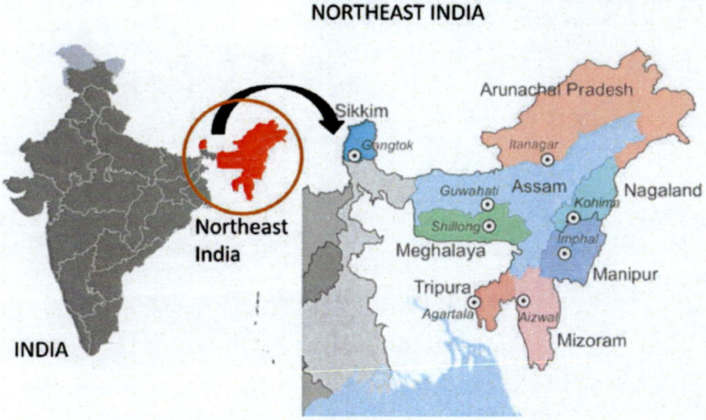

Throughout Meghalaya, there are various tourists spots distributed among various districts. Out of all the districts, the East Khasi Hills District has the most number of tourists spots and destinations followed by other districts (Fig. 8.3).

Meghalaya receives mostly domestic flights from Shillong Airport and a few international flights (indirect). However, most of the international and domestic tourists enter Meghalaya from Guwahati Airport. The government launched additional flights through the "Ude Desh ka Aam Naagrik (UDAN)" scheme of the Centre. Additional flights have been proposed from Shillong with Imphal and Dibrugarh (TimesofIndia, 2020). Airports Authority of India (AAI) is also planning to elevate the runways in seven airports, counting "Shillong Airport," across the country by March 2022 to house Airbus and Boeing aircrafts (Times, 2020). Meghalaya is quite landlocked in terms of air connection. There are several national highways and state highways connecting throughout the State which provides quite an efficient road network across the State (Fig. 8.4). However, several roadways (internal roadways and village roadways) are in need of renovation. World Bank is funding 120 million dollars to propose road projects across Meghalaya to accommodate the "tourism sector, agriculture sector, small-scale industries, and many other sectors" (Bank, 2020).

Even though budget allocation has been increasing in the yesteryears considering the potential of the tourism sector, the expenditure has been meager since the implementation at the grassroot level has been minimal (Refer Table 8.1).

Fig. 8.3 Major tourist destinations across Meghalaya. *Source* Author

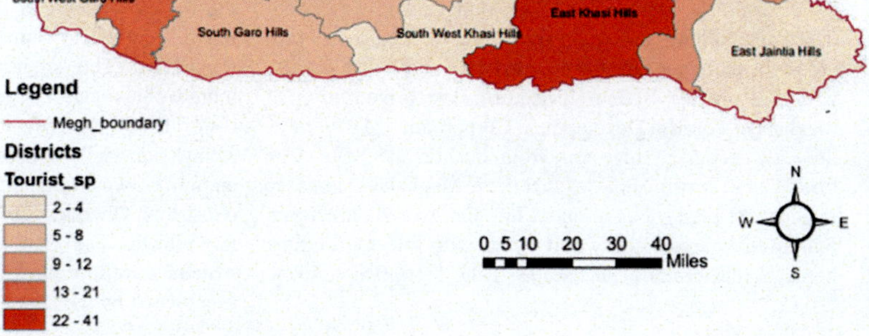

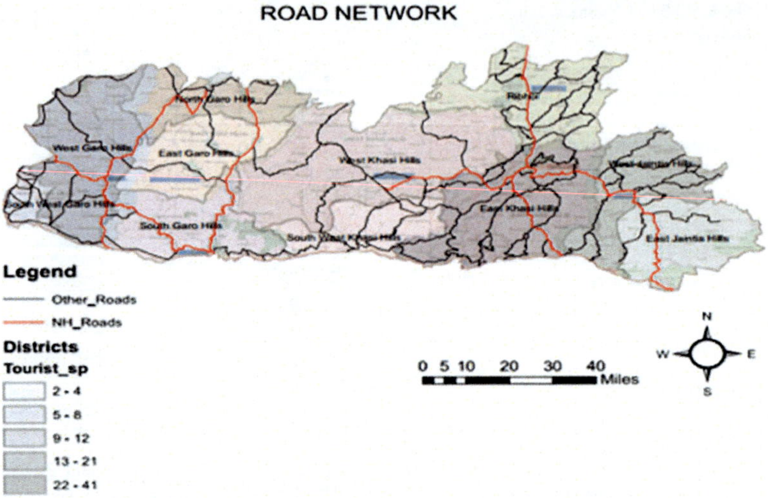

Fig. 8.4 Road network across Meghalaya. *Source* Author

Table 8.1 Budget allocation of Meghalaya tourism

Year	2016–17	2017–18	2018–19	2019–20	2020–2021
Allocation	Rs.2560.00 (in lakhs)	Rs.5119.00 (in lakhs)	Rs. 9329.00 (in lakhs)	Rs. 11,020.00 (in lakhs)	Rs. 11,036.37 (in lakhs)
Expenditure	Rs. 2170.35 (in lakhs)	Rs.1967.80 (in lakhs)	Rs.1903.01 (in lakhs)	Rs.1837.33 (in lakhs)	Rs. 3018.95 (in lakhs)

Source Primary Survey by Author

Revenue generated from tourism across the State has also been increasing through the years proving that tourism in Meghalaya is a promising sector having future scope to improve. Apart from State allocations, there are several cooperative societies in Meghalaya that are helping with the tourism activities of the State. The cooperative societies are able to work at the grassroot level and involve the community as a major stakeholder.

The government adopted a policy resolution declaring tourism as an industry, and tourism promotional activities were initiated. The first Tourism Policy was declared in 2001. Later, the Institute of Hotel Management was established in Shillong and incentives were decided to be provided for entrepreneurs for new projects and house owners who will provide Bed 'n' Breakfast. The privatization of Meghalaya Tourism Development Corporation (MTDC) was done to encourage investors from outside the State. The vision of Tourism Policy, 2011 of Meghalaya aims to facilitate Meghalaya as a tourist hub that would attract various tourists across the world by tapping into the culture, heritage, and beautiful landscape of the State (Department, 2011).

3.1 Tourist Footfall

The total of Indian and Foreign tourists visiting Meghalaya has been increasing steadily through the years except the year 2020 due to the COVID-19 pandemic. The number of domestic tourists visiting Meghalaya is more than the number of foreign tourists (Figs. 8.5 and 8.6).

4 The Traditional Communities of Meghalaya

Khasi, Jaintia, and Garo are the major tribal communities of Meghalaya. The Traditional Institutes of the Khasi Community constitutes a larger territory base (the Hima) which is controlled by the Syiem. The "Rangbah Shnong, Sordar, or Myntri Shnong" which means head of the village council has authority over different types of forestry, its resources (Christoph Oberlack, 2015), and other common properties of the village. The clan chief handles the clan affairs and matters regarding the village. All affairs relating to the forests owned by the clan are watched after and exacted by the

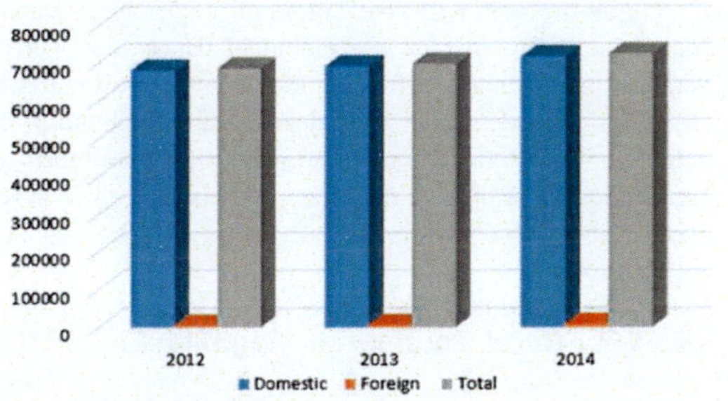

Fig. 8.5 Tourist arrival in Meghalaya from 2012 to 2014. *Source* India, (2014–2015)

clan head and the elders. The family originates from the clan where the father has a very important part in decision-making even though the heredity is through the mother. The family head who is the father or the maternal uncle is represented in the village council (Tiwari, 2019).

The Jaintias define the areas that they agglomerate in as "Elaka (province)." The "Doloi is the head of the province and the "Doloi" must be chosen from the elders of the community. The "Doloi" has to strictly adhere to the laws of no corruption and equality as this could pose a threat to their position who could be removed if any form of misconduct is noticed by the clan. The "Doloi" will also have members called "U Basan (elder)" who would help him with the dealings of the community, and they are usually chosen in numbers based on the requirement of the community and the area the province covers. Forests being one of the most important natural resources of the Jaintia people, the elder, and the head have been given the authority to manage the same with the consensus of the community according to the "Jaintia Hills Autonomous District Council (JHADC)" under the purview of the Sixth Schedule of the Indian Constitutions (Lapasam, 2016).

The Garo tribe is known to be from the Bodo community and is also known to descend from the "Tibeto-Chinese" family (Miah, 2012). The Garo tribe attribute land as a property of the clan, and these lands are known as "A'king." Lands are under the guardianship of the women of the clan known as the Nokma (Tiwari, 2019). Though the women are the guardians of the lands as they follow the matrilineal customs the husband of the Nokma is usually the one to make the decisions on behalf of the women (Khan, 2017). Though the Nokma is the guardian of the lands in the area, they are not in a position to make decisions about the dealings of the land devoid of the consensus of the clan through the representatives known as Chra as a whole. The "Maharis" are the members of the "clan who are from the shared motherhood," and they ensure that decisions regarding guardianship of the lands are done in order (Tiwari, 2019). To date, the traditional institutions are only involved in providing lands to the individuals who would like to establish homestays for tourism. Not all traditional institutions are able to exercise the role as they are mandated to represent and encourage the communities toward sustainable tourism. However, some communities have started realizing the importance of sustainable tourism like the Mawlynnong Village (known as the "cleanest village in Asia" in 2003) which has involved the community to handle the tourism-related activities in the area.

Fig. 8.6 Tourist arrival in Meghalaya from 2016 to 2020. *Source* Primary Survey by Author

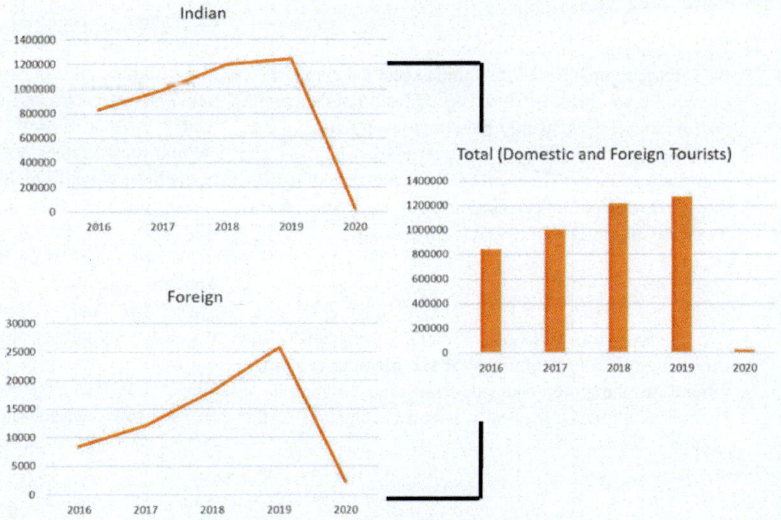

4.1 The Festivals and Traditional Art Forms

The Khasi Tribe, Jaintia Tribe, and the Garo Tribe are involved in various festivals celebrated in connection with their agricultural practices at different times of the year.

Khasi Tribe

- "Shad Suk Mynsiem": The yearly spring dance is executed to rejoice "harvesting and sowing." The dance is performed relative to the agrarian cycle (Culture D. o., Department of Arts and Culture, 2019a, 2019b, 2019c)
- "Ka Pom-Blang Nongkrem": This festival is dedicated to the "Lord Almighty" for a good yield and the members pray for "peace and prosperity" of the community for five days (Culture D. o., Department of Arts and Culture, 2019a, 2019b, 2019c).
- "Ka-Shad Shyngwiang-Thangiap": This dance is performed to lament the death of a "family member." Male musicians play music on the flute, drum, and bamboo pole. The dance commences on the death day, at a place next to the kitchen of the house (called the Rympeiling) and lasts till the last rites are executed on the burning grounds" (Culture D. o., Department of Arts and Culture, 2019a, 2019b, 2019c).
- "Ka-Shad-Kynjoh Khaskain": A dance that is performed when a family shifts to a newly built home. Once the formal ceremonies are done, the dance is performed in the following stages, "Ka Shad Kyuntui, Ka Shad Khalai Miaw, and Ka Shad Brap", and these are continued through the night till dawn of the next day. (Culture D. o., Department of Arts and Culture, 2019a, 2019b, 2019c).

Jaintia Tribe

- "Behdienkhlam": Behdienkhlam is the chief festival in the socioeconomic life of the Jaintias, and the emphasis is on requests for the property and good health of the people and on appealing heavenly blessings for a generous harvest (Culture D. o., 2019a, 2019b, 2019c).
- The "Laho" Dance: This dance form is performed by both men and women folk taking part in the best colorful dresses (Culture D. o., 2019a, 2019b, 2019c).

Garo Tribe

- "Wangala or Drua Wanbola or Wanma Rongchua": The ceremony of the agricultural year is a blessing ceremony offered to the gods and goddesses in the month of October (Culture D. o., Festivals and Ceremonies of the "Garos", 2019a, 2019b, 2019c).

Observing such a diverse culture of Meghalaya, it can be concluded that most of the traditional dances and performances of the major tribes of Meghalaya revolve around their agricultural practices and depicts the close bond of the community with nature and its surroundings. These festivals are either to rejoice or to offer gratitude for the harvest (Department of Tourism, n.d.).

5 Why Rural Tourism in Meghalaya?

Rural tourism can be defined as tourism activities in a "rural area where economic action is carried out, chiefly through agriculture" (Notes, n.d.). The following opportunities can be achieved by applying rural tourism in Meghalaya:

- A grassroot level of working toward sustainable development of villages.
- Promoting sustainable agricultural practices through tourism.
- Promoting public participation.
- A promising way of involving the traditional and State-run governing bodies.
- A sustainable way of promoting the preservation of nature and the environment.
- To promote the untapped tourist spots in the region and sustainably manage them.
- An opportunity to protect the communal lands.

5.1 Future Potential

Meghalaya is endowed with various tourism forms like nature tourism which includes waterfalls, national parks, and parks; adventure tourism which includes cave tourism, trekking, and camping, and cultural and heritage tourism which showcases only historical monuments. However, cultural tourism does not showcase the agrarian community of the State which is highly connected with the culture of the Khasi, Jaintia, and Garo Communities of Meghalaya (Fig. 8.7).

A vast majority of land in Meghalaya is forest land. Though agriculture is the main occupation the extent of agricultural land is limited due to human intervention. Because a significant proportion of the population depends on jhum or shifting cultivation and mining many lands in Khasi Hills, Jaintia Hills, and Garo Hills have wasteland around 15% caused due to forest loss and degradation. Hence, there is a need for sustainable agricultural practices

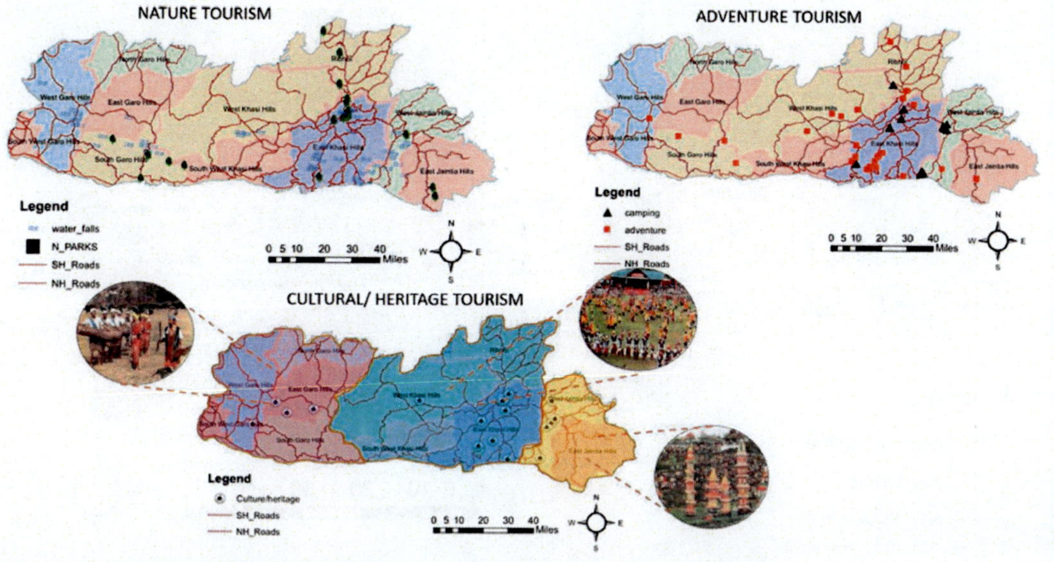

Fig. 8.7 Tourism forms in Meghalaya. *Source* Author

to be promoted to counteract the jhumming cultivation (Department, 2008–2009). The landholding of approximately 52% of people involved in agricultural practices is below 2 ha which makes them small and marginal workers (Department, 2008–2009). The Department of Agriculture, Meghalaya offers quality planting materials including other inputs with a full package of practices at no cost to farmers possessing the land of at least 0.2 ha. However, many are not aware of such schemes provided by the government resulting in the agricultural farmers reaping fewer benefits out of agricultural practices and not being able to tap into the formal market. Therefore, proper awareness from the government side is required and different forms of employment are required for the small and marginal workers.

The Sacred Groves that are present across the State to conserve the forest areas by the communities can also be used to promote indigenous practices of preserving forests in the State (see Fig. 8.8) (Department, 2019). The residents of Meghalaya when asked about the expansion of sacred groves (Ormsby, 2013), about 65% of them responded that they wish to see the sacred groves grow. Tourism can provide an opportunity for the communities to preserve the culture of maintaining the sacred groves sustainably. Major farms are present all around the State which accommodates horticulture, floriculture, and agricultural practices (see Fig. 8.9). Currently, reserves like Nokrek Bioreserve are being promoted to tourists.

The existing tourism forms like waterfall tourism, cave tourism, and adventure tourism, combined with the potential tourism forms like village tourism and agro-tourism in the existing villages across the State will allow the culture of the Garo, Jaintia, and Khasi communities to be promoted through their rural and agricultural practices (Fig. 8.10). Other villages like Kongthong Village in the East Khasi Hills District which is known as the "Whistling Village" can also be included based on their unique characteristics and can be formally added to the existing tourism circuits of the State. Various villages have different types of agricultural activities and other activities that could be used to engage the tourists. There are various produces in Meghalaya that are indigenous and unique. These can be sold to the tourists which will, in turn, promote the agricultural sector and these areas will be recognized by the officials as a potential area for improvement. Produce like the Lakadong Turmeric from the Jaintia Hills, orange blossom honey extracted from the Khasi Mandarin oranges from the Ri-War areas of the Khasi Hills (Zizira, Unique Honey from Meghalaya—Orange Blossom Honey, 2020), plantations like tea and cashew nuts from the Garo Hills (Welfare, Horticulture Crops Plantation, n.d.) can also be chosen to promote rural tourism, to depict the way they are cultivated and thereby showcasing the agrarian nature of the communities. Indigenous fruits like Sohphie and Sohshang could be utilized and introduced in the traditional cuisine to be offered to the tourists (Welfare, Horticulture Crops Indigenous Plants, n.d.).

There are various traditional farming methods that are practiced across Meghalaya which include shifting

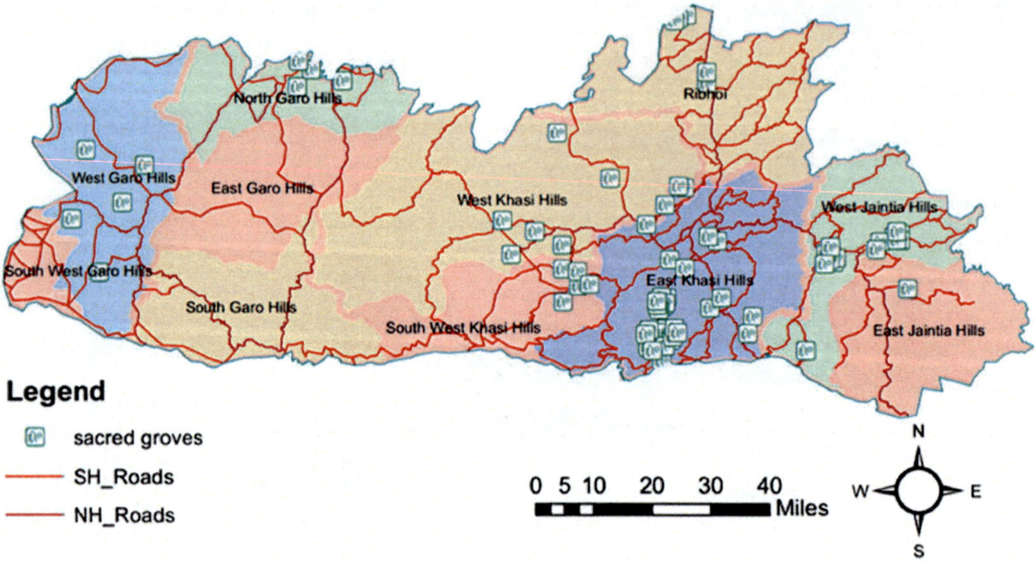

Fig. 8.8 Sacred Groves across Meghalaya. *Source* Author

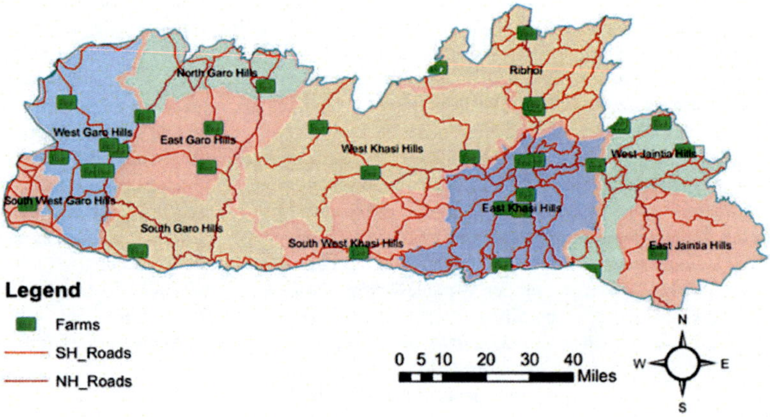

Fig. 8.9 Farms across Meghalaya. *Source* Author

cultivation and bun or terrace cultivation. Irrigation practices include bamboo drip irrigation and bench terrace irrigation (Zizira, Traditional Farming Practices in Meghalaya, Northeast India, 2017). There are unique forms of agricultural methods employed in Meghalaya like "Alder-based plantation practice, Areca nut-coconut-based plantation practice, bamboo-based plantation practice, Khasi pine-based plantation practice, and timber tree-tea-based plantation practice". Incorporating these in the tourism map and itinerary will provide a range of activities for the tourists to engage in and simultaneously promote the culture. This will eventually lead to improvement of agricultural productivity, tourism, and consequently better livelihood of the people.

6 Conclusion: Rural Tourism—The Wayforward

Meghalaya has numerous forms of tourism. However, they do not involve the agrarian community and their practices across the villages. It is possible to include the unique

Fig. 8.10 Potential tourism products in Meghalaya. *Source* Author

agricultural practices, agricultural produces, handicrafts, and community festivals in the scope of tourism. This would consequently promote sustainable agricultural practices in Meghalaya, promote market transaction of agricultural produces, and increase the number of days of stay for the tourists and their opportunities to engage in more activities and explore the Region better. It is to be noted that the Sustainability Development Goals mandated by the United Nations World Tourism Organization (UNWTO) in Agenda of 2030 can be very well accommodated and the goals of the SDGs for tourism can be applied across the State (UNWTO, 2016).

Acknowledgements With great sincerity, the authors acknowledge the officials, faculty, colleagues, and friends who have contributed to this endeavor and place on record their encouragement and support for the study.

References

Bank, T. W. (2020, November 19). *The World Bank*. (The World Bank). https://www.worldbank.org/en/news/press-release/2020/11/19/world-bank-signs-usd120-million-project-improve-connectivity-in-meghalaya-india

Christoph Oberlack, P. L. (2015). Institutions for sustainable forest governance: Robustness, equity, and cross-level interactions in Mawlyngbna, Meghalaya India. *International Journal of the Commons, 9*(2), 670–697.

Culture, D. O. (2019a, June 11). *Department of arts and culture*. (Government of Meghalaya) http://megartsculture.gov.in/festivals-khasi.htm

Culture, D. O. (2019b, June 11). *Department of arts and culture*. (Government of Meghalaya) http://megartsculture.gov.in/festivals-jaintia.htm

Culture, D. O. (2019c, June 11). *Festivals and Ceremonies of the "Garos"*. (Government of Meghalaya) Department of Arts and Culture Ceremonies and Festivals of the Garos: http://megartsculture.gov.in/festivals-garo.htm

Department of Tourism, G. O. (n.d.). *Meghalaya other festivals*. (Department of Tourism, Government of Meghalaya). https://www.meghalayatourism.in/experiences/other-festivals/

Department, G. O. (2011). *The tourism policy of Meghalaya*. Government of Meghalaya.

Department, M. F. (2019). *Meghalaya Forests and Environment Department*. Sacred Groves: http://www.megforest.gov.in/forest_sacredgroves.html#:~:text=In%20Meghalaya%2C%20sacred%20groves%20represent,of%20Khasi%20and%20Jaintia%20Hills

Department, P. (2008–2009). *Meghalaya State Development Report*. Government of Meghalaya.

India), M. O. (2014–2015). *Annual final report of tourism survey for the State of Meghalaya (April 2014–March 2015)*. Ministry of Tourism, Government of India.

Khan, S. (2017, March). *Mission to revive nokma rights*. The Telegraph online: https://www.telegraphindia.com/north-east/mission-to-revive-nokma-rights/cid/1428577

Lapasam, S. B. (2016). Traditional political institutions in Jaintia Hills: Raliang Dolloiship. In C. R. Lyngdoh (Ed.), *Revisiting Traditional Institutions in the Khasi-Jaintia Hills* (pp. 93–111). Cambridge Scholars Publishing.

Miah, S. (2012). Society and culture of the Garo's in Meghalaya. *IJCAES Special Issue on Basic, Applied & Social Sciences*, 376–378.

Notes, T. (n.d.). *Rural Tourism*. February 2021, from Tourism Notes: https://tourismnotes.com/rural-tourism/

Office of Registrar General & Census Commissioner, I. (2011). *Census 2011*. Ministry of Home Affairs, Government of India. Office of Registrar, India. https://censusindia.gov.in/2011-common/censusdata2011.html

Ormsby, A. (2013). Analysis of local attitudes toward the sacred groves of Meghalaya and Karnataka, India. *Conservation and Society, 11* (2), 187–197.

Times, E. T. (2020, September 27). *ET The Economic Times*. (ET The Economic Times). https://m.economictimes.com/industry/transportation/airlines-/-aviation/aai-plans-to-upgrade-runways-at-seven-airports-by-march-2022/articleshow/78347873.cms

Times of India. (2020, September 2). *Timestravel*. https://timesofindia.indiatimes.com/travel/travel-news/udan-4-government-approves-78-new-routes-with-special-focus-on-north-east/as77886581.cms

Tiwari, B. K. (2019, July). Structure and functioning of traditional institutions in Meghalaya.

UNWTO, T. W. (2016). *Tourism in the 2030 Agenda*. (United Nations). https://www.unwto.org/tourism-in-2030-agenda

Welfare, G. O. (n.d.). *Horticulture crops indigenous plants*. Government of Meghalaya department of agriculture & farmers' welfare: http://megagriculture.gov.in/public/crops_indigeneous_plants.aspx

Welfare, G. O. (n.d.). *Horticulture crops plantation*. Government of Meghalaya department of agriculture & farmers' welfare: http://megagriculture.gov.in/public/crops_plantation_crops.aspx

Zizira. (2017, Aug 18). *Traditional farming practices in Meghalaya, Northeast India*. Zizira. https://www.zizira.com/blogs/people-and-process/traditional-farming-practices-meghalaya-northeast-india

Zizira. (2020, February). *Unique honey from meghalaya—orange blossom honey*. Zizira. https://www.zizira.com/blogs/raw-honey/unique-honey-from-meghalaya-orange-blossom-honey

Effects of Market Intelligence Generation, Online Reviews, and Management Response on the Business Performance of Rural Accommodation Establishments in France

António J. D. V. T. Melo, Rosa M. Hernández-Maestro, and Pablo A. Muñoz-Gallego

Abstract

In successful businesses, positive consumer reviews are growing in importance because they can influence consumers' purchasing decisions and, consequently, their performance. The authors propose that market intelligence generation associated with the comments in consumer evaluations and the advantages of management response drive business performance. The researchers analyze data about the rural accommodation (positive online consumer reviews and management response) on the infomediary website, Toprural website (Vrbo), from a sample of 237 French rural tourism accommodations (RTA). An additional survey provided information on the RTA market intelligence generation and business performance (profitability). The researchers adopted a quantitative and demand-driven approach to describe drivers that enhance business performance. The researchers identify a significant direct positive effect of positive volume of online consumer reviews and a non-significant direct effect of management response on business performance. Moreover, those entrepreneurs who are genuinely interested in detecting and understanding shifts in the environment and pursuing best practices are more likely to succeed than those who show less interest. In the presence of market intelligence, online consumer reviews and the management response have stronger effects on business performance. Furthermore, this knowledge appears to be linked with a better use of accessible intelligence in online consumer reviews, as well as a more successful accommodation management response. It also directly influences performance. This variable thus helps explain the influence of online consumer reviews in the hospitality and tourism industry.

A. J. D. V. T.Melo (✉)
University of Maia, Maia, Portugal
e-mail: amello@umaia.pt; ; ajdvtm@gmail.com

R. M. Hernández-Maestro · P. A. Muñoz-Gallego
University of Salamanca, Salamanca, Spain

Keywords

Market intelligence • Online consumer reviews • Management response • Business performance • Services • Tourism • Accommodation

1 Introduction

With the worldwide dissemination of the COVID-19 pandemic virus and the consequent reduction in tourist activity, the need for more professional business management to better serve consumers has become crucial, not only to maintain the performance of companies, but to ensure their survival. With these environmental conditions, online consumer reviews have become even more important as extrinsic clues consumers look for when purchasing experience products, such as tourist services, allowing businesses and consumers, and consumers, to share content and knowledge about consumer products, and activities. Online consumer reviews are one of the main boosters of business performance, have a decisive effect on consumer decision making and bring critical consumer information to business managers (Ahani et al., 2019b; Melo et al., 2017; Oliveira et al., 2020; Phillips et al., 2017).

To deal with consumers effectively, hotels are increasingly using management's response to online consumer reviews as a sign of concern for them, that they actively care about them and improving the service they deliver. This management's response is often grateful for positive guest reviews and pursue to maintain the reputation of the business by promising to resolve or apologizing issues described by guests (Zhang et al., 2020a, 2021b), that produce a greater volume of positive consumer reviews that benefit business performance (Gu & Ye, 2014; Hernández-Maestro, 2020; Li et al., 2017; Melo et al., 2017; Proserpio & Zervas, 2017; Xie et al., 2014). Thus, both online consumer reviews and management response provide signals to potential

A. Mandić et al. (eds.), *Cultural Sustainable Tourism*, Advances in Science, Technology & Innovation,
https://doi.org/10.1007/978-3-031-10800-6_9

consumers. Several studies attributed positive effects to consumer evaluations and management responses, with benefits for business performance (Esmark Jones et al., 2018; Melo, 2020; Xie et al., 2014, 2017; Zhang et al., 2020a; Zhao et al., 2020).

The tourism industry consists of small businesses. In this sense, many RTAs are family businesses that represent all or just an extra income on top of what they earned from agricultural, commercial, or construction activities. In addition, RTA entrepreneurs often lack managerial education. To identify the determinants of success in this industry, researchers must therefore recognize the roles played by entrepreneurs (Hernández-Maestro et al., 2009). For example, the degree to which entrepreneurs are aware of the market may define the nature of their businesses. In this regard, market intelligence generation appertain to the permanent collection of data about guests, competitors, and the environment in general, identifying opportunities and threats to help make effective marketing decisions (Li & Li, 2013). Several works in different sectors, including RTAs, identified positive effects on the business performance of market intelligence generation (Ahearne et al., 2013; Falahat et al., 2020; Hendar et al., 2020; Jogaratnam & Tse, 2004; Katsikea et al., 2019; Melo, 2020; Webb et al., 2011; Yap et al., 2018). Furthermore, there is a need for a sound understanding of the positive effects of market intelligence generation regarding the positive online consumer reviews and management response to those online consumer reviews.

A greater knowledge of these influences among people would be very important for companies in the tourism sector. Accordingly, to identify the factors that improve business performance, the researchers used a quantitative approach. The researchers analyzed how positive online consumer reviews, management response, and market intelligence generation drives the business performance of French RTAs. In addition, the researchers analyzed whether and how this generation of market intelligence moderated the impact of management response and online consumer reviews on business performance. Therefore, the researchers (1) measured the increments of these positive consumer perceptions of business performance, (2) evaluated the contributions of this management response to positive consumer perceptions, to business performance, (3) assessed the contributions of the generation of market intelligence on business performance, and (4) measured the influence of moderation in generation of market intelligence on the relationships between online consumer reviews and management response on business performance. So the researchers collected information about RTAs from Toprural (Vrbo) and conducted a survey to collect data about the generation of market intelligence and business performance of each of the RTAs in the study sample. The study results contribute to fill the research gap and answer calls from RTAs owners/managers, associated with online consumer reviews (comments), management response, market intelligence, and its impact on the commercial performance of accommodations.

In this study, the researchers argued that the business performance benefits of consumer reviews and management response can be enhanced if managers recognize the impact and importance of online consumer reviews and establish effective processes to ensure positive online consumer reviews and management's response work to the business. The researchers considered how the volume of positive consumer reviews, volume of management response, and market intelligence generation together contribute to business performance in the context of French RTAs. In particular, the researchers included market intelligence generation as a moderator of the relationships of positive consumer review volume and management response volume with business performance. Therefore, as the rural tourism sector is also a service industry, the model variables can also be applied to other services industries.

The researchers begin this article with a brief analysis of rural tourism and RTAs, succeeded by a brief debate of online consumer reviews, the emphasis of management response, market intelligence generation and its moderating effect on online consumer reviews relationships and on management response relationships with business performance. Next, based on scrutiny of the literature, the researchers developed specific hypotheses to test the relationships among positive online consumer reviews about an accommodation, management response to the online consumer reviews, and the generation of market intelligence, and its business performance. The researchers specify the data used, the characteristics of the sample, and the measures, then present the results. Finally, the researchers discuss their findings, conceptual contributions, and theoretical and managerial ramifications along with some limitations and avenues for future research extensions.

2 Theoretical Base and Hypotheses

2.1 Literature Review

The internet offers vast amounts of information and, for some, it functions as an "external memory', changing the way travel information is disseminated. Booking platforms, social networking sites, blogs, and forums enable tourists to digitize and share their experiences, knowledge, emotions, and preferences in form of consumer reviews (Ait-Bakrim et al., 2019; Gonçalves et al., 2018; Hur et al., 2017; Matson-Barkat & Robert-Demontrond, 2018; Xiang et al., 2015; Xu, 2020).

2.2 Online Consumer Reviews

In service contexts, consumer feedback is particularly important because services are mainly influenced by intangible attributes, which, by definition, cannot be seen, but can be felt or smelled, and potential consumers cannot evaluate services quality before their purchase, this prominent level of prepurchase information scarcity creates purchase risk (Flanagin & Metzger, 2013; Gottschalk & Mafael, 2017; Hyun & Han, 2015; Melo, 2020; Melo et al., 2017; Wells et al., 2011). To reduce uncertainty by understanding the decision frame, potential consumers make inferences based on signals and suggestions from others' behaviors (Matzat, 2009). Online consumer reviews are a trusted source of knowledge about consumer recommendations and opinions of their own experiences. Available and easy to understand, help potential consumers to estimate the level of quality of accommodation and identify which properties best suit their preferences (Baek & Choe, 2020; Dwidienawati et al., 2020; Melo et al., 2017; Oliveira et al., 2020; Reich & Maglio, 2020). In the hospitality and tourism industry, online consumer reviews are strong influencers, and tourists have become dependent on them to make sound decisions about where to stay (Huertas-Garcia et al., 2014; Perez-Vega et al., 2018; Ren et al., 2015). Consumers consider online feedback carefully before making purchase decisions; they even may be willing to pay more for services and goods that have strongly positive online consumer reviews (Akbarabadi & Hosseini, 2020; Hsieh & Li, 2020; Septianto et al., 2020).

According to previous research, most travelers publish many more reviews to show their perceptions of positive experiences than their negative perceptions (Tontini et al., 2017). Positive reviews provide an general outlook of overall consumer satisfaction with the services provided, which can provide accurate knowledge to attract new consumers (Ahani et al., 2019b). In particular, the volume of positive consumer reviews is important, because it provides an indicator of market awareness and helps to reduce perceived purchasing risk (Neirotti & Raguseo, 2020). Market awareness grants businesses more visibility; the more people know about them, the more they talk about them. When offerings are associated with high volumes of positive consumer reviews, online reputation improves, and more consumers tend to be positively aware of the offerings. Greater volumes of consumer reviews serve as extrinsic, peripheral cues that help consumers justify their purchase decisions, with the reasoning that many others show that they bought the same products (Park et al., 2007; Zhang et al., 2010). Is also a cue to the plausibility of the consumer reviews, increasing the reliability and correctness of their content, potentially driving more purchases (Dellarocas et al., 2007; Salganik & Watts, 2008).

Even with much prior research on online consumer reviews in hospitality and tourism industry (Gursoy, 2019; Zhao et al., 2019), there is little research on its use by companies (Melo, 2020; Melo et al., 2017; Sun et al., 2014; Tran, 2020). Online consumer reviews, such as post-experience consumer feedback, help in understanding the service quality and other aspects like facilities, are of excellent value to know what satisfies their consumers and, as a result, improves their performance (Melo, 2020; Mittra & Khamkar, 2021; Oliveira et al., 2020; Thi et al., 2021). This source of market feedback signals both potential consumers and entrepreneurs about what were the consumers' perceptions about the level of quality of the post-experience service (Ahani et al., 2019a; Melo et al., 2017; Torres et al., 2015). Entrepreneurs can leverage business performance using this information to improve their knowledge of the market, improving the quality of their service, engaging with the consumer through closer communication, and knowing the positioning of each consumer segment (Ladhari & Michaud, 2015; Melo, 2020; Melo et al., 2017).

2.3 Management Response

In general, management response to consumer reviews indicates to both consumer reviewers and readers that businesses care about consumers and offer quality products; response improve companies' images and reduce risk perceptions, and become a vital tool for managers to cooperate with their consumers (Hernández-Maestro, 2020; Zhang et al., 2020a). Reducing the likelihood of potential consumers making potentially erroneous negative inferences can be achieved by management's response (Sparks & Bradley, 2014) and can improve guests' possible clues about hospitality management's concern for them (Sparks et al., 2016).

In face of the positive effect of management response (Hernández-Maestro, 2020; Ye et al., 2009; Zhao et al., 2020), in specific contexts, previous studies also pointed out that there may potentially some negative effects of the management response. For instance, Mauri and Minazzi (2013) study establish a negative impact on consumer expectations and booking intentions in the presence of management response to online consumer reviews, and Xie et al. (2017) also argued that topic repetitions in management response have a negative association with future performance of the hotel. Liu et al. (2021) discover that rote management response reduces the volume of future positive reviews, but not negative ones, and these effects are stronger for independent hospitality business than for chains.

Responding companies gain competitive advantages (Hernández-Maestro, 2020; Rose & Blodgett, 2016; Walker, 2010). Generating more positive online consumer reviews can be helped by management response (Lee & Song, 2010). Additionally, management response to positive reviews magnify consumers' satisfaction, perceived equity, and intentions to buy again (Gu & Ye, 2014).

2.4 Market Intelligence Generation

The conceptualization of market orientation followed two different approaches, but both are similar in driving organizational behavior to acquire more information about the market, consumers, and competitors, and to disseminate this information throughout the organization (Lorentz et al., 2020; Quang-Huy, 2021). First, Kohli and Jaworski (1990) suggested that this variable described organizational behavior, defining market orientation as the generation of information, diffusion, and an adequate response to present and future consumer needs and preferences across the organization. Second, Narver and Slater (1990) considered this variable as a cultural variable, in which the behavior needed to create outstanding value for buyers and, therefore, continuous superior performance for the business created most effectively and efficiently by it. A market intelligence generation strategy addressed to acquire knowledge about the (latent and expressed) needs of consumers, the skills and strategies of competitors (e.g., Kohli & Jaworski, 1990; Narver & Slater, 1990) provides a hub for performance-enhancing efforts, allowing the company the development of stronger relationships with crucial consumers and wisdom in opportunities to develop the market.

For the purposes of this study, market intelligence defined by Talvinen (1995) is all information collected outside the organization to identify opportunities, changes, problems in the market environment in which we operate or wish to operate. Understanding how the organization's markets tend to respond to the value propositions of the marketing mix must be the basis of the organizational decision (Gebhardt et al., 2006; Kohli & Jaworski, 1990). To achieve higher levels of performance, companies must be more market-oriented to better obtain and use market intelligence consistently (Kirca et al., 2005; Kumar et al., 2011).

The accessibility of data and information is in an increasing degree of dynamism, thanks to new media and digital platforms, and companies need to accompany it to improve their market orientation, which is essential to maintain greater business performance. Market intelligence generation, it is part of the market orientation concept that implies the formation and delivery of superior value for consumers (Donavan et al., 2004; Santos, 2006).

2.5 Moderating Effects

Narver and Slater (1990) cultural measurement proposed three aspects: (1) "consumer orientation," which means that, gathering information based on existing and anticipating consumer perceptions of their experience, entrepreneurs can discover the needs, desires and expectations of future consumers; (2) "competitor orientation" looks at perceived short-term advantages and disadvantages, as well as potential long-term development strategies and competencies, of potential and current consumers in other competitors; (3) "inter-functional coordination," illustrates that, by virtue of integrating and applying positive consumer perceptions in management and services, and responding to changes in the market environment, RTA can create value in the future services, which is contributed to the target consumer (Hendar et al., 2020; Li & Li, 2013).

By generating market intelligence on continually changing consumer needs and desires and preparing the organization's response to this, you are helping to increase consumer engagement (Sashi, 2012). In addition, the value creation process shifts for more personalized consumer experiences, with the most knowledgeable, networked, more empowered, and more proactive consumers in generating intelligence about their ever-changing needs and helping the organization to co-create more value in response to these needs (Prahalad & Ramaswamy, 2004).

2.6 Hypothesis Development

High volumes of consumer reviews allow tourism managers to learn more about guests' experiences, attitudes, and preference. Such information helps them improve service quality and enhance consumer value; it results in increased consumer satisfaction and higher margins (Neirotti et al., 2016). Managers must use reviews to constantly measure consumers' demands, shifts in their behavior (Korfiatis et al., 2019).

Further, the researchers confirmed that the volume of positive consumer reviews improves business performance (Melo et al., 2017; Oliveira et al., 2020; Torres et al., 2015). In summary, though some research finds no significant positive effect on business performance (e.g., Filieri & McLeay, 2014), many works in different contexts, including RTAs, show that the volume of positive consumer reviews positively determine performance (Gellerstedt & Arvemo, 2019; King et al., 2014; Melo et al., 2017; Phillips et al., 2017; Purnawirawan et al., 2012; Xiang et al., 2015; Zhang et al., 2016). Therefore, the researchers propose:

H1: Higher volume of positive consumer reviews leads to improved businesses performance.

In previous research, researches that study the drivers consumers look at when choosing services like hospitality and tourism products demonstrate that management response to consumer reviews are important (Park & Allen, 2013; Ye et al., 2008). Especially, response to perceived service issues or service failure recovery can enhance the likelihood that a guest will recommend the accommodation (Levy et al., 2013). Management response is seen by others, because the internet is transparent, so that they show both reviewers and readers that they care about the provision of a good service (Xie et al., 2014). It was found that answer to positive online consumer reviews has become a key strategy in the hospitality industry (Oliveira et al., 2020). Xie et al. (2017) recommend that managers, to redirect consumer attention to other positive aspects of goods or services, should provide more different topics of information. Compared with the generic response, the hotel's sincerity toward consumer issues is demonstrated by the specific response from management and allows for higher quality of communication and level of trust (Lee & Hu, 2005), RTAs need to implement adequate response strategies for online consumer reviews. Potential consumers are more motivated to get involved looking for additional information (Chen & Chaiken, 1999) and in evaluating the response of the management of an RTA.

Management's response to online consumer reviews affects the perception of consumers who receive the answer, but it also influences the behavior of potential consumers who are attentive to the interaction.

Management response volume will likely have significant effects on the performance of RTAs (Xie et al., 2014; Zhang et al., 2022). Responding to positive reviews allows for a warmer human relationship with reviewers, allowing management acquire competencies from this interaction and build a stronger goodwill from their consumers (Park & Allen, 2013; Proserpio & Zervas, 2017; Zhang et al., 2020a). We have to consider that this opportunity to intensify the relationship with potential and current consumers will increase the probability that consumers will recommend the RTA more, be more willing to book, build stronger goodwill toward RTA and, for these arguments, can provide the most positive reviews closing this performance virtuous circle (Melo, 2020).

When businesses respond to consumer reviews, the cost for consumers to produce negative reviews increases, because consumers anticipate that their reviews will be examined in detail, which may prompt less-negative reviews. At the same time, when consumers feel that companies are paying attention to them, they are more motivated to express positive reviews (Melo, 2020; Proserpio & Zervas, 2017). Higher numbers of positive of reviews and lower numbers of negative reviews should enhance business performance (Chevalier & Mayzlin, 2006; Nieto et al., 2014; Pathak et al., 2010). Accordingly, the researchers propose that,

H2: Higher volume of management response positively affects business performance.

Entrepreneurs can complement existing knowledge by checking various sources of information (Kaish & Gilad, 1991); in the process, they become aware of consumers' needs, wants, and value assessments, as well as develop knowledge about competitors and environmental trends. By examining market information, entrepreneurs can identify cues that enable them to create competitive advantages and stay ahead of the market (Donavan et al., 2004; Narver & Slater, 1990; Santos, 2006). Interest in the market and the collection of market information usually is linked to responsiveness to the information (Jaworski & Kohli, 1993; Kohli & Jaworski, 1990; Narver & Slater, 1990); thus, developing market intelligence is the first step in developing new products that are adapted to consumer preferences, in developing stronger relationships with crucial consumers and highlights of opportunities for market development (Heusinkveld et al., 2009; Slater & Narver, 2000).

Moreover, with market knowledge, companies not only adapt to but also shape business environments, through their innovation and collaboration with various stakeholders (e.g., other companies and institutions). Therefore, the researchers analyze the impact of market intelligence generation in the context of French RTAs, where many entrepreneurs lack managerial education and the business often represents just as a supplementary income, according to the following prediction:

H3: Market intelligence generation has a positive influence on business performance.

Compared to other entrepreneurs, it is assumed that those who regularly collect accurate market information and use it appropriately also value the collected information more highly, as they know it is a decisive component to the prosperity of their business (Ahearne et al., 2013; Hall et al., 2017; Katsikea et al., 2019; Talvinen, 1995; Wood, 2001). Congruently, they are expected to make better use of the information contained in consumer reviews and more likely to effectively manage the consumer reviews feedback (Katsikea et al., 2019; Melo, 2020). By making decisions that guarantee a united, consistent, and integrative implementation of the quality elements that are adjusted to consumers' expectations of quality or by providing corrections to service delivery when needed (Ahlin et al., 2012; Heusinkveld et al., 2009). The interaction between consumer and seller and between consumers themselves, as well as with potential consumers, allows the co-creation of value with, later, content generation and feedback provision, dissemination of knowledge, and making them advocates of the services delivered by the organization among their peers creating a virtuous cycle of consumer more relational and emotional engagement (Sashi, 2012).

As previously noted, management response to consumer reviews produce positive effects on consumer perceptions and behavior and thus on business performance (Proserpio & Zervas, 2017). However, according to Mauri and Minazzi (2013), management response could have negative consequences, because consumers may consider them advertising and not completely credible.

In this context, both signaling theory (Connelly et al., 2011; Rao et al., 1999; Spence, 1974) and attribution theory (Kelley, 1971) provide support for the contradictory about the effects of management response. The effectiveness of a response should be greater or lesser (or even negative), depending on consumers' perceptions of management's reason for responding. The content and style of response provide signals to consumers; for negative consumer reviews, for example, the effects of specific response (e.g., providing solutions) or of immediate versus lagged response may differ. In general, if consumers perceive a response to be inappropriate, the credibility of the responder may be reduced (Lee & Blum, 2015; Ullrich & Brunner, 2015; Wei et al., 2013; Xie et al., 2014).

On this subject, the researchers expect entrepreneurs' profiles to determine the content, style, and timing of their response. Entrepreneurs who embark in market intelligence generation are expected to be aware of the need to provide appropriate response, due to their contact with the market; they also may be more likely to add positive elements to their response, because of their knowledge of what constitutes appropriate response. Accordingly, to understand the full effects of market intelligence generation, the researchers consider its moderating role, such that market intelligence generation may enhance the positive effect of the number of consumer online consumer reviews and management response volume on business performance:

H4a: Market intelligence generation moderates the relationship between the volume of positive online consumer reviews and business performance in a positive way.
H4b: Market intelligence generation moderates the relationship between management response volume and business performance in a positive sense.

3 Methodology

3.1 Sample

Authors data selection consisted of two phases. First, the researchers gathered information on consumer review volume, management response volume, and accommodations' characteristics from Toprural.com, a leading location for rural accommodation in Southern Europe. Second, the researchers surveyed owners'/managers' perceptions of business performance and their market intelligence generation. To obtain a higher response rate, the researchers reduced the number of questions to the essential of the model proposed in the research.

Initially, researcher Toprural.com data included about 2275 French RTAs with 10,047 consumer online consumer reviews. The researchers eliminated all accommodations without complete information, duplicates, or blank guest comments to filter the data. The researchers loaded the final short, structured, online questionnaire into a website that the researchers created and provided respondents with a password that allowed respondents to access the website. In line with Jordan and Troth (2020), to the extent possible, the researchers, to minimize the ex-ante bias of the common method, developed good research information and a set of instructions. To this end, they kept the questions concise and simple, avoided double elements, eliminated common scaling properties, used information from multiple sources for predictors and variables, and performed separate data collection.

The researchers solicited responses from 1618 (71.12% of the total) accommodations, sending a reminder e-mail to obtain authors total. Table 1 contains the sample characteristics. The researchers received responses from 237 RTAs (3490 consumer reviews and 196 management responses). Based on previous literature (Hu et al., 2009; Melo, 2020; Melo et al., 2017; Nieto et al., 2014; Racherla et al., 2013; Stringam & Gerdes, 2010), the researchers access consumer perceptions by the volume of online consumer reviews. Then, the researchers concentrated on 4- and 5-star reviews (the positive ones, only 4.6% of reviews were negative between 1 and 3 stars). The mean value of consumer review volume per establishment was 14.73, with 0.83 management responses per establishment, nightly prices per person of 26.96€, average number of accommodations per establishment of 1.17, and average capacity (beds) per establishment of 10.44.

3.2 Measurement

Table 2 presents the variable measures. From the information provided by Toprural.com, the researchers selected the

Table 1 Sample characteristics

Number of RTAs	237
Consumer review volume	3490
Consumer review volume per RTA	14.73
Average night price per person (€)	26.96
Number of accommodations (average)	1.17
Capacity (average, beds)	10.44
Management response	196
Management response volume per RTA	0.83

Table 2 Variable definitions

Variable		Measure
Consumer review volume		For each RTA, number of reviews on the Toprural website
Management response volume		For each RTA, number of management responses to consumer reviews
Business performance		Owner's perception about the status of the rural accommodation establishment, measured on a 7-point Likert scale (1 = "very bad," 7 = "excellent"), in terms of: • Reservation growth • Profitability Owner's perception, measured on a 7-point Likert scale (1 = "strongly disagree," 7 = "strongly agree"): • I am satisfied with the business income
Market intelligence		Owner's perception, measured on 7-point Likert scales (1 = "strongly disagree," 7 = "strongly agree"): • I always attend fairs or important meetings on rural tourism • I am very attentive to news media and specialized publications on rural tourism • I am always aware about available online information of other rural accommodations
Control variable	Price	Average price per night and person (€)

volume of online consumer reviews and management response volume.

The researchers adapted a business performance scale from Melo et al. (2017), Nieto et al. (2014), and Polo Peña et al. (2012). The researchers asked owners about the status of their accommodations (1 = "very bad" to 7 = "excellent") in terms of reservation growth and profitability. The researchers also asked them to complete an agreement scale (1 = "strongly agree" to 7 = "strongly disagree") in response to the item "I am satisfied with the business income," keeping in mind that it is difficult to obtain objective business performance for RTAs (Hallak et al., 2015; Hernández-Maestro et al., 2009; Kropp et al., 2006; Lee et al., 2016; Melo et al., 2017).

According Chandler (1962); Dess and Davis (1984) subjective measures of performance are common in research related to small enterprises and correlate strongly with objective measures. Moreover, prior evidence related to RTAs shows that entrepreneurs have accurate perceptions of their performance. Melo et al. (2017) and Nieto et al. (2014) establish a positive relationship between consumers' online reviews and owners'/managers' perceptions of consumer satisfaction, business reputation, and profitability.

In accordance with authors literature review, the researchers used a measure of market intelligence generation. The scale, based on the works of Melo et al. (2017), Nieto et al. (2011), and Polo Peña et al. (2012), contained three items. The researchers asked owners to indicate their agreement (1 = "strongly disagree" to 7 = "strongly agree") with three items: "I always attend fairs or important meetings on rural tourism," "I am very attentive to news media and specialized publications on rural tourism," and "I am always aware about available online information of other RTAs".

The researchers also included the control variable of nightly price per person, based on average price. In the literature, price is a proven determinant of consumer behavior (Zeithaml et al., 1988), even in modern purchasing environments that are high extraordinarily complex specially since the emergence of social media and the internet (Book et al., 2015). Thus, the researchers included it as a control variable with the intention of controlling for its potential effect on business performance. With these inputs, in line with Farahani et al. (2010), after testing the model with PLS and OLS regressions, the researchers concluded that the OLS regression model is less robust than the PLS regression model. The researchers analyzed these data with SPSS 27 and SmartPLS 2.0.

4 Results

4.1 Main Results

In the sample analysis, there was no single factor or overall factor responsible for most of the covariance between measures, therefore, it was concluded that there was not a significant amount of common method bias (Jordan & Troth, 2020).

The authors began by analyzing the quality of the two focal constructs, that of generating market intelligence and that of business performance. In Table 3, we find the results of the construct assessment, which reveals composite reliability ($\rho_c > 0.80$), convergent validity ([AVE] > 0.5, significant outer loadings), and discriminant validity (Hair et al., 2014). Thus, the reliability and validity of the construct measures were confirmed by the researchers, confirming their suitability to the model (Hair et al., 2014).

Next, the researchers introduced the two constructs into Model 1 (Fig. 1). The results show that positive consumer review volume has a significant positive influence of 0.152 (2.689) on business performance, consistent with H1. Surprisingly though, management response volume exhibits a non-significant relationship with business performance, so the researchers must reject H2. Regarding the effect of market intelligence generation, it helps rural businesses achieve better business performance, at a value of 0.254 (4.272), in support of H3. Nevertheless, price does not present a significant effect on business performance. And Model 1 presents a R^2 with the value of 8.3%. This value is satisfactory considering that many other factors have a potential influence on business performance and that two different data sources (Toprural platform and a survey) are used in this study.

4.2 Moderating Effects

Regarding the moderating effects, the results for Models 2 and 3 in Table 4 show that the generation of market intelligence positively moderates the relationship between the volume of consumer reviews and business performance, at a value of 0.232 (1.968), consistent with H4a. It also slightly and positively moderates the relationship between management response volume and business performance, at a value of 0.196 (1.682), in line with H4b. Thus, when market intelligence generation is greater, the influences of consumer review volume and management response volume on business performance increase.

5 Conclusions

The internet and social media provide new opportunities for entrepreneurs to create value. Online consumer reviews, management response, and market intelligence can improve relationships between companies and consumers and between consumers themselves (Hendar et al., 2020; Melo, 2020; Melo et al., 2017; Thi et al., 2021; Xie et al., 2014, 2017; Zhang et al., 2020b). They also provide a cost-effective way to monitor consumers' voices; such monitoring can provide competitive advantages to even the smallest RTAs.

It is a challenge to define appropriate management strategies for service businesses such as hospitality. Most research point out to positive impacts of online consumer reviews quantity (e.g., Ahani et al., 2019b), of positive response quantity (e.g., Zhao et al., 2020), and of market intelligence (e.g., Wood, 2001; Melo, 2020) on consumer

Table 3 Reflective constructs' assessment

Latent variable	Indicators	Outer loadings	*t*-value	AVE	Composite reliability (ρ_c)	Indicator reliability	Cronbach's alpha	Discriminant validity
Business performance	Income	0.863	27.528	0.800	0.923	0.745	0.875	Yes
	Reservation growth	0.885	27.267			0.783		
	Profitability	0.934	76.576			0.872		
Market intelligence generation	Fairs	0.846	16.381	0.664	0.855	0.716	0.749	
	News media/publications	0.886	25.071			0.785		
	Online information/competitors	0.701	7.207			0.491		

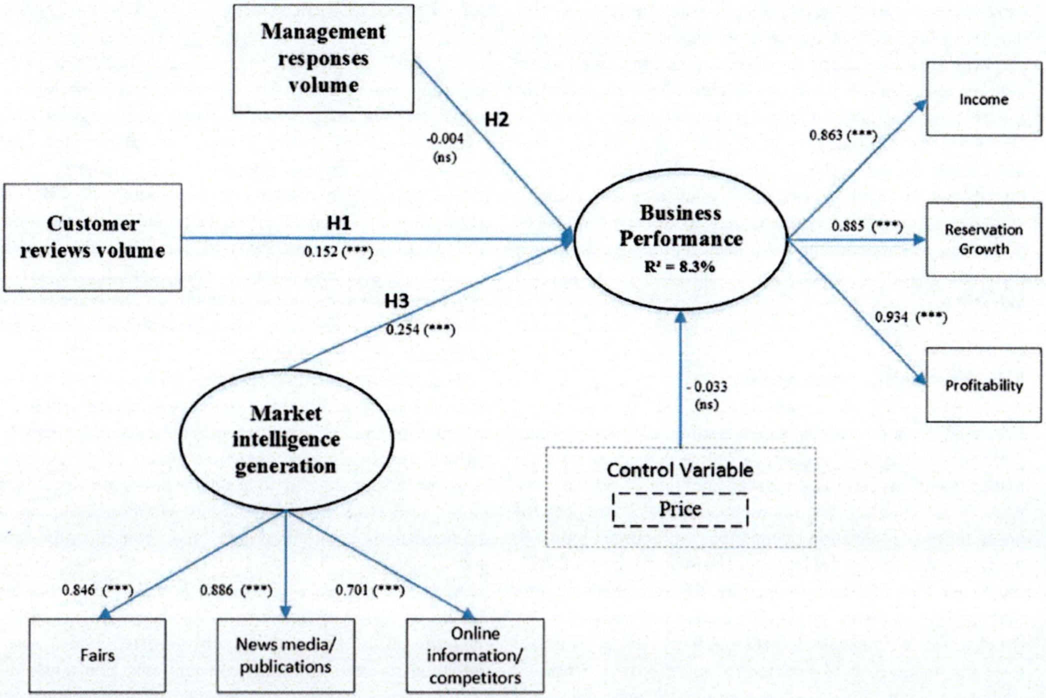

Fig. 1 Model 1 direct effects on business performance, with standardized path estimates

Table 4 Moderating effects

	Business performance	
	Model 2	Model 3
Consumer online review	0.320 (***)	0.143 (***)
Management response	−0.010 (ns)	0.123 (ns)
Market intelligence	0.245 (***)	0.245 (***)
Consumer review volume × market intelligence generation	0.232 (**)	–
Management response volume × market intelligence generation	–	0.196 (*)
R^2	10.8%	10.5%

*** $p < 0.01$; ** $p < 0.05$; * $p < 0.1$; (ns) non-significant

perception and, consequently, on their behaviors. However, some studies have not found positive significant effects on consumer perceptions and behavior from online consumer reviews (e.g., Zhang et al., 2021a), and management response (e.g., Mauri & Minazzi, 2013). Past research seems to suggest that including thank you statements in management response could have positive effects (Mate et al., 2019). Accordingly, further investigation is needed, especially in this pandemic situation for an tourism industry that is one of the main drivers of rural development in a sustainable way. However, research on the specificity of hospitality management strategies is still in its initial phase.

In this sense, through quantitative methods with a demand-driven perspective, this study promotes the conscious of the effects that the positive online consumer reviews, management's response, and market intelligence generation have on business performance within the scope of RTAs. It affirms a positive effect on business performance of positive online consumer reviews and a non-significant effect of management response volume. Moreover, those entrepreneurs who are genuinely interested in detecting and understanding shifts in the environment and pursuing best practices are more likely to succeed than those who show less interest. Online consumer reviews and management response have stronger effects on business performance in the presence of market intelligence generation. This knowledge seems to be associated with better uses of information available in consumer reviews, as well as more effective management response. It also directly influences performance. This variable helps shed light on the influence of online consumer reviews on the hospitality and tourism industry.

5.1 Theoretical Indications

Providing a more holistic understanding of how business performance responds to positive online consumer reviews, management response and market intelligence are the purpose of this research. The researchers' model guarantees the possibilities of replication beyond the rural tourism industry, because it is a service industry with similar variables. While the larger part of online consumer reviews research only touched on the merits of positive online consumer reviews (Mathwick & Mosteller, 2017), Zhang et al. (2021a) research points to evidence that positive online consumer reviews are not always advantageous for companies, this research highlights evidence that positive online consumer reviews improve business performance. Due to this research objective, the present research expands the theoretical understanding of sharing economy, hospitality industry, and the influence of other consumers in several main aspects. Early, this research brings unique improvements to the online consumer reviews marketing literature, especially on business performance impact of online consumer reviews, management response, and market intelligence generation in rural tourism setting (e.g., Katsikea et al., 2019; Mariani & Borghi, 2020; Palese et al., 2021; Quang-Huy, 2021; Uslu Cibere et al., 2020; Xie et al., 2017). Second, the results of this research advance the knowledge of market orientation theory that provides a structure to study market orientation measures and investigating their impact on business performance (Kohli & Jaworski, 1990; Narver & Slater, 1990). It also contributes to the body of organizational learning theory as a process that aims to gather information and disseminate acquired knowledge, related to knowledge management (e.g., Nonaka, 1994). Still, this approach to organizational learning has been used in hospitality and tourism industry research (e.g., Lemmetyinen & Go, 2009; Thomas & Wood, 2015).

5.2 Practical Indications

In the field of practice, this research highlights several important managerial notes, the conclusions of this study follow the line of previous research that suggests review volume positively predicts business performance. More reviews tend to increase awareness and popularity of businesses and reduce consumers' perceptions of risk. Consumers use the volume of positive online consumer reviews as a cue to find out how many people visited certain tourist accommodations and were satisfied with the experience. Thus, businesses should promote online consumer feedback to reduce uncomfortable risk perceptions and increase intentions to decide to buy.

This research also indicates that the volume of consumer reviews has a greater impact on business performance in the presence of market intelligence generation. Managers who actively and continuously generate market intelligence seem to know how to use it to produce better perceptions and more confidence in accommodations, represented in recommendations, new bookings, and, consequently, better performance.

Market intelligence generation is strategically meaningful, because it implies access to relevant information about industry shifts and focuses on tourists' evaluations of delivered services. Owners/managers must constantly monitor consumer reviews to understand what consumers think about the services they provide and what drives consumer perception; so, they can adjust their services to meet consumers' needs and wants. Considering the dynamism of tourists' perceptions, owners/managers must track these factors continuously and attempt to engage consumers in deeper relationships.

However, the volume of management response does not influence business results directly. Nevertheless, researcher results indicate that management response volumes can influence outcomes if the response come from entrepreneurs with market knowledge. These findings suggest that it is not response volume that is important; rather, it may be the content or style of response that is relevant. Response by entrepreneurs with market knowledge may contain more appealing content or styles, such that they offer credible signals that are appreciated by potential consumers. For example, if a consumer review identifies a problem, the response could include an explanation and an apology, as well as an indication of the actions taken to resolve the problem. Although responses on the internet are effortless ways to interact with consumers, especially for small businesses, it is important to devote effort to responding in a personalized professional way.

Permanently collecting market information requires an attitude of openness to relevant information generated by the market. It appears worthwhile for entrepreneurs to be in contact with their markets, for example, to attend rural tourism events, stay abreast of rural tourism news, and seek out available online information about other RTAs. It is likely that the positive consequences the researchers identify reflect the proximity of entrepreneurs to their markets, as well as their abilities to transform market information into good business decisions.

Nonetheless, this research has some limitations that could potentially indicate ideas for further studies. In the first place, it draws data from only one platform (Toprural.com) and from only a country (France), which could limit the generalizability of authors results. Second, does not analyze the content of online consumer reviews and, at the same time, management response, which could provide additional insights.

Acknowledgements This work was supported by the Junta de Castilla y León under SA369A11-1 and the Spanish Ministry of Economy and Competitiveness under ECO2014-53060-R.

References

Ahani, A., Nilashi, M., Ibrahim, O., Sanzogni, L., & Weaven, S. (2019a). Market segmentation and travel choice prediction in Spa hotels through TripAdvisor's online reviews. *International Journal of Hospitality Management, 80*, 52–77. https://doi.org/10.1016/j.ijhm.2019.01.003

Ahani, A., Nilashi, M., Yadegaridehkordi, E., Sanzogni, L., Tarik, A. R., Knox, K., Samad, S., & Ibrahim, O. (2019b). Revealing customers' satisfaction and preferences through online review analysis: The case of Canary Islands hotels. *Journal of Retailing and Consumer Services, 51*, 331–343.https://doi.org/10.1016/j.jretconser.2019.06.014

Ahearne, M., Lam, S. K., Hayati, B., & Kraus, F. (2013). Intrafunctional competitive intelligence and sales performance: A social network perspective. *Journal of Marketing, 77*(5), 37–56. https://doi.org/10.1509/jm.11.0217

Ahlin, B., DrnovŠEk, M., & Hisrich, R. D. (2012). Exploring moderating effects of proactivity on the relationship between market information and innovation performance. *Economic & Business Review, 14*(2), 121–146. https://doi.org/10.15458/2335-4216.1206

Ait-Bakrim, A., Attouch, H., Guerreiro, M., & Perez-Aranda, J. (2019). Examining the relationship between hoteliers' review site use intensity and hotel characteristics—Moroccan hotels. *Journal of Spatial and Organizational Dynamics VI, I*(3), 226–239.

Akbarabadi, M., & Hosseini, M. (2020). Predicting the helpfulness of online customer reviews: The role of title features. *International Journal of Market Research, 62*(3), 272. https://doi.org/10.1177/1470785318819979

Baek, J., & Choe, Y. (2020). Differential effects of the valence and volume of online reviews on customer share of visits: The case of US casual dining restaurant brands. *Sustainability, 12*(13), 11. https://doi.org/10.3390/su12135408

Chandler, A. (1962). *Strategy and structure: Chapters in the history of the industrial enterprise*. MIT Press.

Chen, S., & Chaiken, S. (1999). In S. Chaiken, & Y. Trope. (Eds.), *Dual-process theories in social psychology* (pp. 73–96). Guilford.

Chevalier, J. A., & Mayzlin, D. (2006). The effect of word of mouth on sales: Online book reviews. *Journal of Marketing Research, 43*(3), 345–354. https://doi.org/10.1509/jmkr.43.3.345

Connelly, B. L., Certo, S. T., Ireland, R. D., & Reutzel, C. R. (2011). Signaling theory: A review and assessment. *Journal of Management, 37*(1), 39–67. https://doi.org/10.1177/0149206310388419

Dellarocas, C., Zhang, X., & Awad, N. F. (2007). Exploring the value of online product reviews in forecasting sales: The case of motion pictures. *Journal of Interactive Marketing, 21*(4), 23–45. https://doi.org/10.1002/dir.20087

Dess, G. G., & Davis, P. S. (1984). Porter's (1980) generic strategies as determinants of strategic group membership and organizational performance. *Academy of Management Journal, 27*(3), 467–488. https://doi.org/10.2307/256040

Donavan, D. T., Brown, T. J., & Mowen, J. C. (2004). Internal benefits of service-worker customer orientation: Job satisfaction, commitment, and organizational citizenship behaviors. *Journal of Marketing, 68*, 128–146. https://doi.org/10.1509/jmkg.68.1.128.24034

Dwidienawati, D., Tjahjana, D., Abdinagoro, S. B., Gandasari, D., & Munawaroh. (2020). Customer review or influencer endorsement: which one influences purchase intention more? *Heliyon, 6*(11). https://doi.org/10.1016/j.heliyon.2020.e05543

Esmark Jones, C. L., Stevens, J. L., Breazeale, M., & Spaid, B. I. (2018). Tell it like it is: The effects of differing responses to negative online reviews. *Psychology and Marketing, 35*(12), 891–901. https://doi.org/10.1002/mar.21142

Falahat, M., Ramayah, T., Soto-Acosta, P., & Lee, Y.-Y. (2020). SMEs internationalization: The role of product innovation, market intelligence, pricing and marketing communication capabilities as drivers of SMEs' international performance. *Technological Forecasting and Social Change, 152*, 119908. https://doi.org/10.1016/j.techfore.2020.119908

Farahani, H. A., Rahiminezhad, A., Same, L., & Immannezhad, K. (2010). A comparison of partial least squares (PLS) and ordinary least squares (OLS) regressions in predicting of couples mental health based on their communicational patterns. *Procedia—Social and Behavioral Sciences, 5*, 1459–1463. https://doi.org/10.1016/j.sbspro.2010.07.308

Filieri, R., & McLeay, F. (2014). E-WOM and accommodation: An analysis of the factors that influence travelers' adoption of information from online reviews. *Journal of Travel Research, 53*(1), 44–57. https://doi.org/10.1177/0047287513481274

Flanagin, A. J., & Metzger, M. J. (2013). Trusting expert- versus user-generated ratings online: The role of information volume, valence, and consumer characteristics. *Computers in Human Behavior, 29*(4), 1626–1634. https://doi.org/10.1016/j.chb.2013.02.001

Gebhardt, G. F., Carpenter, G. S., & Sherry, J. F., Jr. (2006). Creating a market orientation: a longitudinal, multifirm, grounded analysis of cultural transformation. *Journal of Marketing, 70*(4), 37–55. https://doi.org/10.1509/jmkg.70.4.37

Gellerstedt, M., & Arvemo, T. (2019). The impact of word of mouth when booking a hotel: Could a good friend's opinion outweigh the online majority? *Information Technology & Tourism, 21*(3), 289–311. https://doi.org/10.1007/s40558-019-00143-4

Gonçalves, H. M., Silva, G. M., & Martins, T. G. (2018). Motivations for posting online reviews in the hotel industry. *Psychology and Marketing, 35*(11), 807–817. https://doi.org/10.1002/mar.21136

Gottschalk, S. A., & Mafael, A. (2017). Cutting through the online review jungle—Investigating selective eWOM processing. *Journal of Interactive Marketing, 37*, 89–104. https://doi.org/10.1016/j.intmar.2016.06.001

Gu, B., & Ye, Q. (2014). First step in social media—Measuring the influence of online management responses on customer satisfaction. *Production and Operations Management, 23*(4), 570–582. https://doi.org/10.1111/poms.12043

Gursoy, D. (2019). A critical review of determinants of information search behavior and utilization of online reviews in decision making process (invited paper for 'luminaries' special issue of International Journal of Hospitality Management). *International Journal of Hospitality Management, 76*, 53–60. https://doi.org/10.1016/j.ijhm.2018.06.003

Hair, J. F. J., Hult, G. T. M., Ringle, C. M., & Sarstedt, M. (2014). *A primer on partial least squares structural equation modeling (PLS-SEM)*. SAGE Publications, Inc.

Hall, Z. R., Mullins, R. R., Syam, N., & Boichuk, J. P. (2017). Generating and sharing of market intelligence in sales teams: An economic social network perspective. *Journal of Personal Selling & Sales Management, 37*(4), 298–312. https://doi.org/10.1080/08853134.2017.1393342

Hallak, R., Assaker, G., & Lee, C. (2015). Tourism entrepreneurship performance: The effects of place identity, self-efficacy, and gender. *Journal of Travel Research, 54*(1), 36–51. https://doi.org/10.1177/0047287513513170

Hendar, H., Ratnawati, A., Razak, W. M. W. A., & Abdullah, Z. (2020). Market intelligence on business performance: The mediating role of specialized marketing capabilities. *Journal of Intelligence Studies in Business, 10*(1), 42–58.

Hernández-Maestro, R. M. (2020). Management response to eWOM in rural lodging establishments. *Spanish Journal of Marketing— - ESIC, 24*(2), 263–278. https://doi.org/10.1108/SJME-11-2019-0093

Hernández-Maestro, R. M., Muñoz-Gallego, P. A., & Santos-Requejo, L. (2009). Small-business owners' knowledge and rural tourism establishment performance. *Journal of Travel Research, 48*(1), 58–77. https://doi.org/10.1177/0047287508328794

Heusinkveld, S., Benders, J., & Van den Berg, R.-J. (2009). From market sensing to new concept development in consultancies: The role of information processing and organizational capabilities. *Technovation, 29*(8), 509–516. https://doi.org/10.1016/j.technovation.2009.02.003

Hsieh, J.-K., & Li, Y.-J. (2020). Will you ever trust the review website again? The importance of source credibility. *International Journal of Electronic Commerce, 24*(2), 255. https://doi.org/10.1080/10864415.2020.1715528

Hu, N., Zhang, J., & Pavlou, P. A. (2009). Overcoming the J-shaped distribution of product reviews. *Communications of the ACM—A View of Parallel Computing CACM, 52*(10), 144–147. https://doi.org/10.1145/1562764.1562800

Huertas-Garcia, R., Laguna García, M., & Consolación, C. (2014). Conjoint analysis of tourist choice of hotel attributes presented in travel agent brochures. *International Journal of Tourism Research, 16*(1), 65–75. https://doi.org/10.1002/jtr.1899

Hur, K., Kim, T. T., Karatepe, O. M., & Lee, G. (2017). An exploration of the factors influencing social media continuance usage and information sharing intentions among Korean travellers. *Tourism Management, 63*, 170–178. https://doi.org/10.1016/j.tourman.2017.06.013

Hyun, S. S., & Han, H. (2015). Luxury cruise travelers: Other customer perceptions. *Journal of Travel Research, 54*(1), 107–121. https://doi.org/10.1177/0047287513513165

Jaworski, B. J., & Kohli, A. K. (1993). Market orientation: Antecedents and consequences. *Journal of Marketing, 57*(3), 53–70.

Jogaratnam, G., & Tse, E. C. Y. (2004). The entrepreneurial approach to hotel operation: Evidence from the Asia-Pacific hotel industry. *Cornell Hotel and Restaurant Administration Quarterly, 45*, 248–259. https://doi.org/10.1177/0010880404266192

Jordan, P. J., & Troth, A. C. (2020). Common method bias in applied settings: The dilemma of researching in organizations. *Australian Journal of Management, 45*(1), 3–14. https://doi.org/10.1177/0312896219871976

Kaish, S., & Gilad, B. (1991). Characteristics of opportunities search of entrepreneurs versus executives: Sources, interests, general alertness. *Journal of Business Venturing, 6*(1), 45–61. https://doi.org/10.1016/0883-9026(91)90005-X

Katsikea, E., Theodosiou, M., & Makri, K. (2019). The interplay between market intelligence activities and sales strategy as drivers of performance in foreign markets. *European Journal of Marketing, 53*(10), 2080–2108. https://doi.org/10.1108/EJM-06-2017-0402

Kelley, H. H. (1971). *Attributions in social interactions*. General Learning Press.

King, R. A., Bush, V. D., & Racherla, P. (2014). What we know and don't know about online word-of-mouth: A review and synthesis of the literature. *Journal of Interactive Marketing, 28*(3), 167–183. https://doi.org/10.1016/j.intmar.2014.02.001

Kirca, A. H., Jayachandran, S., & Bearden, W. O. (2005). Market orientation: A meta-analytic review and assessment of its antecedents and impact on performance. *Journal of Marketing, 69*(2), 24–41. https://doi.org/10.1509/jmkg.69.2.24.60761

Kohli, A. K., & Jaworski, B. J. (1990). Market orientation: The construct, research propositions, and managerial implications. *Journal of Marketing, 54*(2), 1–18. https://doi.org/10.1177/002224299005400201

Korfiatis, N., Stamolampros, P., Kourouthanassis, P., & Sagiadinos, V. (2019). Measuring service quality from unstructured data: A topic modeling application on airline passengers' online reviews. *Expert Systems with Applications, 116*, 472–486. https://doi.org/10.1016/j.eswa.2018.09.037

Kropp, F., Lindsay, N. J., & Shoham, A. (2006). Entrepreneurial, market, and learning orientations and international entrepreneurial business venture performance in South African firms. *International Marketing Review, 23*(5), 504–523. https://doi.org/10.1108/02651330610703427

Kumar, V., Jones, E., Venkatesan, R., & Leone, R. P. (2011). Is market orientation a source of sustainable competitive advantage or simply the cost of competing? *Journal of Marketing, 75*(1), 16–30. https://doi.org/10.1509/jm.75.1.16

Ladhari, R., & Michaud, M. (2015). eWOM effects on hotel booking intentions, attitudes, trust, and website perceptions. *International Journal of Hospitality Management, 46*, 36–45. https://doi.org/10.1016/j.ijhm.2015.01.010

Lee, C., Hallak, R., & Sardeshmukh, S. R. (2016). Innovation, entrepreneurship, and restaurant performance: A higher-order structural model. *Tourism Management, 53*, 215–228. https://doi.org/10.1016/j.tourman.2015.09.017

Lee, C. C., & Hu, C. (2005). Analyzing hotel customers' e-complaints from an internet complaint forum. *Journal of Travel & Tourism Marketing, 17*, 167–181. https://doi.org/10.1300/J073v17n02_13

Lee, H., & Blum, S. C. (2015). How hotel responses to online reviews differ by hotel rating: An exploratory study. *Worldwide Hospitality and Tourism Themes 7*(3), 242–250. https://doi.org/10.1108/WHATT-03-2015-0016

Lee, Y. L., & Song, S. (2010). An empirical investigation of electronic word-of-mouth: Informational motive and corporate response strategy. *Computers in Human Behavior, 26*(5), 1073–1080. https://doi.org/10.1016/j.chb.2010.03.009

Lemmetyinen, A., & Go, F. M. (2009). The key capabilities required for managing tourism business networks. *Tourism Management, 30*(1), 31–40. https://doi.org/10.1016/j.tourman.2008.04.005

Levy, S. E., Duan, W., & Boo, S. (2013). An analysis of one-star online reviews and responses in the Washington, D.C., lodging market. *Cornell Hospitality Quarterly, 54*(1), 49–63. https://doi.org/10.1177/1938965512464513

Li, C., Cui, G., & Peng, L. (2017). The signaling effect of management response in engaging customers: A study of the hotel industry. *Tourism Management, 62*, 42–53. https://doi.org/10.1016/j.tourman.2017.03.009

Li, Y.-M., & Li, T.-Y. (2013). Deriving market intelligence from microblogs. *Decision Support Systems, 55*(1), 206–217. https://doi.org/10.1016/j.dss.2013.01.023

Liu, S., Wang, N., Gao, B., & Gallivan, M. (2021). To be similar or to be different? The effect of hotel managers' rote response on subsequent reviews. *Tourism Management, 86*, 104346. https://doi.org/10.1016/j.tourman.2021.104346

Lorentz, H., Aminoff, A., Kaipia, R., Pihlajamaa, M., Ehtamo, J., & Tanskanen, K. (2020). Acquisition of supply market intelligence—an information processing perspective. *Journal of Purchasing and Supply Management, 26*(5), 100649.

Mariani, M. M., & Borghi, M. (2020). Online review helpfulness and firms' financial performance: An empirical study in a service industry. *International Journal of Electronic Commerce, 24*(4), 421–449. https://doi.org/10.1080/10864415.2020.1806464

Mate, M. J., Trupp, A., & Pratt, S. (2019). Managing negative online accommodation reviews: Evidence from the Cook Islands. *Journal of Travel & Tourism Marketing, 36*(5), 627–644. https://doi.org/10.1080/10548408.2019.1612823

Mathwick, C., & Mosteller, J. (2017). Online reviewer engagement: A typology based on reviewer motivations. *Journal of Service Research, 20*(2), 204–218. https://doi.org/10.1177/1094670516682088

Matson-Barkat, S., & Robert-Demontrond, P. (2018). Who's on the tourists' menu? Exploring the social significance of restaurant experiences for tourists. *Tourism Management, 69*, 566–578. https://doi.org/10.1016/j.tourman.2018.06.031

Matzat, U. (2009). A theory of relational signals in online groups. *New Media & Society, 11*(3), 375–394. https://doi.org/10.1177/1461444808101617

Mauri, A. G., & Minazzi, R. (2013). Web reviews influence on expectations and purchasing intentions of hotel potential customers. *International Journal of Hospitality Management, 34*, 99–107. https://doi.org/10.1016/j.ijhm.2013.02.012

Melo, A. J. D. V. T. (2020). *Ewom effects on business performance in rural tourism accommodation* (Doctor of Philosophy—Business Economics Scientific research dissertation). University of Salamanca. https://www.educacion.gob.es/teseo/mostrarSeleccion.do

Melo, A. J. D. V. T., Hernández-Maestro, R. M., & Muñoz-Gallego, P. A. (2017). Service quality perceptions, online visibility, and business performance in rural lodging establishments. *Journal of Travel Research, 56*(2), 250–262. https://doi.org/10.1177/0047287516635822

Mittra, S., & Khamkar, P. R. (2021). To understand customers perception towards hotel amenities, services quality and online reviews while booking hotels. *ANWESH: International Journal of Management & Information Technology, 6*(1), 38–42.

Narver, J. C., & Slater, S. F. (1990). The effects of a market orientation on business profitability. *Journal of Marketing, 54*, 20–35. https://doi.org/10.1177/002224299005400403

Neirotti, P., & Raguseo, E. (2020). Mitigating resource dependence on internet visibility providers: Exploring complementarity effects in the positioning of small hotels on online intermediaries. *Information & Management* 103279. https://doi.org/10.1016/j.im.2020.103279

Neirotti, P., Raguseo, E., & Paolucci, E. (2016). Are customers' reviews creating value in the hospitality industry? Exploring the moderating effects of market positioning. *International Journal of Information Management, 36*(6, Part A), 1133–1143.

Nieto, J., Hernández-Maestro, R. M., & Muñoz-Gallego, P. A. (2011). The influence of entrepreneurial talent and website type on business performance by rural tourism establishments in Spain. *International Journal of Tourism Research, 13*, 17–31. https://doi.org/10.1002/jtr.794

Nieto, J., Hernández-Maestro, R. M., & Muñoz-Gallego, P. A. (2014). Marketing decisions, customer reviews, and business performance: The use of the Toprural website by Spanish rural lodging establishments. *Tourism Management, 45*, 115–123. https://doi.org/10.1016/j.tourman.2014.03.009

Nonaka, I. (1994). A Dynamic Theory of Organizational Knowledge Creation. *Organization Science, 5*(1), 14–37. https://doi.org/10.1287/orsc.5.1.14

Oliveira, A. S., Renda, A. I., & Correia, M. B. (*2020*). Online reviews: A pathway to improve hotel management. *dosalgarves: A Multidisciplinary e-Journal, 36*, 108–126. https://doi.org/10.18089/DAMeJ.2020.36.7

Palese, B., Piccoli, G., & Lui, T.-W. (2021). Effective use of online review systems: Congruent managerial responses and firm competitive performance. *International Journal of Hospitality Management, 96*.https://doi.org/10.1016/j.ijhm.2021.102976

Park, D. H., Lee, J., & Han, I. (2007). The effect of on-line consumer reviews on consumer purchasing intention: The moderating role of involvement. *International Journal of Electronic Commerce, 11*(4), 125–148. https://doi.org/10.2753/JEC1086-4415110405

Park, S.-Y., & Allen, J. P. (2013). Responding to online reviews: Problem solving and engagement in hotels. *Cornell Hospitality Quarterly, 54*(1), 64–73. https://doi.org/10.1177/1938965512463118

Pathak, B., Garfinkel, R., Gopal, R. D., Venkatesan, R., & Yin, F. (2010). Empirical analysis of the impact of recommender systems on sales. *Journal of Management Information Systems, 27*(2), 159–188.

Perez-Vega, R., Taheri, B., Farrington, T., & O'Gorman, K. (2018). On being attractive, social and visually appealing in social media: The effects of anthropomorphic tourism brands on Facebook fan pages. *Tourism Management, 66*, 339–347. https://doi.org/10.1016/j.tourman.2017.11.013

Phillips, P., Barnes, S., Zigan, K., & Schegg, R. (2017). Understanding the impact of online reviews on hotel performance empirical analysis. *Journal of Travel Research, 56*(2), 235–249. https://doi.org/10.1177/0047287516636481

Polo Peña, A. I., Jamilena, D. M. F., & Molina, M. Á. R. (2012). Validation of a market orientation adoption scale in rural tourism enterprises. Relationship between the characteristics of the enterprise and extent of market orientation adoption. *International Journal of Hospitality Management, 31*(1), 139–151. https://doi.org/10.1016/j.ijhm.2011.06.005

Prahalad, C. K., & Ramaswamy, V. (2004). Co-creation experiences: The next practice in value creation. *Journal of Interactive Marketing, 18*(3), 5–14. https://doi.org/10.1002/dir.20015

Proserpio, D., & Zervas, G. (2017). Online reputation management: Estimating the impact of management responses on consumer reviews. *Marketing Science, 36*(5), 645–665. https://doi.org/10.1287/mksc.2017.1043

Purnawirawan, N., Pelsmacker, P. d., & Dens, N. (2012). Balance and sequence in online reviews: How perceived usefulness affects attitudes and intentions. *Journal of Interactive Marketing, 26*(4), 244–255. https://doi.org/10.1016/j.intmar.2012.04.002

Quang-Huy, N. (2021). How market orientation induces small business performance: The role of strategic fits. *Gadjah Mada International Journal of Business, 23*(1), 55–75. https://doi.org/10.22146/gamaijb.60623

Racherla, P., Connolly, D. J., & Christodoulidou, N. (2013). What determines consumers' ratings of service providers? An exploratory study of online traveler reviews. *Journal of Hospitality Marketing & Management, 22*(2), 135–161. https://doi.org/10.1080/19368623.2011.645187

Rao, A. R., Qu, L., & Ruekert, R. W. (1999). Signaling unobservable product quality through a brand ally. *Journal of Marketing Research, 36*(2), 258–268. https://doi.org/10.1177/002224379903600209

Reich, T., & Maglio, S. J. (2020). Featuring mistakes: The persuasive impact of purchase mistakes in online reviews. *Journal of Marketing, 84*(1), 52–65. https://doi.org/10.1177/0022242919882428

Ren, L., Zhang, H. Q., & Ye, B. H. (2015). Understanding customer satisfaction with budget hotels through online comments: Evidence from home inns in China. *Journal of Quality Assurance in Hospitality & Tourism, 16*(1), 45–62. https://doi.org/10.1080/1528008X.2015.966299

Rose, M., & Blodgett, J. G. (2016). Should hotels respond to negative online reviews? *Cornell Hospitality Quarterly, 57*(4), 396–410. https://doi.org/10.1177/1938965516632610

Salganik, M. J., & Watts, D. J. (2008). Leading the herd astray: An experimental study of self-fulfilling prophecies in an artificial cultural market. *Social Psychology Quarterly, 71*(4), 338–355. https://doi.org/10.1177/019027250807100404

Santos, M. (2006). Strategic information management profiles in brazilian companies. *Brazilian Business Review (english Edition), 3* (1), 116–134. https://doi.org/10.15728/bbr.2006.3.1.9

Sashi, C. M. (2012). Customer engagement, buyer-seller relationships, and social media. *Management Decision, 50*(2), 253–272. https://doi.org/10.1108/00251741211203551

Septianto, F., Kemper, J. A., & Choi, J. (2020). The power of beauty? The interactive effects of awe and online reviews on purchase intentions. *Journal of Retailing and Consumer Services, 54*, 102066. https://doi.org/10.1016/j.jretconser.2020.102066

Slater, S. F., & Narver, J. C. (2000). Intelligence generation and superior customer value. *Journal of the Academy of Marketing Science, 28*(1), 120. https://doi.org/10.1177/0092070300281011

Sparks, B. A., & Bradley, G. L. (2014). A "Triple A" typology of responding to negative consumer-generated online reviews. *Journal of Hospitality & Tourism Research, 41*(6), 719–745. https://doi.org/10.1177/1096348014538052

Sparks, B. A., So, K. K. F., & Bradley, G. L. (2016). Responding to negative online reviews: The effects of hotel responses on customer inferences of trust and concern. *Tourism Management, 53*, 74–85. https://doi.org/10.1016/j.tourman.2015.09.011

Spence, A. (1974). *Market signaling: Informational transfer in hiring and related screening processes*. Harvard University Press.

Stringam, B. B., & Gerdes, J. (2010). An analysis of word-of-mouse ratings and guest: Comments of online hotel distribution sites. *Journal of Hospitality Marketing & Management, 19*(7), 773–796.

Sun, T., Viswanathan, S., & Zheleva, E. (2014, August 05). *Impact of message design on online interactions: An empirical investigation.* Paper presented at the ICEC '14: Sixteenth International Conference on Electronic Commerce, Philadelphia, PA, USA.

Talvinen, J. M. (1995). Information systems in marketing. *European Journal of Marketing, 29*(1), 8–26.

Thi, N. V., Hung, V. V., Tuckova, Z., Pham, N. T., & Nguyen, L. H. L. (2021). Guest online review: An extraordinary focus on hotel users' satisfaction, engagement, and loyalty. *Journal of Quality Assurance in Hospitality & Tourism* 1–32.

Thomas, R., & Wood, E. (2015). The absorptive capacity of tourism organisations. *Annals of Tourism Research, 54*, 84–99.

Tontini, G., Bento, G. d. S., Milbratz, T. C., Volles, B. K., & Ferrari, D. (2017). Exploring the nonlinear impact of critical incidents on customers' general evaluation of hospitality services. *International Journal of Hospitality Management, 66*, 106–116.

Torres, E. N., Singh, D., & Robertson-Ring, A. (2015). Consumer reviews and the creation of booking transaction value: Lessons from the hotel industry. *International Journal of Hospitality Management, 50*, 77–83.

Tran, L. T. T. (2020). Online reviews and purchase intention: A cosmopolitanism perspective. *Tourism Management Perspectives, 35*, 100722.

Ullrich, S., & Brunner, C. B. (2015). Negative online consumer reviews: Effects of different responses. *Journal of Product & Brand Management, 24*(1), 66–77.

Uslu Cibere, G., Basaran, M. A., & Kantarci, K. (2020). Evaluation of hotel performance attributes through consumer generated reviews: The case of Bratislava. *Advances in Hospitality & Tourism Research (AHTR): An International Journal of Akdeniz University, Tourism Faculty, 8*(1), 48–75.

Walker, K. (2010). A systematic review of the corporate reputation literature: Definition, measurement, and theory. *Corporate Reputation Review, 12*(4), 357–387.

Webb, J. W., Ireland, R. D., Hitt, M. A., Kistruck, G. M., & Tihanyi, L. (2011). Where is the opportunity without the customer? An integration of marketing activities, the entrepreneurship process, and institutional theory. *Journal of the Academy of Marketing Science, 39*(4), 537–554.

Wei, W., Miao, L., & Huang, Z. (2013). Customer engagement behaviors and hotel responses. *International Journal of Hospitality Management, 33*, 316–330.

Wells, J. D., Valacich, J. S., & Hess, T. J. (2011). What signal are you sending? How website quality influences perceptions of product quality and purchase intentions. *MIS Quarterly, 35*(2), 373–396.

Wood, E. (2001). Marketing information systems in tourism and hospitality small- and medium-sized enterprises: A study of Internet use for market intelligence. *International Journal of Tourism Research, 3*(4), 283–299.

Xiang, Z., Wang, D., O'Leary, J. T., & Fesenmaier, D. R. (2015). Adapting to the internet: Trends in travelers' use of the web for trip planning. *Journal of Travel Research, 54*(4), 511–527.

Xie, K. L., So, K. K. F., & Wang, W. (2017). Joint effects of management responses and online reviews on hotel financial performance: A data-analytics approach. *International Journal of Hospitality Management, 62*, 101–110.

Xie, K. L., Zhang, Z., & Zhang, Z. (2014). The business value of online consumer reviews and management response to hotel performance. *International Journal of Hospitality Management, 43*, 1–12. https://doi.org/10.1016/j.ijhm.2014.07.007

Xu, X. (2020). Examining the role of emotion in online consumer reviews of various attributes in the surprise box shopping model. *Decision Support Systems, 136*.

Yap, C. S., Cheng, B. L., Mohamad Hussain, N., & Ahmad, R. (2018). Innovativeness, market intelligence practices, and firm performance of small- and medium-sized tour operators. *Tourism & Hospitality Research, 18*(2), 143–151.

Ye, Q., Gu, B., & Chen, W. (2008). *Measuring the influence of managerial responses on subsequent online customer reviews: A natural experiment of two online travel agencies.* Paper presented at the Twenty Ninth International Conference on Information Systems, Paris.

Ye, Q., Law, R., & Gu, B. (2009). The impact of online user reviews on hotel room sales. *International Journal of Hospitality Management, 28*(1), 180–182. https://doi.org/10.1016/j.ijhm.2008.06.011

Zeithaml, V. A., Varadarajan, P. R., & Zeithaml, C. P. (1988). The contingency approach: Its foundations and relevance to theory building and research in marketing. *European Journal of Marketing, 22*(7), 37–64.

Zhang, L., Wei, W., Line, N. D., & Cheng, Y. (2021a). When positive reviews backfire: The effect of review dispersion and expectation disconfirmation on Airbnb guests' experiences. *International Journal of Hospitality Management, 96*, 102979.

Zhang, L., Wu, L., & Mattila, A. S. (2016). Online reviews: The role of information load and peripheral factors. *Journal of Travel Research, 55*(3), 299–310.

Zhang, X., Qiao, S., Yang, Y., & Zhang, Z. (2020a). Exploring the impact of personalized management responses on tourists' satisfaction: A topic matching perspective. *Tourism Management, 76*, 103953. https://doi.org/10.1016/j.tourman.2019.103953

Zhang, X., Yang, Y., Qiao, S., & Zhang, Z. (2020b). Responsive and responsible: Customizing management responses to online traveler reviews. *Journal of Travel Research* 1.

Zhang, X., Yang, Y., Qiao, S., & Zhang, Z. (2022). Responsive and responsible: Customizing management responses to online traveler reviews. *Journal of Travel Research, 61*(1), 120–135. https://doi.org/10.1177/0047287520971046

Zhang, Z., Li, H., Yang, Y., & Xu, Y. (2021b). Not all words are beneficial: The impact of management response contents on customer engagement behavior. *International Journal of Hospitality Management, 93*.

Zhang, Z., Ye, Q., Law, R., & Li, Y. (2010). The impact of e-word-of-mouth on the online popularity of restaurants: A comparison of consumer reviews and editor reviews. *International Journal of Hospitality Management, 29*(4), 694–700. https://doi.org/10.1016/j.ijhm.2010.02.002

Zhao, Y., Wen, L., Feng, X., Li, R., & Lin, X. (2020). How managerial responses to online reviews affect customer satisfaction: An empirical study based on additional reviews. *Journal of Retailing and Consumer Services, 57*. https://doi.org/10.1016/j.jretconser.2020.102205

Zhao, Y., Xu, X., & Wang, M. (2019). Predicting overall customer satisfaction: Big data evidence from hotel online textual reviews. *International Journal of Hospitality Management, 76*, 111–121.

The Fortress of Santa Catarina de Ribamar (Portimão) as a Proposal for Good Practices of Military Heritage Preservation

Bruna Ramalho Galamba

Abstract

Early modern fortresses by the sea in Portugal are an historical mark of protection during the maritime expansion. Pirates and privateers were a constant threat to the economy and the safety of the populations who were targets. Nowadays, many of these fortresses have been requalified for tourism, especially in Oeiras and Cascais, near Lisbon, that have been blessed with investments and projects and, either by land or sea, can be visited or communicated by careful planning of circuits. In the Algarve region, many military buildings are lacking the same investment, projects and policies. The Fortress of Santa Catarina de Ribamar is a part of a much bigger defense system, that is rarely acknowledged, communicated and invested in, even if it is visited by many people, as it also represents a shortcut to the beach (Praia da Rocha). In the prior decades, the fortress has been mismanaged by many entities and had a small cafeteria and a restaurant that have been closed for a long time. Even though it has been a platform for a wine tasting festival in recent years, nothing else has been done in the location, either taking advantage from its history or the proximity to the beach. In 2016, it was included in a project of concession to private entities in exchange of requalification, as the fortress is in a state of abandonment and near ruin. In 2020, good news finally came to this historical landmark, as it might be managed by the municipality, that has longed for changes, a proper management and to communicate its story. This is a moment of transition and of planning. In virtue of a dissertation in Archaeology of Architecture, this fortress is being studied and will benefit from a project of requalification as a military building that accommodates heritage preservation and history and can promote tourism (not only in benefit of the visitor, but in benefit to the local community).

Keywords

Early modern fortress • Algarve region • Space modification • Conservation • Communication of heritage

1 Introduction

The Fortress of Santa Catarina de Ribamar, located in Praia da Rocha, in Portimão (Algarve) is an important landmark in the city landscape and history. Since the Modern Age, this fortress dominates the entrance to the Arade River, it has resisted the forces of nature, and after extensive works in the twentieth century, it has come to us, disfigured.

This defensive system was built in 1633 and was a symbol of protection for those who lived in the old town and feared looting and attacks by Muslim pirates and European privateers, as well as a symbol of recovery for those who survived the three violent earthquakes of the eighteenth century (including the most famous 1755 earthquake and tsunami). In the twentieth century, this fortress, after ceasing to be the property of the former Ministry of War, withstood a violent earthquake and was requalified for tourism with the construction of an access to the beach.

In 1977, it was classified as a Property of Public Interest (Imóvel de Interesse Público) by decree-law, which should have granted it additional protection and investments, but this was not implemented. Several walls of the former garrison barracks were demolished to open a restaurant, with its kitchen in the bastion to the West. In the Captain's barracks, to the East, doors were opened to explore a small cafeteria. In the early 2000s, both establishments had already closed.

This fortress is the object of study in a master's thesis in archaeology. It is essential to understand the design of this

B. R. Galamba (✉)
Department of History, NOVA FCSH, Lisbon, Portugal
e-mail: bruna.r.galamba@gmail.com

A. Mandić et al. (eds.), *Cultural Sustainable Tourism*, Advances in Science, Technology & Innovation,
https://doi.org/10.1007/978-3-031-10800-6_10

building, its changes over almost four centuries and its use so far. For the last eighty years, this fortress has been almost exclusively used for tourism. However, no project was successful in the most favorable place for tourism in Portugal.

This article will explore the modifications and uses of the Fortress of Santa Catarina de Ribamar during the twentieth and twenty-first centuries to accommodate and promote tourism, especially using the files in the Army Infrastructure Directorate (Direção de Infraestruturas do Exército—DIE) and the Information System for the Architectural Heritage (Sistema de Informação para o Património Arquitetónico—SIPA) and using the evolution of space modification of this building. As it is the first study done for this location, the bibliography is scarce, and the analysis always starts from the primary sources (the archives and the building itself). Finally, I will offer a holistic conservation perspective for the future, how to communicate its history, and provide local people and tourists with a place that can and should be enjoyed, always with a focus on the past.

2 Brief History of the Fortress of Santa Catarina de Ribamar

The Algarve region, always peripheral to the kingdom of Portugal, gained prominence in the fifteenth century by the action of Infante D. Henrique and D. João I, who made their ventures from Lagos to Ceuta, in order to conquer it, in the year 1415 (Barreto, 1972). With the death of the Infante, the Algarve progressively began to lose the prominence it had gained in the process of overseas expansion. In this century and until the sixteenth century, the Algarve held a geostrategic position of importance compared to the strongholds that were conquered in North Africa. Almost the entire coast of that territory was under Portuguese possession or surveillance, and the Algarve was the closest place for supplies and aid. Unfortunately, over the course of the sixteenth century, there was a progressive abandonment of the fortresses, from the South of Morocco toward the North, which led the Algarve to lose its functions (Magalhães, 1999).

The Algarve coast then became the weakest point in the kingdom, facing the attacks of Berber, Turkish and European pirates and privateers. With the abandonment of North African strongholds, Muslim pirate frequently came to devastate the Algarve's localities (Coutinho, 1991).

It was only with the disappearance and death of King Sebastião, in 1578, that the Algarve became chronically vulnerable to pirates and privateers (Magalhães, 1999). Two years later, with the beginning of the Iberian Union and the reign of the Filipes, serious problems arose given the rivalries between the Spanish monarch and some European nations (namely the Dutch, English and French), so the coast was nearly always in danger (PT/TT/CC/1/114/136; Coutinho, 1999).

In 1617, the Italian military engineer, Alexandre Massaii, in this juncture of constant enemy attacks, carried out a preliminary inspection report on the existing fortresses in the Algarve and subsequently drew up a proposal for the fortification of strategic points (PT/TT/CCDV/29). However, it was only in 1621 that Massaii delivered his final report (MC. DES.1402). In the 1617 report, Massaii offers the proposal to build a fortress on the cliff of Santa Catarina (Praia da Rocha), where it existed a small chapel for that saint (Saint Catherine of Alexandria). In 1621, Massaii consolidates its proposal with three plans for the site. The fortress started to be built in 1630 and was completed in 1633 (PT/TT/CC/2/364/00141). The fortress had a dry moat and a drawbridge (and later a dormant bridge); there were loopholes (an arrow slit) in the walls to defend the moat and several gun emplacements on the bastions, that were accessible with two staircases and several cannons facing the sea. The fortress encompassed the small chapel, and it was built with barracks for the Captain of the fortress and for the garrison.

In the eighteenth century, the Algarve was hit by three earthquakes, one in 1719, another in 1722 and finally, together with the tsunami, in 1755 (ANPC, 2010). The fortress was in a state of ruin and unable to defend Portimão and undergoes two reconstructions, an immediate one in 1758 and another in 1794, under the direction of Lieutenant-Colonel José de Sande Vasconcelos (Ventura & Marques, 1993).

Reaching the nineteenth century, contrary to very brief studies on the fortress, especially carried out by Coutinho (1991, 1999, 2001), the fortress had not been deactivated for being incapable of defense. The fortress was considered essential for the preservation of the great defensive system and was constantly inspected and underwent construction works in 1864 (PT/AHM/DIV/3/09/88/22).

3 The Twentieth Century: Adapting to Tourism

Reaching the twentieth century, the Fortress of Santa Catarina de Ribamar housed the Fiscal Guard (Guarda Fiscal—GNR), the Shipwreck Aid Institute (Instituto de Socorro a Náufragos—ISN) and the Captaincy of the Port of Portimão (Capitania do Porto de Portimão), among others. As far as archives and correspondence show, in 1927, the fortress was already visited by locals and foreigners that enjoyed its views (6-Oct-1927 PT/DIE/RGP/PM001/PTM[T-85/2A]). We can infer that, at least for 94 years, the fortress was already used for tourism, in a very rudimentary way.

The City Council of Portimão (Câmara Municipal de Portimão), in 1934, understanding the clear advantage of using the fortress for tourism, tried to classify the building as a National Monument, the highest classification for heritage in Portugal (15-May-1934 PT/DIE/RGP/ PM001/PTM [T-85/2A]). In that same year, the opinion of the Ministry of War was not favorable, concluding that there was no plausible justification for such an important classification to be attributed to a building that had no outstanding history or architecture and, not only are these obstacles named, the Ministry of War confirmed that the budget allocated to the conservation of National Monuments under its charge is scarce (Major Francisco Goulão, 23-Jun-1934 PT/DIE/RGP/ PM001/PTM[T-85/2A]).

The most significant changes to its adaptation to tourism began in the 1940s, when the fortress was handed over to the Ministry of Finances (SOP Militares Directorate in Évora, 18-Jan-1940 PT/DIE/RGP/PM001/PTM [T-85/2A]). Thus, it is communicated to the Directorate-General for Buildings and National Monuments (Direção-Geral dos Edifícios e Monumentos Nacionais—DGEMN) that the Municipal Tourism Commission of Portimão had the intentions of "beautifying the place that constitutes a beautiful viewpoint" and, being much visited by tourists, requested authorization to landscape and put benches on the terrace (SIPA TXT.01619339), the project was approved by the Directorate-General.

In 1943, the first major project for tourism began, when the Ministry of Finances concessioned the fortress to the City Council of Portimão, through the intermediary of the Municipal Tourism Commission of Portimão. The objective was to convert the fortress into a viewpoint and terrace for tea (SIPA TXT.01619345). The project foresaw the conversion of the old lower war square (Praça Baixa) to a tea terrace, protected by parasols; benches would be built on the wall (which would be excavated for this purpose) and covered with blue and white tiles with Algarvian motifs. In one of the annexes, a kitchen would be built (in the western bastion), an arcade would be built separating the upper war square (Praça Alta) from the lower war square (Fig. 1), and, finally, the cistern (water well) would be equipped with traditional ironwork (Fig. 2) (SIPA TXT.01619386). The DGEMN would eventually agree with the project, but would not authorize the excavation of the walls, taking into account the various conservation problems that would arise (SIPA TXT.01619389 and SIPA TXT.01619390).

In 1963, a work to consolidate the cliff was approved, due to the erosion it suffered, in order to avoid any kind of collapse, thus preserving the integrity of the building, inhabitants and visitors. In other words, they carried out an entire construction of reinforced concrete, lined with limestone and mortar, with several levels and platforms where small shops for tourism purposes were being installed (Figs. 3 and 4) (SIPA TXT.01619446).

In 1964, new works were authorized in the fortress, in order to modify it even more to accommodate the tourist demand. The interior of the West bastion was demolished, and the construction of reinforced concrete slabs was carried out. At the East bastion, the escarpment was aggressively

Fig. 1 Perspective of the arcade built in 1943, with six of its seven arches visible. Photograph taken by the author in August 2021

Fig. 2 Traditional ironwork in the water well. Photograph taken by the author in August 2021

Fig. 3 Construction of platforms and a concrete wall to preserve the cliff from erosion in the early 60s. Photograph in SIPA (FOTO.00174545)

Fig. 4 The artificial cliff as it is in present day. It is in an advanced state of erosion. Photograph taken by the author in August 2021

excavated, compromising its integrity, a problem that has not yet been resolved. Finally, the bastions were excavated in order to install ventilation in the kitchen and warehouse (SIPA TXT.01619454 and SIPA TXT.016194689).

On February 28th, 1969, a strong earthquake ruined the building. The cliff was in imminent danger of collapsing and needed urgent maintenance, it needed protection from the action of the sea and showed cracks inside the rooms and roofs, and the tea terrace needed to be consolidated. The works were carried out almost immediately (SIPA TXT.01619476).

From the 70s onwards, few reports exist in the archives, but it is known that from 1973 onwards the classification of the fortress as a Property of Public Interest began, which would be implemented in 1977 by decree-law nº129/77 of 29 September (SIPA TXT.01619503). The last news of the fortress in the twentieth century is from 1982, when the Fiscal Guard deactivates the post in the fortress.

4 The Twenty-First Century: Progressive Abandonment and the Current State

The first records of the fortress in the present century appear in the year 2002. The fortress housed kennels that had to be demolished, as they contributed to the degradation of the site (it is never specified in which annex they were located). The next news about the building would only come twelve years later, through the local and national media. The fortress as a classified monument and prestigious access to the beach was visited by hundreds of people a day; however, it was in an advanced state of abandonment, and the elevator was damaged and with an accumulation of garbage, which allowed the propagation of animals (especially rats). The news reveals that, despite belonging to the State, no entity assumed responsibility for managing the space. Shortly afterward, it was published in Diário da República that the Administration of the Ports of Sines and Algarve (APS) would manage the fortress (S.n., 2014).

In 2016, the fortress was included in the Revive project, promoted by Tourism of Portugal. A survey was carried out the following year and is available on the project's official Web site. Thus, the fortress would be leased to private individuals from thirty to fifty years, who would be obliged to carry out the necessary restoration works (Eusébio, 2017). Also in 2016, the Left Block party (Bloco de Esquerda) denounced that the situation of abandonment and degradation of the fortress was unsustainable and dangerous (Algarve-Primeiro, 2016).

In 2017, the City Council of Portimão stated that it wanted to recover the management of the Fortress of Santa Catarina de Ribamar, in order to recover and dynamize the monument which, after three years under the management of the APS, was still degraded. Mayor Isilda Gomes would eventually reveal that a technical feedback had been commissioned from the National Laboratory of Civil Engineering (Laboratório Nacional de Engenharia Civil—LNEC), in order to understand the state of the structure. The opinion, which is confidential, denounces, in a very brief way, that the property is unstable and needs intervention with a large investment (Lusa, 2017).

In the summers from 2017 to 2019, the fortress was used as the venue for the Portimão Wine Tasting Festival (Paulo, 2019). In some specific years, it was also used as part of the routes of the Ciência Viva center (to observe the stars), of some tourist routes and even as a meeting point for Narcotics Anonymous (Ciência Viva, 2013; Região Portuguesa Narcóticos Anónimos, 2009).

In 2018, emergency works were ordered. The North wall and especially the West bastion, already very degraded, had falling stones and mortar, which needed to be repaired and represented an investment of twenty thousand euros (Eusébio, 2018). In 2019, the City Council of Portimão reinforced, once again, that it wanted to recover the abandoned fortress, and the transfer of management would take place soon. The City Council also denounced that, despite the APS having applied the fortress to the Revive project, in three years it was never granted concession to another entity (Eusébio, 2019).

In 2020, weeks before the Sars-Cov-2 virus was declared a pandemic, the City Council ordered a survey to the company 3D Modelling Studio, in order to document the building and be able to grant a concession for its study. It is with all this bibliographical, archival, photographic and photogrammetric survey that a dissertation in archaeology of architecture is created and its future is considered in terms of conservation, musealization and communication.

The twenty-first century has not been, so far, a time of splendor for the fortress. According to the blueprint (Fig. 5), starting with the set to the West, the kitchen is disabled, dirty and covered with grease, dead rodents and cockroaches. The restaurant is an area that is rarely accessible, due to the fact that when it was abandoned, the doors were reinforced so as not to be vandalized.

In an annex next to the restaurant, a small house had been illegally occupied by a person with addiction problems (the floor being covered with bottles of beer and wine and some syringes). Finally, in the same segment, there are public bathrooms that are closed to the public.

On the opposite side, to the East, the bastion had been converted into a warehouse. It is where chairs and tables are

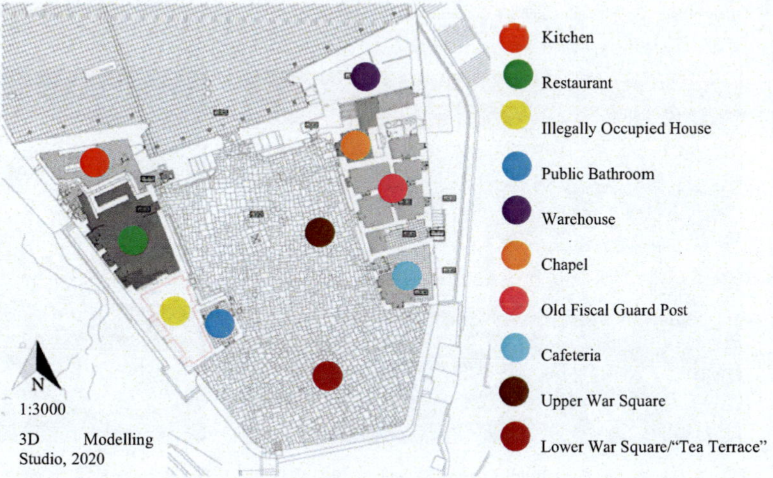

Fig. 5 Blueprint of the fortress in 2020

kept for some activities that take place in the fortified area. The only equipment in the fortress that, at the time the pandemic started, still functioned was the chapel, in which small liturgical ceremonies were held. With the pandemic, these ceremonies were held in the parish church.

The divisions attached to the chapel represent the former Fiscal Guard post, closed in 1982. It has not been used since. What is verified is that the false ceiling has already collapsed in some places and the plaster has also fallen to the ground (Fig. 6). Finally, in this segment, there is the cafeteria, which closed more recently and is in the best condition of all the annexes.

In the exterior, the construction works of the 1940s, such as the flowerbeds for the garden (Fig. 7) that due to the force of the trees and plants growing are broken and could fall and the construction of gunboats, are in an extremely poor state of conservation, with the construction device destroyed and exposed (Fig. 8). The reinforced concrete applied in some places, on the other hand, is degraded, exposed and corroded and is an example of works from the 60s in the worse state of conservation (Fig. 9).

Fig. 6 In the old post of the Fiscal Guard, it is possible to see some fallen plaster. Photograph taken by the author in August 2021

Fig. 8 Gunboats constructed in the 40s are destroyed and exposed. Photograph taken by the author in August 2021

Fig. 7 One of the flowerbeds is destroyed and in risk of falling. Photograph taken by the author in August 2021

Fig. 9 Reinforced concrete applied in the 60s is already broken and eroded. Photograph taken by the author in August 2021

5 Proposal for Good Practices of Military Heritage Conservation

Until the time of writing of this paper (September 2021), the management of this fortress is not yet under the authority of the City Council of Portimão. What can be seen from the brief description of the fortress, especially its history in the twentieth and twenty-first centuries and its alterations and state of conservation, is that the situation is unstable and unsustainable. For all intents and purposes, it is an area of great cultural and tourist potential that has been underutilized for years.

The fortress, as an historical landmark, has to be properly preserved. The arrival of the new millennium also brought new interpretations and care for cultural heritage. Not only does the Base Law on Cultural Heritage (nº107/2001) insist that it is a fundamental duty of the State and its citizens to protect and enhance their heritage (article nº12), a policy of strategic orientations to conserve and recover must be defined, investigate and publicize the heritage (article nº13).

What is certain is that in twenty years, no plans or guidelines have been defined for this space. In fact, although article nº13 does not define that the order should be the one mentioned, this study began almost at the end, with the investigation first. However, with the municipal elections and the emergence of the Portugal 2027 Tourism Strategy, this appears as the ideal opportunity to think about the future of the fortress.

Mayor Isilda Gomes, in 2017, said that the monument could house a museum linked to the sea, due to the connection of Portimão to Portuguese expansion, not excluding the activation of the restaurant, which is an essential equipment for tourism. The mayor ends up saying that the investment for construction works and its new purpose would entail a large investment, which would have to be acquired by applying for community funds (Lusa, 2017). Although the author understands this vision for the monument in question, the author cannot agree that this is the best option, considering that many other fortresses, namely those of Cascais and Oeiras, have already considered the same hypothesis.

The fortresses of Cascais and Oeiras, perhaps municipalities with other investments and accessibilities, are for the most part preserved and closed, only open by appointment, but enjoy well-established tourist routes, whether by land or sea (Município de Oeiras, 2021). These municipalities have a greater advantage in the face of competition for community funds and a "museum linked to the sea" had already been devised for these places, according to one of my advisors.

At another point, in much more recent news, the larger Fortress of Juromenha, in Alentejo, with a greater historical evolution, will receive an investment of nearly five million euros, both by community funds and by private action. The mayor says that it can be leased for hotel purposes, which I do not think is correct (Lusa, 2021).

Strongholds, forts, fortresses and castles are defensive devices and should be characterized as such. Therefore, the vast majority of current visions for fortresses are hard to agree on. So I propose a different vision for the Fortress of Santa Catarina de Ribamar. I advocate a conversion of the fortress to the seventeenth century, a stage for a journey through time. Historical monuments, be they palaces, castles and fortresses, and historic villages should be a place where people can visit the past, but here I defend that people live the past.

Thus, starting on the outside and ending on the inside, I defend the following aspects (in addition to those already implicit in major works of conservation and stabilization of the structure). The moat should be fully excavated and musealized. Likewise, to be able to cross it, it can be constructed either a drawbridge or a dormant bridge. The loopholes should be reopened and equipped with replica weapons aimed at the bridge, giving the image of defending the entrance to the building.

The entire North wall, still in masonry, should be preserved in accordance with the principles of reversibility (which implies the removal of some cement) and should be plastered in order to protect this masonry and prevent erosion. The gunboats on the bastions and on the square facing the sea should be equipped with cannon replicas, and the mast with the country's flag should be placed.

The flowerbeds for trees and plants must be removed, as plants not only ruined these twentieth-century constructions, but it also caused instability in the seventeenth-century structure. Finally, the access stairs to the bastions must be equipped with wooden handrails that adhere to the principles of safety and the conversion of one of the stairs into a ramp should also be considered, for access for people with reduced mobility (since the stairs are a reconstruction of the twentieth century). Lastly, all doors and windows should be replaced by wood, so as not to decharacterize it further.

Inside, the cafeteria should be leased for exploration and recharacterized for a more rustic appearance, as should its equipment (such as tables, chairs, parasols), so as not to contrast with an old fortress. The chapel should resume its functioning, with the possibility of leaving its doors open in order to be visited. In addition, the public bathroom must be rebuilt and regularly cleaned to be used by visitors, just as the elevator must be repaired to give access to the platforms and the beach for all people with reduced mobility.

I believe that the best use for the old restaurant, kitchen and annex, and for the old Captain barrack, would be the conversion to seventeenth-century barracks, with furniture made by local companies and the possibility for local theater companies to do historical reconstructions. Various historical episodes in Portimão have always had the fortress in their background or as the center stage, and it would be interesting to see the theatrical imagination combined with history. I found out on a visit to Portimão that there is a replica of a pirate ship named Santa Bernarda, that travels along the river and the sea with tourists, and I believe that good historical reenactments with visitors on board and visitors to the fortress could be excellent.

Finally, it should never be forgotten that screens and/or explanatory signs for each location (the signs should be resistant to the elements) should always be added, in order to report the history, studies and evolution of knowledge. Passive communication of heritage should always be contemplated and should always be written, first in Portuguese and then in English. The possibility of Braille being included should also be considered.

The security of the site and the possibility of paying entrance fees in certain places or on certain days must be considered, especially the various days that exist for the celebration of heritage, such as the European Archaeology Days and the European Heritage Days, the International Day of Archaeology, the International Day of Monuments and Sites and even the International Day of Museums. Also, on the topic of spreading the history, guided tours by appointment should be considered and scientific events can be held on site. Likewise, scheduled and publicized activities must be regular throughout the year, the hosting of local festivals must be held in this monument, and an agenda full of initiatives in the high season must be taken into account.

6 Final Considerations

One of the main purposes of this paper is to communicate the conversion of the Fortress of Santa Catarina de Ribamar to tourism, during the twentieth century. Although the best options were considered and implemented on this location (regarding its time frame), the restaurant and the cafeteria did not last (the cafeteria, the last one to close, probably due to the crisis in the late 2000s). Furthermore, the history of this fortress was never considered as a means to promote tourism or to communicate local history.

As of right now, with the municipal elections and new strategies for tourism, there is an opening window to perceive local cultural heritage in a new light. All elements of history and the landmarks they constitute should be used to communicate history and to promote tourism. Also, all vestigies of local and national history should always be of use to the residents, should be promoted for them and should count with local help to attain success.

I strongly believe that the promotion of local industries (such as woodworking and theaters), combined with a cafeteria operated by locals, the management and support of activities by the City Council of Portimão and the Portimão Municipal Museum, represents an added value for the locals, which could enjoy the whole year of a monument that is representative and explored by the City Council and its inhabitants, as it would be an added value for visitors to be able to enjoy an interactively preserved and musealized historical monument.

The Fortress of Santa Catarina de Ribamar, barely used so far, could receive a new vision. Due to the master's thesis in arcaheology that will be submitted this year, a plan for the conservation, musealization and communication of the site was advised, which makes perfect sense, as the knowledge produced must be disseminated and discussed and a perspective for the future must always be considered. It is my belief that this proposal adheres to all good principles of conservation and musealization of historical monuments, especially military monuments, because it invokes a different view of what has been done to adapt Modern Age fortresses for tourism. Also, this proposal is sustainable and interesting for the locals and visitors.

I also conclude that studies should be commissioned taking into account architecture, conservation and restoration, engineering and even musealization. This fortress needs all the technical and multidisciplinary support for its hypothetical candidacy for community funds or private initiatives (as patrons), and all studies should be subject to public consultation for the population to give their opinion based on how the fortress should be explored.

As this paper is a presentation of one of many possibilities for the strategies in the future, my proposal is based on what I believe is the best and most original way for the fortress to be musealized and explored, offering service and employment opportunities to the local population and a different experience for visitors. I have not seen a similar proposal for any other place in this chronology, especially for the Algarve region.

Acknowledgements I would like to thank the City Council of Portimão and APS for all the help and access to the fortress, the 3D Modelling Studio for giving me access to the photogrammetry survey and my dissertation advisors, Dr. Leonor Medeiros (NOVA FSCH) and Colonel José Berger (Chief of GEAEM-DIE).

References

ANPC. (2010). *Estudo do Risco Sísmico e de Tsunamis do Algarve*. Carnaxide: Autoridade Nacional de Proteção Civil.

Barreto, J. M. (1972). *O Algarve*. Lisboa: Direção-Geral da Educação Permanente.

Base Law for Cultural Heritage nº107/2001.

Coutinho, V. (1991). *Castelos, Fortalezas e Torres*. Faro: Algarve em Foco. D.L.

Coutinho, V. (1999). As Fortalezas da Costa Algarvia durante o Período das Economias-Mundo Centradas em Amsterdão e em Londres. In *O Algarve: da Antiguidade aos nossos Dias. Elementos para a sua História*. Lisboa: Colibri, pp. 263–268.

Coutinho, V. (2001). *Dinâmica Defensiva da Costa do Algarve: do Período Islâmico ao Século XVIII*. Portimão: Instituto de Cultura Ibero-Atlântica.

Decree-Law nº129/77 of 29 of September, classifying the Fortress as a Property of Public Interest.

Magalhães, J. R. (1999). A Conjuntura Económica. In: *O Algarve: da Antiguidade aos nossos Dias. Elementos para a sua História*. Lisboa: Colibri. P.253–262.

Ventura, Maria de Graça A. Mateus. Marques, Maria de Graça Maia (1993). *Portimão*. Coleção Cidades e Vilas de Portugal. Lisboa: Presença.

Archives in Torre do Tombo, Arquivo Histórico Militar, SIPA and others:

PT/TT/CC/1/114/136—Carta de Diogo de Menezes para o Vice-Rei D. Pedro de Castilho, informando que as Cidades de Tavira, Faro e Portimão se encontravam sem artilharia, muito mal reparadas de muros e falta de munições, solicitando que as enviasse para o caso de invasão de armada inimiga.18 de Janeiro de 1606.

PT/TT/CCDV/29—Alexandre Massaii—Descrição e Plantas da Costa, dos Castelos e Fortalezas, desde o Reino do Algarve até Cascais, da Ilha Terceira, da Praça de Mazagão, da Ilha de Santa Helena, da Fortaleza da Ponta do Palmar na Entrada do Rio de Goa, da Cidade de Argel e de Larache. 1617.

PT/TT/CC/2/364/00141—Cópia das Diligências feitas pessoalmente pelo Dr. Francisco Soares de Albergaria, Corregedor da Comarca, pelo Capitão Mor de Vila Nova de Portimão, pelo Sargento Mor, e pelos Três Capitães de Infantaria Nela Existentes, em Cumprimento de Ordem Régia emanada pelo Tribunal do Governo de Portugal e do Mandado do Governador do Reino do Algarve. 20 de Julho de 1633.

PT/AHM/DIV/3/09/88/22—Reconhecimento a Diversos Fortes do Algarve. Pós-1864.

PT/DIE/RGP/PM001/PTM[T-85/2A]—Process of the Fortress of Santa Catarina, Military Building nº1 of Portimão.

PT DGEMN:DSID-001/008-0909/2—Administrative process of the Fortress of Santa Catarina. Available online in SIPA: http://www.monumentos.gov.pt/site/app_pagesuser/SIPA.aspx?id=1286. Each page of the documents are identified by the extention TXT and/or FOTO, and that is how they are cited in this paper, according to their guidelines.

MC.DES.1402—Alexandre Massaii—Descripção do Reyno do Algarve. 1621.

In the Media:

Algarve-Primeiro. (December 22, 2016). Portimão: Bloco de Esquerda recomenda ao Governo urgente requalificação da Fortaleza de Santa Catarina. *Algarve Primeiro*. Available online: https://m.algarveprimeiro.com/d/portimaobloco-de-esquerda-recomenda-ao-governo-urgente-requalificacao-da-fortaleza-de-santa-catarina/15017-82

Ciência Viva. (2013). *Observação Astronómica Nocturna em Portimão*. Available online: https://arquivo.pt/wayback/20170607155822// http://www.cienciaviva.pt/veraocv/comum/2013/actividadeshoje.asp?accao=pesqbyid&id_accao=4754

Eusébio, J. C. (April 1, 2017). Fortaleza ao abandono vai ser concessionada. *Correio da Manhã*. Available online: https://www.cmjornal.pt/portugal/cidades/detalhe/fortaleza-ao-abandono-vai-ser-concessionada

Eusébio, J. C. (June 7, 2019). Câmara de Portimão quer recuperar fortaleza ao abandono. *Correio da Manhã*. Available online: https://www.cmjornal.pt/portugal/cidades/detalhe/camara-de-portimao-quer-recuperar-fortaleza-ao-abandono

Eusébio, J. C. (April 9, 2018). Fortaleza na Praia da Rocha está a ser alvo de obras. *Correio da Manhã*. Available online: https://www.cmjornal.pt/portugal/cidades/detalhe/fortaleza-na-praia-da-rocha-esta-a-ser-alvo-de-obras

Lusa. (January 17, 2017). Portimão quer gerir Fortaleza de Santa Catarina. *Diário de Notícias*. Available online: https://www.dn.pt/portugal/portimao-quer-gerir-fortaleza-de-santa-catarina-5610079.html

Lusa. (September 6, 2021). Restauro das muralhas de antiga fotaleza em Alandroal arranca este mês. *Público*. Available online: https://www.publico.pt/2021/09/06/local/noticia/restauro-muralhas-antiga-fortaleza-alandroal-arranca-mes-1976480

Municipio de Oeiras. (August 1, 2021). *Passeios "Com História" e Vela Latina*. Available online: https://www.oeiras.pt/-/passeios-com-historia-vela-latina

Paulo, A. (July 20, 2019). 5 coisas para fazer no Algarve, já neste fim de semana. *Visão*. Available online: https://visao.sapo.pt/visaose7e/sair/2019-07-20-5-coisas-para-fazer-no-Algarve–ja-neste-fim-de-semana/

Região Portuguesa Narcóticos Anónimos. (2009). Reuniões Zona Sul. Available online: https://arquivo.pt/wayback/20090624064824//http://www.na-pt.org/reuniao/reuniao-sul.php

S.n. (March 21, 2014). Fortaleza turística de Portimão com rato e lixo. *Correio da Manhã*. Available online: https://arquivo.pt/wayback/20160120053357//http://www.cmjornal.xl.pt/nacional/portugal/detalhe/fortaleza-turistica-de-portimao-com-ratos-e-lixo.html

Sustainable Tourism in Natural Territories that Have Suffered from Catastrophes: The Perception of Public and Private Stakeholders of the Alva Great Route

Manuel António Brites Salgado, José Alexandre dos Santos Vaz Martins, and Ricardo Jorge da Costa Guerra

Abstract

This study analyses the current meaning of the sustainable tourism model, especially for natural and rural areas, with the aim of understanding the new development scenarios in a post-COVID-19 era. The interpretation of the value of water and landscape for tourism and leisure is our main motivation for studying the different types of resources associated with the Alva River in the central region of Portugal. This region has been recently affected by two catastrophes that affect the tourist attraction, first the large forest fires in 2017 and then the pandemic calamity of COVID-19 that still affects tourism worldwide. The objective of this study is to interpret the tourism potential and the appropriate development model for this region, which leads us to evaluate the model of sustainability and competitiveness of this area with low population density. Considering the impact of these disruptive events, both regional and global, it is important to understand what are the associated threats and ultimately the opportunities. The methodology adopted for this exploratory study is essentially quantitative, and, in addition to the literature and secondary data review, questionnaires were sent to the main public and private stakeholders of the municipality of Oliveira do Hospital to obtain responses to the objectives of the study and to understand their opinions and expectations regarding the recent creation of the Alva Great Route (AGR). The main results indicate that new and innovative actions are needed to better articulate the development of tourism in the region, the majority of whose tourist situation is considered adequate or negative. In fact, there is a great lack of knowledge in the region itself about the existence and purpose of the AGR. However, those who know about this route consider it very important and agree on the strategy to be followed, which is an integrated management structure at the level of all the municipalities along the Alva River. The sustainability of this area, whose main attributes are the landscape/nature and the river, and which is therefore very vulnerable to disasters, requires a model of shared management between public and private entities. Their involvement in a network performance is important both for the planning and development of the tourist offer and for the promotion and communication with the tourist demand.

Keywords

Sustainability • Tourism development • Catastrophes • Nature and rural tourism • Hiking

1 Introduction

The World Tourism Organisation (UNWTO, 2005, 2017), among many organisations around the world, for example Biosphere Tourism (2021), points out at the importance of more sustainable tourism practices to safeguard the cultural and natural resources till 2030. This will ensure the economic growth of the activity, by being able to meet the needs of present and future generations. There must be an interest in reconciling the wishes of tourists with the regions' residents, ensuring environmental protection and stimulating the

M. A. B. Salgado
GOVCOPP—Unidade de Investigação em Governança, Competitividade E Políticas Públicas, CITUR and UDI/IPG, Polythecnic Institute of Guarda, Guarda, Portugal
e-mail: manuelsalgado@ipg.pt

J. A. dos Santos Vaz Martins
CITUR—Centro de Investigação, Desenvolvimento E Inovação Em Turismo and UDI/IPG, Polythecnic Institute of Guarda, Guarda, Portugal
e-mail: jasvm@ipg.pt

R. J. da Costa Guerra (✉)
CEGOT—Centro de Estudos de Geografia e Ordenamento do Território, CITUR, and UDI/IPG, Polythecnic Institute of Guarda, Guarda, Portugal
e-mail: ricardoguerra@ipg.pt

A. Mandić et al. (eds.), *Cultural Sustainable Tourism*, Advances in Science, Technology & Innovation,
https://doi.org/10.1007/978-3-031-10800-6_11

development of the tourism activity, in line with the interests of the local society.

A sustainable tourism development model in rural areas requires established organisations to share objectives, activities and resources in order to maximise benefits with a minimum of cost and effort (Augusto et al., 2010; Kastenholz et al., 2014; Inskeep, 1991; Salgado, 2010). The sustainability of tourism should be designed to meet the needs of local populations and also tourists and tourism industry stakeholders. The growing importance of tourism in rural areas is crucial for the appreciation of natural, human and cultural resources, which in most cases is a gap that needs to be addressed, as we can observe in the Alva Valley region (CIM Região de Coimbra, 2020; ERTCP, 2008, Salgado & Leitão, 2011).

Tourism development requires more effective planning and management, considering a more rational use of resources, in order to ensure their preservation for future generations in terms of sustainable development, but also to promote a socio-economic environment that is necessarily more attractive and competitive, as postulated in these two principles in Portuguese Law (Ministério da Economia e da Inovação, 2009). Participative and integrated management is a necessary tool to involve all tourism stakeholders in this important process.

Tourism destination characterisation is essential to gain regional awareness and consistent endorsement at both national and international levels, based on a strategy of recognising and valuing the potential of emerging tourism products (Silva, 2013; Salgado & Leitão, 2011; Salgado et al., 2015). Networking in the regional context can contribute to the coherence of local strategies and the improvement of the region as a tourism destination. Tourism can be a profitable investment for local investors and generate employment for residents (Augusto et al., 2010; Klimek, 2013; Silva, 2013).

The research developed intends to assume the importance of developing a sustainable tourism governance model in inland territories of Portugal, in which it is essential to value the rurality and nature aspects, in order to promote the adequate resumption of activities in the post-COVID-19 period. Therefore, this article also emphasises the interest and importance of interpreting the value of water resources and the natural landscape for tourism and recreation in relation to the Alva Valley region. In this area, there are tangible and intangible cultural assets of inestimable value associated with the ancient human settlement of this area, especially with the natural, rural and cultural activities and lifestyles associated in particular with the primary and secondary sectors. Although the evident potential, it is not fully developed and lacks better articulated intervention between stakeholders who have a fundamental role in the territory's tourism development.

The tourism products to be developed in this territory can be differentiated according to the quality of the endogenous resources. It is therefore necessary to define the best sectoral policies and develop the most appropriate strategies for sustainable and competitive tourism development, both at local and regional level.

In this study, therefore, the following objectives are pursued: characterise the tourism offer and the main public and private stakeholders of the Alva Valley that are part of the Great Alva Route, especially in the municipality of Oliveira do Hospital; identify the perceptions of stakeholders regarding the tourism development of the Alva Valley, especially the existing public policies and strategies and the actions implemented, as well as the cooperation and network participation between them; identify the level of knowledge of stakeholders about the Great Alva Route and the importance attributed to it; Identify the major problems and the greatest potentials that can be improved in the Great Alva Route; recognise the heritage resources that can determine the image of this region as a tourism destination; sort out the different elements of the tourism offer that contribute most to the enhancement of the Great Alva Route; and analyse the complementarity of this "new" tourism product with the established network.

To this end, the methodology is quantitative in nature, as it takes into account the perspective of local development of the Alva Valley region by private and public representatives of a municipality (Oliveira do Hospital) of this region, as they are responsible for managing the sites that have a socio-economic importance for tourism, but also promote environmental and cultural benefits.

In terms of the structure of the article, the first part addresses the importance of an effective governance model for regional tourism planning and management, which should help to reduce the impact and risks of catastrophes in terms of a more rational use of resources. In this sense, it is appropriate to discuss the main concepts around sustainable tourism development, the creation of routes and also the catastrophes of disasters on tourism. The geographical and tourism framework of the Great Route of Alva is then presented, highlighting the main characteristics of this tourism destination in order to identify regional potential and assess the potential of strategic tourism products. This will help municipalities to prioritise the various tourism development interventions. The section also looks at the importance that nature tourism currently has in the Alva Valley region and, in parallel, the possibility of promoting hiking as a strategy to enhance this tourism product, both internally and internationally, is addressed. The second part of this paper focuses on the methodology used and the empirical analysis of the results obtained from the questionnaires sent to the public and private stakeholders of the Alva Great Route. Finally, the conclusions as well as the limitations of the study and possibilities for future research are presented.

The purpose of this research focuses on interpreting the meaning of the necessary paradigm shift in tourism development opportune after a global pandemic (COVID-19) that lasted for 2 years and exposed the serious weaknesses of the model of tourism growth, in addition to other natural disasters affecting the central region of Portugal. Is this change coming at the right time for the rural and natural areas, or should it? Indeed, it has yet to be proven in Portugal that tourism is a key factor in the sustainable development of rural areas. So, this study aims to prove that there are more sustainable ways for regional development, which in this case can include a tourism strategy based on hiking and activities that are more connected to the resources of the rural world, which in this case concerns the Alva Valley region.

2 Sustainable Tourism Development

Tourism and leisure are of increasing importance in terms of their relationship with environmental, social, economic and cultural systems and many others with which the tourism system establishes fundamental interactions (Cunha & Abrantes, 2013, p. 104). Among the nine systems mentioned by Cunha and Abrantes, we would like to highlight the environmental system. This has a growing influence on tourism due to the importance that attractions (especially historical and natural heritage) have on tourism activities. Tourism can help to preserve attractions or damage them through an excessive concentration of tourists in certain areas. Tourism is based on the presence of natural and cultural resources, the knowledge and management of which are increasingly an essential basis for tourism development.

Simões and Ferreira (2009) assume that niche tourism groups together a numerous of tourism products whose limited demand tends to be one of the main criteria for identification. In many territorial contexts, these niche products are a new opportunity for innovation and leverage on the development process, not only of tourism but also of the territory, as it allows the commercialisation of specific resources. In fact, the ORTE project (Kastenholz et al., 2014) shows that the tourism experience in rural areas can be understood as a global experience of the destination, experienced by visitors to rural areas, involving a large amount and variety of resources, attractions, services, people and environments. In this context, the potential of the tourism experience in these rural destinations is very much dependent on the resources and heritage (tangible and intangible) present in the villages and surrounding areas, as is also recognised in the Alva Valley destination.

The Portuguese Tourism law (Decreto-Lei nº 191/2009, p. 5337) establishes the principle of sustainability, i.e. the adoption of policies that promote: "the enjoyment and the use of environmental resources, with respect for ecological processes that contribute to the conservation of nature and biodiversity; respect for the cultural authenticity of local communities, aiming to preserve and promote their traditions and values; the economic viability of businesses as a basis for job creation, better facilities and entrepreneurship opportunities for local communities".

Working to promote a cohesive and competitive territory implies determining the model that best helps to achieve this goal. This can be either an exogenous or endogenous approach, or, as it is also said, a top-down or bottom-up approach, or finally a mixture of both forms of governance (top-down/bottom-up), supported by a strategic planning methodology applied to territorial planning (Augusto et al., 2010). It seems that this methodology, combining both approaches, is suitable to improve policies and strategies, thus enabling their implementation in the Alva territory. In this context, Ahmed et al. (2010) suggest an integrated and territorial approach in low-density rural areas.

The Alva Valley is a predominantly rural territory that should consider the complementarity between the two territorial planning approaches. It should integrate strong relationships between the stakeholders of the territory, but to do so, it is important to assume a new model for the design and implementation of territorial/industrial policies based on the intrinsic nature of each territory as well as on the capacity to enable integration as a whole. In this way, the development process is oriented towards sustainable development with an ascending/descending and descending/ascending orientation in territorial terms, which the government defines as the guidelines that allow the country to develop abroad (Augusto et al., 2010). These authors advocate a governance model based on the bottom-up model to allow the existence of the principle of citizenship and on the top-down model to ensure discussion and coordination between the strategies defined at the different territorial levels. The participation of the local population is also a fundamental strategy for involving stakeholders who are directly interested in the sustainable tourism development of everyday life in their territory.

Academic studies (e.g. Leitão, 2004) and legislation (e.g. Ministério da Economia e da Inovação, 2009) have increasingly focused on the principle of sustainability, which has become one of the most important issues for tourism. The social and environmental consequences of tourism development required the consideration of environmental, social and cultural aspects alongside economic ones. Increasing attention was paid to the competitiveness of regional tourism destinations in order to secure economic growth and employment through long-term strategies.

According to Burnay (2006, p. 168), pursuing a sustainable tourism policy in a given region requires: "a strong political will; the existence of an appropriate legal framework and sectoral activities; the existence of resources and financial incentives to support the necessary investments, whether in the public or private sector; the training and qualification of investors; the certification and monitoring of destinations and products; and committed cooperation with local stakeholders".

In the process of territorial development, tourism activity generates new relationships and new activities resulting from the spatial structuring of services and attractions that ensure tourism production (Cunha & Abrantes, 2013; Queirós, 2014). Tourism is a complex web of relationships and interconnections that require a systemic view in which territory is a central element. From this perspective, the analysis of tourism activities must adopt a multidimensional approach. As it combines different elements that interact with each other and with the environment, the tourism system is characterised by the relationships between its players and by the influence of the various factors of tourism activity. In this sense, only a holistic view of tourism allows for a systemic analysis.

The tourism destination is the result of an area that has gone through the process of developing tourism activities, creating new relationships and a new economic reality. Buhalis (2000) refers to destinations as an amalgamation of tourism products that provide consumers with an integrated experience. Tourists consume destinations through their brand/name based on subjective personal constraints such as itinerary, reasons for visiting, level of education and culture and previous experiences.

The tourism product is characterised by being a complex variety of services. From the consumer's point of view, the tourism product is a subjective concept and depends heavily on the image and expectations of the place/destination. The tourism product according to Kotler et al. (2021) is physical objects, services, destinations, ideas and organisations that fulfil a desire or need. The idea of meeting the needs of tourists comes from the development of integrated tourism products that meet consumers' desires and build a positive image of the destination. These authors mention that the spatial boundary of a destination can be real or perceived, with boundaries created by the market or by the elements that build the destination. Destinations are often divided by physical and administrative barriers that do not take into account tourists' preferences or the positioning of the tourism industry.

Buhalis (2000) gives the example of the Alps, which are shared by France, Switzerland, Italy and Austria and which are perceived and consumed as a unique tourism product despite the borders between the four countries. This example can also be applied to the case of the Alva River, which is shared by seven municipalities. So, based on the idea of the value chain, upstream activities, research and product development, supporting activities, operations and logistics are successful activities for cooperation between competitors.

In analysing the sustainability practices of tourism destinations in seven European countries, Klimek (2013) noted the development of "green" and eco-labelled tourism products are important elements of the marketing mix of nature destinations. In Switzerland in particular, mountain destinations market "green" tourism through integrated product packages. In Polish tourism destinations, each service provider sells its non-integrated service as an integrated "green" product. Klimek (2013) justifies the lack of tourism product packages in Polish destinations with the lack of cooperation between the public and private sectors and with the strong competitive pressure.

The development of new products associated with exploration and observation of natural or urban spaces, cultural and natural heritage, is part of the goal of endogenous development, based on the preservation of resources and their subsequent tourist valorisation. Silva (2013) states that nature tourism will always be a key product linked to preservation and sustainability. But, there are other activities that can be included in more than one product. Therefore, the segmentation of tourism products and the classification of activities are a complex and evolving process that depends on many factors. In this sense, the creation of tourist routes can be one of the possible solutions for the sustainable development of tourism destinations.

The bottom-up governance model is very important in rural areas and small villages and is supported by the community-based participatory tourism development model, as we can see in several projects in the central region of Portugal (as Historical, Schist or Mountain Villages). The need for tourism development planning is widely discussed in Portugal, as it is a Mediterranean country that, like other destinations in this region, has had a mass growth model that urgently needs to be replaced by a sustainable development model (Silva & Umbelino, 2017).

3 Tourism Routes

Tourism can play a central role in the new socio-economic and environmental development model of several inland areas, as in our case, the Alva Valley. The definition of a tourism anchor product, like for example a distinguished tourism route, can be a lever for the integrated development of the region by creating a common denominator for the development and marketing of tourism that allows for better coordination of the efforts of the different stakeholders.

Queirós (2014, p. 111) defines the tourism route as "an organised set of discovery trips and enjoyment of the entire cultural heritage with its unique identity, based on the metaphysics of ecology and landscape, accessible to all, but with differentiated products according to their segments, promoting the organisation and development of value chains of the tourist activity". It is therefore important to understand the concept of the tourist circuit, which in turn is defined as "a short route linking several heritage sites (it should not take more than a journey/a day), accessible to all but segmented, with an autonomous and distinctive identity, organised from the perspective of discovery and enjoyment of landscape ecology (an interdisciplinary scientific contribution to its understanding) and landscape metaphysics (imaginary classical and vernacular intangible cultural landscape), and according to communicative/emotional attractions capable of sustaining and developing tourism value chains".

Indeed, it is possible to find common elements in different circuits, such as a watercourse, in our case the Alva River, but the sum of the different types of heritage sites is expected to produce a unique offer. It is precisely in this context that, according to Queirós (2014, p. 111), "the activity of tourism differs from other scientific domains, because the selection and valuation is determined by the differentiation of the tourist product", because its "methodological construction consists in the reappropriation of concepts traditionally used in other scientific fields for a new study". Queirós (2014, p. 116) considers the hermeneutics of the cultural landscape important in its application to the study of the corpus of tourism and tourist activity, aiming to explore the concepts of the route and the tourist circuit, which are "based on the need to apply an interdisciplinary and multidisciplinary scientific methodology to organise and guide the visit to the territory, allowing its cultural landscapes, the material and immaterial heritage of the humanised landscape to be read and interpreted". There is a growing interest of travellers in the authentic reality of places, whether natural or built (or cultural), which determines important motivations even for simple walks through the territory (Leitão, 2004; Rodrigues & Kastenholz, 2010; Tovar, 2011). The routes and circuits integrated into the cultural landscapes and their tourism destinations generate the most important capital gains.

Figueira (2013) states the importance of the creation of tourism routes, which is inherent with the planning and tourism development of territories and is an essential component of the factors for destination differentiation and competitiveness. In the context of "economic valuation of resources, the creation of routes aims to make better use of tourist attractions and optimise the cost–benefit ratio" (Figueira, 2013, p. 122). This strategy affects the actors within the tourism value chain, implies the organisation and optimisation of resources and places, and, if properly applied, it provides profits to the tourism system. This also represents the added value that needs to be developed in the Alva Valley. Figueira (2013, p. 132) concludes that an interdisciplinary and multidisciplinary approach to route planning is essential and that routes are "territorial and productive factors used at different scales and themes, they are productive elements of thematic differentiation in the universe of tourism products, they promote the movement of people, mobilise new factors and means of production, create cooperation and competition networks, promote regions and are, in fact, the pillar of the tourism trip".

Rodrigues and Kastenholz (2010) consider that the demand for natural areas for the development of tourism activities continues to increase, especially for landscapes where people live and which have water resources. These authors consider that activities such as hiking in natural and rural spaces are increasingly in demand for recreational and tourism purposes. In this context, the characteristics of the area in question make it a very interesting destination for leisure, recreation and tourism, as it has a water surface that offers a unique tourist experience. Tovar (2011, p. 158) concludes that "hiking is an activity with great possibilities to grow, as is hiking tourism within nature tourism".

The tourism products are quite diverse in Portugal, and due to the capital of resources of interest, they become promoters of sustainable development strategies, whether at national, regional or even national level (Salgado, 2010). In this context, it is important to implement a rigorous inventory of endogenous resources in order to know them and structure the tourism offer to promote quality products with international projection. As a result of this study, it is necessary to define the tourism resources that are crucial for the tourism image of this region, thus promoting the development of new tourism products of excellence, working as facilitators of policies and sectoral strategies of tourism at the regional level.

4 Impact of Catastrophes in Tourism

In this context, we would first like to refer to the report presented by the UNWTO (2008), which aims to identify the necessary global responses to the challenges of climate change and tourism. After a series of meetings, conferences and declarations, the UNWTO adopted a resolution on tourism and climate change in 2007, mainly due to the need to ensure measures that contribute to the sustainability of tourism. Following the declaration of a global pandemic COVID-19 and its impact on tourism, much literature was also published from 2020 onwards (Rodrigues et al., 2021; Bauzá Martorell & Melgosa Arcos, 2020; Figueira & Oosterbeek, 2021). The Alva Valley region has suffered strong impact after two disruptive events, namely the major forest fires of 2017 in the central region of Portugal and the declaration of the global pandemic in 2020.

Estevão and Costa (2020) outline a framework for future tourism destination catastrophe management systems for researchers and tourism planners, concluding that there are very few studies that address strategies and operational guidelines for tourism destination management. They consider that the impacts and negative consequences for tourism cluster businesses and the surrounding tourist destination community clearly demonstrate the importance of strategic planning for managing such issues. After a natural disaster, there is usually a great loss of life, damage to infrastructure and material losses.

Tourism flows have decreased sharply worldwide in 2020 and 2021, as a result of the pandemic triggered by COVID-19, with significant negative impacts that require optimal planning, management and development of tourism activities at the regional level, especially in predominantly rural and natural territories. Among the challenges for rural areas, tourism is perceived as a structuring sector as it allows for diverse and balanced development at regional level, but above all, tourism can counteract the depopulation of these areas. Rodrigues et al. (2021), for example, point out that the pandemic situation had a significant impact on global tourism activity in many regions of the world. Their study aims to analyse the tourism planning and management carried out through the implementation of sanitary measures, social distancing and other policies to mitigate the impacts of the pandemic COVID-19 and their applicability in the reality of the municipalities of Belem, Soure and Salvaterra. They conclude that the pandemic had a significant impact on the tourism sector in 2020.

Tourism activity tends to be, increasingly, affected by diversified events, whether of human origin or of nature. In fact, according to Almeida (2017, p. 105), the complexity and diversity of this sector and the high level of interdependencies "on the one hand, can result in the amplification of the effects of an event and, on the other hand, make it difficult to act to increase its resilience". The sources of risk in tourism are diverse, including "extreme natural phenomena, climate change, environmental disasters, public health problems, disturbances associated with the energy and oil sectors, political crises, economic and financial crises, terrorism" (Almeida, 2017, p. 105), which may jeopardise the safety of visitors to a destination. It is precisely in the context of resilience and risk management that identifying the trends of climate change and their implications for tourism must be studied, because there will be opportunities but, above all, major challenges for mankind, hence the need to incorporate them into tourism planning towards sustainable development.

Resilience, risk management and crisis planning, in the context of tourism development (Almeida, 2017), have gained prominence in the literature, especially after the 2009 economic crisis and the COVID-19 pandemic, which resulted in a collapse of global tourism. In fact, according to Almeida (2017, p.105), the complexity and diversity of this sector and the high level of interdependencies, which are global, "can, on the one hand, lead to a widening of the impact of an event and, on the other, make it more difficult to act in terms of increasing its resilience". It seems that the sources of risk in tourism are multiple, including "extreme natural phenomena, climate change, environmental disasters, public health problems, disruptions related to the energy and oil sectors, political crises, economic and financial crises, terrorism", which can threaten the safety of visitors to a destination. It is in the context of resilience and risk management that we need to study and better understand the trends in health problems and their impact on tourism, as they lead to major challenges for humanity. Therefore, it is essential to include them in global planning, but also in tourism to ensure a more sustainable development.

Almeida (2017, p. 108) recognises the need to put in place appropriate spatial and tourism planning, design and management systems, which are "essential to improve the resilience of destinations and avoid uncontrolled growth of the sector, which can have a negative impact on the resources that sustain it, and contribute to the valorisation of these resources, which is often only possible because of tourism interest". Therefore, it is considered necessary to interpret the risk typologies of natural phenomena, which can be classified into 4 main categories (Almeida, 2017): geophysical; meteorological, climatic or maritime; hydrological; and biological. The second category includes, among others, strong winds, lightning discharges, storms and heat waves, which together lead to extreme weather conditions that can have unpredictable consequences, especially in the case of forest fires that can spread uncontrollably, as was observed in Portugal and the state of California (USA) in 2017.

Figueira and Oosterbeek (2021) argue for an open and participatory debate to help countries and regions cope with the dilemmas that are evident in the context of the main impact of the pandemic COVID-19 in these times of uncertainty. In the same direction, Bauzá Martorell and Melgosa Arcos (2020) argue that the situation caused by the global pandemic that the Sars-COVID-19 virus has triggered throughout the world is critical in many ways, especially in terms of people's health, but also in terms of social welfare, employment and economic growth. In fact, the pandemic caused by COVID-19 has led the many governments to declare a state of alert, restricting the free movement of people and taking a series of measures to curb various economic activities and especially tourism activities, such as stopping the hotel and restaurant industry. But, tourism is a strategic sector, and both the government and local authorities have enacted regulations to reduce the harsh impact on businesses and workers. In the study by Saramago (2021), it was noted that the data shows that there are measures and strategies in place in Portugal to address the disruption caused by COVID-19 in the tourism sector. Nevertheless, stakeholders show that the measures and policies that had to be taken were not able to reverse the losses caused by this pandemic. It is therefore necessary to rethink and restructure the entire tourism system in order to resume the activity in a more appropriate and harmonious way, which should also be more aligned with the new paradigm of sustainability.

5 Characterisation of the Alva Great Route

The Alva Valley is an area in the central region of Portugal that includes the administrative boundaries of seven municipalities where the Alva river spans: from the Fervença River on the southwestern hillside of the Serra da Estrela Natural Park (which extends from the village of Sabugueiro—the highest mountain village in the municipality of Seia) to Porto da Raiva in the municipality of Penacova, where it flows into the Mondego River. The Alva River is about 106 km long, creating a valley of natural beauty that is much appreciated by tourists. The attraction of this area is characterised by the remarkable diversity and richness of the existing tourist resources, as well as by its geographical location, central position and easy accessibility, the hospitality of the residents, competitive prices and safety, which are some of the reasons that highlight the high potential of this region.

The Alva is a mountain river and is famous for its river beaches, green fields, willows and ferns that soak in the fresh waters and the schist houses. There are also a number of villages that have coexisted side by side with the river for a long time, as well as some granite "mountain" villages in the middle of the Serra da Estrela Mountains, which makes this territory unique. There are also very slim passages due to the narrow canals built over the centuries, mainly for agricultural irrigation practices.

The Alva Great Route, promoted by the Intermunicipal Community of the Region of Coimbra (CIM-RC), is a long-distance walking route in Central Portugal. It has 76.8 km from the village of São Gião (325 m in altitude) Oliveira do Hospital, Tábua, Arganil, Vila Nova de Poiares and Penacova (45 m in altitude). It is one of the routes of the Great Route approved by the FCMP—Federation of Camping and Mountaineering in Portugal.

This geographic area (Fig. 1), like much of inland Portugal, has some structural problems resulting from its peripheral location, both geographically and developmentally. This is due to poor accessibility, the inability of cities to grow and drive the development of surrounding areas, the lack of skilled labour due to the depopulation of rural areas and out-migration (Augusto et al., 2010). This area also shows signs of a certain degradation, destruction of places and natural resources as a result of conditions that provide for increasing desertification. However, the formerly unfavourable conditions can now act as a catalyst for the development of integrated tourism. A whole range of structural, economic and social changes that benefit these regions include: the cultural heritage, such as monuments and historical sites, festivals and fairs and ethnic heritage; the proximity of population clusters and centres of local

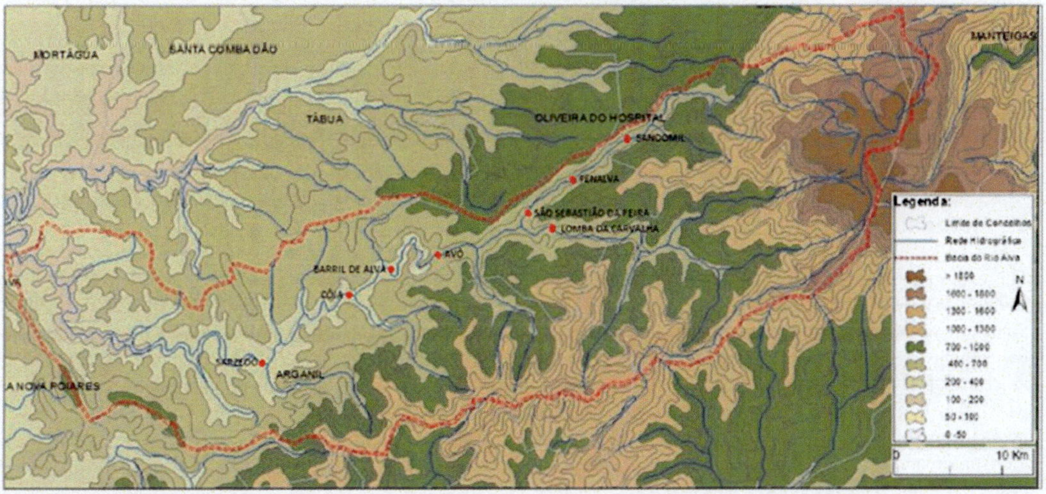

Fig. 1 Location of the Alva River watershed and its framework. *Source* Lourenço et al. (2014)

businesses; the conditions for sporting activities or leisure (excursions, hunting, fishing); the active intervention of local public authorities as well as local development associations; the competence and effectiveness in promoting the region and marketing the existing units; the quality of hospitality and accommodation facilities, as well as the competence of the services offered; the preservation of traditional rural architecture; the traditionalism of agricultural practises; the peculiarities of the flora and fauna; the proximity to the border; the mountains, the rugged terrain and the landscape diversity, among other characteristics (Brito, 2015; Queirós et al., 2008; Salgado et al., 2015). Unfortunately, until today, there are still enormous weaknesses and constraints in the development of tourism in the region.

The presence of abundant resources should suggest a tourism vocation to every municipality and therefore encourage the development of outstanding tourism products that act as policy promoters and sectoral strategies of tourism at the local level. This territory is facing very serious problems resulting from forest fires and the pandemic crisis, some of which may jeopardise the future of the analysed region. Therefore, the AGR project must now adopt an ambitious plan that takes into account the necessary strategy to guarantee the protection of people and goods, especially from the disruptive events that will become more frequent due to climate change.

According to Salgado et al. (2019), tourism is very important for rural and mountain regions characterised by low population density in their territories, which can transform it into a structuring phenomenon at the spatial and socio-economic levels, supported by the richness and diversity of natural and cultural resources. In this context, the necessary dynamics at the sub-regional level are considered, as the different municipalities in the Alva River region need to combine their efforts to promote sustainable and competitive development, as tourism in the area is still very incipient in this area. The area has been affected by the two aforementioned catastrophes: forest fires and COVID-19.

6 Methodology

This explorative study is essentially quantitative in nature. In addition to literature and secondary data research, a questionnaire was used as the primary data collection tool and was administered in the municipality of Oliveira do Hospital in central Portugal. This questionnaire consisted of three parts, the first relating to the tourism situation of the municipality of Oliveira do Hospital, the second to the tourism situation of the Alva Valley and the third to the characterisation of the respondent. The first part consists of 9 questions, including 7 evaluation questions with a scale of 5 values and 2 open questions. The second part consists of 9 questions, including 3 closed dichotomous questions, one of which is a philtre question, 4 evaluation questions with a scale of 5 values and 2 ordering questions with six options. The third part consists of 6 questions.

The target population includes, in the context of local government, 16 villages and the municipality of Oliveira do Hospital and, in the context of private businesses, according to the survey conducted on 12 August 2020 through the National Tourism Registry (RNT), 23 hotels and rural accommodation, 6 tourism animation companies, 5 travel agencies and 61 other local accommodation and 28 restaurants, according to a cross-search on the websites of the municipality of Oliveira do Hospital, Tripadvisor and Lifecooler.

The defined questionnaire was implemented in a pdf file version and in a Google Forms version. The file and link were sent by email to the members of the population who had an email to contact, i.e. all 16 villages and the municipality of Oliveira do Hospital, as well as 5 (22%) hotels and rural accommodation, 58 other local accommodation (95%), 10 (91%) tourist animation companies and travel agencies and 10 (36%) restaurants, representing 100 (71%) of the 140 elements of the target population. This initial contact took place from 14 September 2020, with further contact by telephone and/or email. Data collection ended on 20 November 2020, and a sample of 24 individuals was identified through a randomised, non-probabilistic process. Of the 24 responses received, representing a response rate of 24% and 17% of the target population, 11 (46%) were from local government, representing 65% of the respective population, and the remaining 13 (54%) were from private businesses, representing 16% of the respective population contacted and 11% of the reachable population.

The data was processed using Microsoft Excel 2016 software to conduct descriptive statistical analysis and IBM SPSS software to conduct inferential statistical analysis. As part of the inferential statistical analysis and to investigate whether the distributions of an ordinal variable and a dichotomous nominal variable are similar, the Kolmogorov–Smirnov test was applied, assuming that the distributions have the same shape, i.e. with similar asymmetries and similar variability, and since, the limitations of the t-test were not met. The non-parametric Mann–Whitney U test was also applied to these cases. In addition, a qualitative analysis of the collected qualitative data was also conducted.

7 Empirical Study

The sample of 24 respondents consists of 16 men (67%), and this percentage is 82% for public sector respondents and 54% for private sector respondents. The average number of years in the profession is 12.6 years, with a standard deviation (SD) of 9.7, with a lower average in the public sector (7.6 years, with SD 3.4), which is normal due to the characteristic of positions with a limit on duration, while in the private sector, the average is higher (16.8 years, with SD 11.4). Regarding educational qualifications, 55% of respondents reported having a bachelor's or master's/Ph.D. degree, with the same percentage in both the public and private sectors. In terms of the private sector, 54% of responses were from restaurants, 23% from hotels and rural accommodation, 15% from other local accommodation and 8% from travel agencies.

Regarding integrated planning strategies at the regional level, 87% of respondents indicated that these are fairly important or very important, which corresponds to an average (on a Likert scale of 1 to 5) of 4.42. When looking at the sectors separately, the percentage of those who said they were fairly important or very important was 91% in the public sector and 84% in the private sector, with averages of 4.73 and 4.15, respectively. There is thus a unanimous sense of the great importance of integrated planning strategies at the regional level, regardless of the sector.

Regarding the policies of sustainable development of the territory for the municipality of Oliveira do Hospital, there is also a clear assumption about their importance, in which 75% stated that they are quite important or very important, which corresponds to an average of 4.17. When analysed by sector, the percentage of those who stated that they were fairly important or very important was 82% in the public sector and 69% in the private sector, with averages of 4.55 and 3.85, respectively. The importance of the sustainable development policy of the territory for the municipality of Oliveira do Hospital is thus clearly expressed in both sectors, but more so in the public sector and less intense in relation to the integrated planning strategies at regional level.

Regarding the evaluation of public policies and strategies for the development of tourism in this region and public measures for the development of tourism in the municipality of Oliveira do Hospital, only 25% and 21% of the respondents respectively indicate a negative weighting (Table 1). There is also a parallel between the public and private sectors on these two issues, with the private sector showing a less positive/optimistic tendency. In order to strengthen the analysis and determine whether the distributions of the evaluation of public policies and strategies for the development of tourism in Oliveira do Hospital are similar in the groups of private organisations (private sector) and public administrative units (public sector), and considering that one variable is an ordinal variable and the other a dichotomous nominal variable, the Kolmogorov–Smirnov test was applied, assuming that the distributions are uniform, i.e. have similar asymmetries and variability. This assumption is made because the degree of asymmetry for both categories is low (0 and −0.244, where the absolute value is not greater than 2), both medians belong to the respective confidence intervals and the means also do not differ greatly (3 and 3.64) and are not near the maximum value (5), and the values for the variance and the coefficients of variation do not differ greatly (variances 1.167 and 1.255 and coefficients of variation 0.36 and 0.31). The value of the test statistic was therefore 0.58 and $p = 0.889 > 0.05$, meaning that the variables are independent, i.e. the scores do not differ between the public and private sector groups, with no significant differences between the groups at a 95% confidence level. Reinforcing this conclusion, the non-parametric Mann–Whitney U test with a statistical value of 93.5 also yields a $p = 0.188 > 0.05$, so exactly the same conclusion can be drawn. The Kolmogorov–Smirnov test was also applied to the evaluation of the public measures for the development of tourism in Oliveira do Hospital in the groups of private organisations (private sector) and public administrative units (public sector). It is assumed that the distributions have the same shape, since the degree of asymmetry for both categories is low (−0.76 and −0.65, where the absolute value is not greater than 2), both medians are within the respective confidence intervals, and the mean values also do not differ greatly (3.08 and 3.91) and are not close to the maximum value (5), and the values for the variance and coefficient of variation are not very different (variances 1.577 and 1.091 and coefficients of variation 0.41 and 0.27). Thus, the value of the test statistic was 0.70 and $p = 0.711 > 0.05$, meaning that the variables are

Table 1 Assessment of public policies and strategies and public measures for the development of tourism in the municipality of Oliveira do Hospital

	Public policies and strategies for the development of tourism			Public measures for the development of tourism		
Appreciation	Global	Public	Private	Global	Public	Private
Very negative (%)	4	0	8	8	0	15
Negative (%)	21	18	23	13	9	15
Neither negative nor positive (%)	33	27	38	25	27	23
Positive (%)	25	27	23	33	27	38
Very positive (%)	17	27	8	21	36	8
Average	3.29	3.64	3.00	3.46	3.91	3.08

independent, i.e. the scores do not differ between the public and private sector groups, with no significant differences between the groups at a 95% confidence level. This conclusion is confirmed by the non-parametric Mann–Whitney U test with a statistical value of 93.5 and a $p = 0.134 > 0.05$, so exactly the same conclusion can be drawn.

Regarding the strategies of participation and collaboration of public sector actors in the development of tourism in the Alva Valley region, 33% of respondents gave a rating of little or no participation and an average of 3. This percentage is 27% in the public sector and 39% in the private sector, with averages of 3.27 and 2.77, respectively (Table 2). Regarding the strategies of collaboration among private sector stakeholders related to tourism in the Alva Valley region, 21% of the respondents indicated that they rated collaboration as little or not at all participatory, with an average of 3.21, with the corresponding percentage being 0% in the public sector and 39% in the private sector, with average values of 3.73 and 2.77, respectively. That is, considering the average values obtained, overall the respondents consider the strategies mentioned to be only somewhat participatory or collaborative, with this rating being less positive and more critical of the private sector actors and, on the other hand, the rating for the participation and collaboration of public sector actors in the development of tourism in the Alva Valley region is less positive than for the collaboration between private sector actors in the context of tourism in the Alva Valley region.

The Kolmogorov–Smirnov test was applied to strengthen the analysis and to find out whether the distributions of the evaluation of participation and cooperation strategies of public sector representatives are similar in the groups of private organisations and public administrative bodies. The assumption that the distributions have the same shape is made because the degree of asymmetry for both categories is not high (−0.22 and −0.52, where the absolute value is not greater than 2), both medians belong to the respective confidence intervals, and also the mean values do not differ much (2.77 and 3.27) and are not close to the maximum value (5), and the values for the variance and the coefficients of variation hardly differ (variances 2.026 and 1.255 and coefficients of variation 0.51 and 0.46). The value of the test statistic was therefore 0.48 and $p = 0.976 > 0.05$, meaning that the variables are independent, i.e. the ratings do not differ between the public and private sector groups, with no significant differences between the groups at a 95% confidence level. Reinforcing this conclusion, also the non-parametric Mann–Whitney U test with a statistical value of 85.5 also yields a $p = 0.405 > 0.05$, so exactly the same conclusion can be drawn. The Kolmogorov–Smirnov test was also used to assess the cooperation strategies between private sector actors in the groups of private organisations and public administrative units. It is assumed that the distributions have the same shape, since the degree of asymmetry for both categories is low (0.47 and 0.44, where the absolute value is not greater than 2), both medians lie within the respective confidence intervals and the mean values, despite some deviation from each other (2.77 and 3.73) do not lie near the maximum value (5), and although the values of the variance and coefficient of variation also show some differences (variances 2.359 and 0.418 and coefficients of variation 0.55 and 0.17). The value of the test statistic was 0.99 and $p = 0.281 > 0.05$, meaning that the variables are independent, i.e. the scores do not differ between the public and private sector groups, with no significant differences between the groups at a 95% confidence level. This conclusion is further strengthened by the Mann–Whitney non-parametric U test with a statistical value of 101.5 and a $p = 0.082 > 0.05$, so exactly the same conclusion can be drawn.

Respondents from the public sector focused essentially on a strategy with greater integration of restaurants, tourist animation and accommodation, with particular attention to the camping parks, on the potential of culture and especially nature, with emphasis on river beaches and pedestrian routes, and on the need to pay attention to promotion. In the private sector, the focus was on a better connection between the policy and the entrepreneurs, on better promotion, on increasing tourism activities in order to increase the attractiveness and permanence of tourists in the region, with a stronger focus on walking routes and cycling trails, also with

Table 2 Assessment of the participation and collaboration strategies of public sector agents and collaboration between private sector agents in the development of tourism in the Alva Valley region

	Participation and collaboration strategies of public sector agents			Collaboration strategies between private sector agents		
	Global	Public	Private	Global	Public	Private
Not participatory (%)	25	18	31	17	0	31
Little participatory (%)	8	9	8	4	0	8
Reasonably participatory (%)	25	27	23	38	36	38
Quite participatory (%)	25	18	31	25	55	0
Very participatory (%)	17	27	8	17	9	23
Average	3.00	3.27	2.77	3.21	3.73	2.77

reference to reforestation and cleaning of the forest and the river.

Regarding the main problems for the development of tourism in the region, the public sector highlighted poor accessibility, lack of incentives and investment in the sector, ageing and depopulation of the region, lack of promotion and lack of spatial and supply planning. The private sector, on the other hand, mentioned as problems the lack of innovative and highly attractive initiatives, the insensitivity of policy makers and the lack of networking between the municipality and between them and the private sector, the lack of financial support, the poor accessibility and signage, the lack of dissemination and information to tourists, the lack of staff to work in the region and the lack of investment in reforestation.

When rating the tourism situation in the Alva Valley on a scale of 1 to 5, with 1 being very negative and 5 being very positive, 63% of respondents gave a score of 3 or less, with an average of 3.25 (3.27 in public and 3.23 in private) and a standard deviation of 0.83.

In order to strengthen the analysis and find out if there is a similarity between the distributions of the ranking of the tourist situation of Alva Valley in the groups of private organisations and public administrative units, the Kolmogorov–Smirnov test was applied. It is assumed that the distributions have the same shape, since the degree of asymmetry for both categories is not high (−0.88 and −0.44, where the absolute value is not greater than 2), both medians belong to the respective confidence intervals, and also the mean values do not differ much (2.77 and 3.27) and are not close to the maximum value (5), as are the values for the variance and the coefficients of variation, which differ but not very much (variances 1.026 and 0.418, and coefficients of variation 0.31 and 0.2). Thus, the value of the test statistic was 0.188 and $p = 1.000 > 0.05$, which means that the variables are independent, i.e. the ratings do not differ between the public and private sector groups, and there are no significant differences between the groups at a 95% confidence level. This conclusion is confirmed by the non-parametric Mann–Whitney U test with a statistical value of 71.5 and a *p*-value of $p = 1.000 > 0.05$, so exactly the same conclusion can be drawn.

Regarding specifically the knowledge of the Alva Great Route (AGR) in global terms, 75% of respondents say they are not aware of the AGR project, and among public sector respondents, ignorance is 45%, while in the private sector, it is 100%. Looking now at the 25% of respondents who said they were aware of the AGR project (all from the public sector, 55%), all (100%) considered the level of importance to be 5, i.e. "very important", all (100%) considered it necessary to promote the AGR internally, all (100%) considered it necessary to promote the internationalisation of the AGR in external tourism markets, and 83% considered it important to extend the AGR to the municipality of Seia.

In terms of rating the various elements of tourism in the Alva Valley on a scale of 1, which is very bad, to 5, which is very good, Fig. 2 shows that private sector respondents' perceptions of the order of the elements of tourism provision in the Alva Valley are not very different from those of public sector respondents, although public sector respondents generally give higher scores in terms of absolute rating. In particular, the most valued elements (average equal to or greater than 3.5) were River Beaches, Cultural Heritage and Accommodation, with public sector respondents also indicating Natural Patrimony. Among the least valued elements of the Alva Valley's tourism offer (average equal to or less than 2.5), Tourism Animation stands out as the only one with an overall average of less than 2.5 (2.33). Private sector respondents also earn scores below 2.5 for the elements of tourist services, Cycling and Pedestrian Routes, Accessibility to Resources and Integration of New Technologies.

To strengthen the analysis and find out if there are similarities between the distributions of the different elements of the tourism offer of the Alva Valley in the groups of private organisations and public administrative units, the Kolmogorov–Smirnov test was applied (Table 3), which confirms, with a confidence level of 95%, that the variables are independent in all cases, i.e. the ratings do not differ according to the public sector and private sector groups, with no significant differences between the groups. The non-parametric Mann–Whitney U test was also applied and found exactly the same for all analysed elements of the tourism offer, except for accessibility of resources and tourism services, where differences were found ($p = 0.047 < 0.05$ and $p = 0.022 < 0.05$, respectively), with private sector respondents rating these elements worse.

Having also been asked to rank some attributes in order of importance for tourism development in the Alva Valley, with 1 being the least important and 6 being the most important, it can be seen in Fig. 3 that overall the most important attribute is Landscape/Nature (5.21), followed in order of importance by River (4.08), Culture/Traditions (3.67), People/Services (3.0), Cultural Heritage (2.58) and least important Gastronomy/Wines (2.46). It should also be noted that only for the attributes Gastronomy/Wines and River, there was a higher average among respondents from the private sector than those from the public sector.

In order to deepen the analysis and find out if there are similarities between the distributions of the different elements of the tourism offer in the Alva Valley in the groups of private organisations and public administrative units, the Kolmogorov–Smirnov test was applied (Table 4), which confirms with a confidence level of 95% that the variables are independent in all cases, i.e. the ratings do not differ

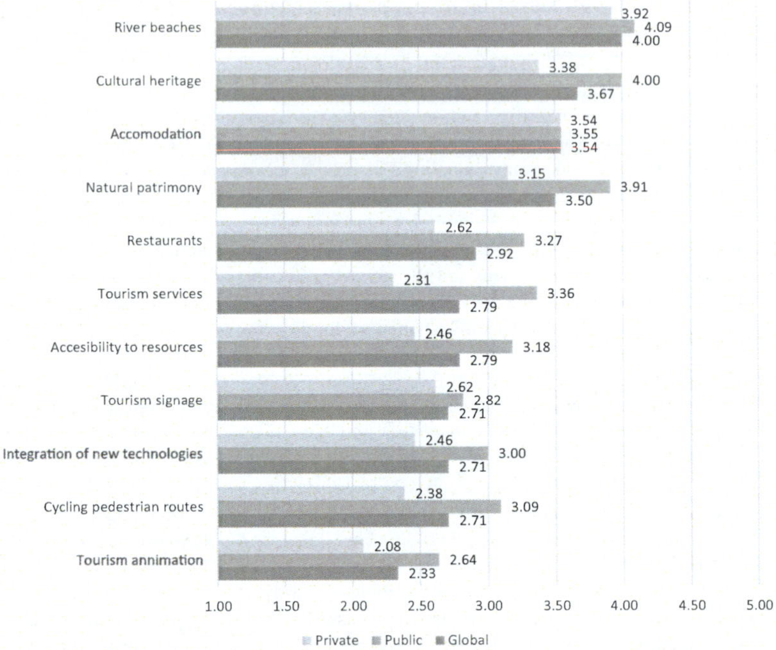

Fig. 2 Average level of various elements of the tourism offer in the Alva Valley

Table 3 Summary of the hypothesis test for independent samples with the Kolmogorov–Smirnov test for the elements of tourism supply in the Alva Valley

Null hypothesis	Sig.[a, b]	Decision
The distribution of cycling pedestrian routes is equal in the private/public categories	0.794	Accept null hypothesis
The distribution of natural patrimony is equal in the private/public categories	0.213	Accept null hypothesis
The distribution of accessibility to resources is equal in the private/public categories	0.075	Accept null hypothesis
The distribution of Integration of new technologies is equal in the private/public categories	0.213	Accept null hypothesis
The distribution of integration of restaurants is equal in the private/public categories	0.740	Accept null hypothesis
The distribution of accommodation is equal in the private/public categories	0.625	Accept null hypothesis
The distribution of tourism services is equal in the private/public categories	0.341	Accept null hypothesis
The distribution of river beaches is equal in the private/public categories	1.000	Accept null hypothesis
The distribution of cultural heritage is equal in the private/public categories	0.213	Accept null hypothesis
The distribution of tourism signage is equal in the private/public categories	0.976	Accept null hypothesis
The distribution of tourism animation is equal in the private/public categories	0.386	Accept null hypothesis

[a]The level of significance is 0.05
[b]The asymptotic significance is displayed

between the public and private sector groups, with no significant differences between the groups. The non-parametric Mann–Whitney U test was also applied and yielded exactly the same conclusion for all analysed elements of the tourism offer, except in the cases of Cultural Heritage and Gastronomy/Wines where differences were found ($p = 0.035 < 0.05$ and $p = 0.026 < 0.05$, respectively).

From the ranking of the importance of some tourism products to be developed in Alva Valley, where 1 is the least important and 6 is the most important, it can be seen (Fig. 4) that the most important product indicated is River Beach Tourism (4.71), followed in order of importance of the products Nature and Rural Tourism (4.54), Health and Wellness Tourism (3.38), Sports and Adventure Tourism

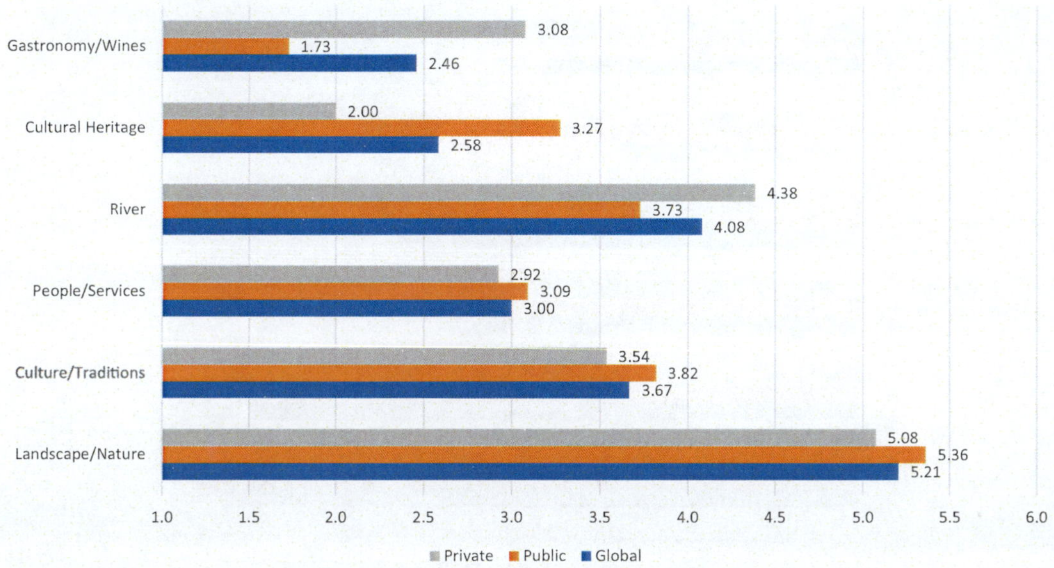

Fig. 3 Degree (average) of importance of some attributes for the tourism development of the Alva Valley

Table 4 Summary of the hypothesis test for independent samples with the Kolmogorov–Smirnov test for the various elements of tourism offer in the Alva Valley

Null hypothesis	Sig.[a,b]	Decision
The distribution of landscape/nature is equal in the private/public categories	1.000	Accept null hypothesis
The distribution of culture/traditions is equal in the private/public categories	0.889	Accept null hypothesis
The distribution of people/services is equal in the private/public categories	1.000	Accept null hypothesis
The distribution of river is equal in the private/public categories	0.486	Accept null hypothesis
The distribution of cultural heritage is equal in the private/public categories	0.281	Accept null hypothesis
The distribution of gastronomy/wines is equal in the private/public categories	0.125	Accept null hypothesis

[a]The level of significance is 0.05
[b]The asymptotic significance is displayed

(3.17), Wine and Gastronomic Tourism (2.83) and the least important Cultural and Religious Tourism (2.38). It should also be noted that only for the products of Nature and Rural Tourism and Wine and Gastronomic Tourism, there was the average of private sector respondents higher than that of public sector respondents, with no second product mentioned, the difference being more marked (3.46 and 2.09, respectively).

In order to deepen the analysis and find out whether the distribution of the different tourism products to be developed in the Alva Valley is similar in the groups of private organisations and public administrative units, the Kolmogorov–Smirnov test was applied (Table 5), which confirms with a confidence level of 95% that the variables are independent in all cases, i.e. the evaluations do not differ between the public and private sector groups, with no significant differences between the groups. Also, the application of the non-parametric Mann–Whitney U test leads to the conclusion that all tourism products to be developed in Alva are evaluated in exactly the same way, with the exception of Cultural and Religious Tourism and Wine and Gastronomic Tourism where differences were found ($p = 0.030 < 0.05$ and $p = 0.013 < 0.05$, respectively), i.e. in the case of Cultural and Religious Tourism, the respondents from the private sector rate this product worse, and in the case of Wine and Gastronomic Tourism, the situation is reversed.

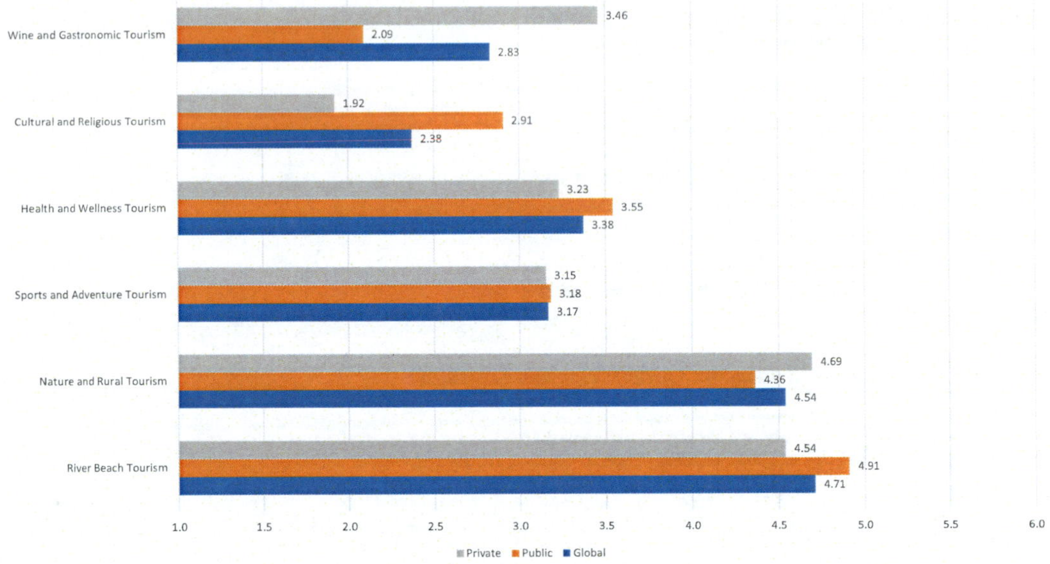

Fig. 4 Degree (average) of importance of some tourism products to be developed in the Alva Valley

Table 5 Summary of the hypothesis test for independent samples with the Kolmogorov–Smirnov test

Null hypothesis	Sig.[a,b]	Decision
The distribution of river beach tourism is equal in the private/public categories	0.998	Accept null hypothesis
The distribution of nature and rural tourism is equal in the private/public categories	0.984	Accept null hypothesis
The distribution of sports and adventure tourism is equal in the private/public categories	1.000	Accept null hypothesis
The distribution of health and wellness tourism is equal in the private/public categories	0.998	Accept null hypothesis
The distribution of cultural and religious tourism is equal in the private/public categories	0.075	Accept null hypothesis
The distribution of wine and gastronomic tourism is equal in the private/public categories	0.058	Accept null hypothesis

[a]O nível de significância é 0.05
[b]A significância assintótica é exibida

8 Conclusions

For some rural and natural regions characterised by low population density, tourism could be a structural phenomenon at the spatial and socio-economic level, based on the capital and diversity of natural, cultural and human resources present in a territory. Therefore, territorial characteristics are essential for tourism products, which can help to make policy choices and pursue sectoral strategies. A sub-regional strategy with joint efforts between municipalities, in this case the municipalities around the Alva River, could help to develop and promote a sustainable and competitive approach for the region. The data shows us that there is a strong awareness of the importance of integrated planning strategies at a regional level, which is confirmed by around 87% of respondents who indicated it as quite or very important and also by 75% of respondents who consider it quite or very important the existence of sustainable development policies in the Alva Valley. Like Buhalis (2000), we believe that the strategy to be followed in the Alva Valley should consider the complementarity between the two approaches, while there is a relational logic between the living forces of the territory. It can be assumed that the strategies should be implemented with the involvement of all stakeholders and implementing integrated management models. In this context, we believe that the sustainability of tourism will develop if it responds to the needs of the local population and simultaneously to the tourists.

The public policies and strategies for the development of tourism in this region and the public measures taken for the development of tourism in the municipality of Oliveira do Hospital are also considered important to promote an appropriate development model for the tourism area, but the strategies of participation and cooperation of the public and private sectors are positive, but should be improved for the development of tourism in the Alva Valley region. Unfortunately, this study has shown that the existence of the Alva Great Route (AGR) is unknown to a large number of respondents (75%), with the public sector accounting for 45% and the private sector 100%. This fact is limiting from the outset and should be the subject of deep reflection on the part of the entities involved.

There are weaknesses and threats that can partially jeopardise the future of the region, especially considering the impact of catastrophes that change the conditions of development in a destination. Namely, the main problems for the development of tourism in this region are the poor accessibility, the lack of incentives and investment in the sector, the ageing and depopulation of the region, the lack of promotion, spatial and offer planning, innovative and highly attractive initiatives and sensitivity of political decision makers. Regarding the tourism situation in the Alva Valley, the interviewees pointed out the need to promote the internationalisation of AGR locally and in external tourism markets.

The most valued elements were river beaches, cultural heritage and accommodation and natural heritage. The most important attributes for tourism development in the Alva Valley are Landscape/Nature (5.21), River (4.08), Culture/Traditions (3.67). The strategic tourism products could be Beach and River Tourism (4.71), Nature and Rural Tourism (4.54) and Health and Wellness Tourism. The development of the Alva Valley region must therefore be based on sustainable territorial development strategies, on the significant natural, rural and historical resources, on the promotion of competitiveness by creating jobs and by retaining and attracting the younger population, by enabling the development of tourism activities and, at the same time, by diversifying economic activity in the region due to the strong multiplier effect of tourism on regional economies. In this context, the AGR project related to the Alva Valley and its tributaries would be a key investment by promoting and linking the hiking trails mentioned here. According to some of the main trends in the tourism market, especially in the national market, and as we show, nature and rural tourism and especially hiking could be the strategic development product of this destination.

Reflections on the concept of sustainability in tourism made it possible to highlight the need for integration of the interests of local people, tourists and the tourism industry, as well as the need to preserve the natural and cultural heritage when developing an area such as the Alva Valley. The identification of natural resources by stakeholders as some of the most important elements of the tourism offer, such as the river beaches and natural patrimony (as seen in Fig. 2), demonstrates the potential for sustainable rural and nature tourism and also highlights the constraints to be considered that arise from situations caused by natural catastrophes. The development of natural and rural areas is difficult to manage, and we need to take into consideration the local development with the current and future needs of residents and also the interests of tourists. In terms of what the region has to offer tourists, it can be said that the AGR provides conditions that allow the development of a specialised tourism offer in accordance with the current needs of various niche markets. Regarding the aspects to be improved, the public sector emphasised the river beaches and pedestrian paths, as well as the need to pay attention to promotion, and the private sector referred to the hiking and cycling trails, mentioning also the reforestation and the cleaning of the forest and the river.

Given the characteristics of the Alva Valley, we believe, like Simões and Ferreira (2009) and Kastenholz et al. 2014, that the natural and cultural resources can generate several specialised products to meet the needs of niche market tourists, especially those seeking more direct contact with nature through hiking. Investing in this niche product requires a rational use of existing resources so that future generations can enjoy them as part of a sustainable and competitive development. This is promoted through participative and integrated management, as it is a necessary tool to involve all tourism stakeholders in the Alva Valley region. This study aims to provide a useful contribution to stimulate all stakeholders, especially at the local level (municipalities), which can be reinforced by educational institutions, as Silva and Umbelino (2017) say, so that together they share responsibilities and duties to achieve sustainable and competitive development.

With regard to the limitations of the study, it can immediately point out the fact that only one of the municipalities that are part of the AGR was analysed. In addition, in contrast to the public sector agents, less attention was paid to the private tourism offer of the area. Finally, we can also point out the limited amount of data available on AGR.

In terms of future studies, the same study can be extended to the other municipalities, taking into account the demand side as well. It would also be interesting to develop a study that identifies and supports the need to create a management unit for the development of AGR.

References

Almeida, M. C. (2017). Resiliência e Gestão do Risco no Turismo. In F. Silva & J. Umbelino (Eds.), *Planeamento e Desenvolvimento Turístico* (pp. 105–120). Lidel.

Augusto, D. D., Pinho, J. C., & Rodrigues, C. J. (2010). Análise Integrada Multissectorial e Pluridimensional no Território. *Revista Turismo e Desenvolvimento, 13/14* (2), 499–511. https://doi.org/10.34624/rtd.v2i13/14.12137

Bauzá Martorell, F. J., & Melgosa Arcos, F. J. (Eds.) (2020). *Turismo post COVID-19. El turismo después de la pandemia global. Análisis, perspectivas y vías de recuperación*. Ediciones Universidad de Salamanca. https://doi.org/10.14201/0AQ0296

Biosphere Tourism. (2021). *The Definitive Guide for Sustainable Tourism*. Comunidade Intermunicipal da Região de Coimbra. https://www.biospheretourism.com/en/download-the-definitive-guide-for-sustainable-tourism/113

Brito, A. S. (2015). *Vila Cova (Seia): Uma História Multissecular*. Simões & Linhares.

Buhalis, D. (2000). Marketing the competitive destination of the future. *Tourism Management, 21*(1), 97–116. https://doi.org/10.1016/S0261-5177(99)00095-3

Burnay, M. J. (2006). O Turismo Sustentável e o Turismo de Natureza - Constrangimentos e Oportunidades, Turismo de Montanha. *Revista Turismo e Desenvolvimento, 6*, 167–179.

CIM Região de Coimbra (2020). *GR51—Grande Rota do Alva*. Retrieved October 13, 2021, from https://pt.wikiloc.com/trilhas-trekking/gr51-grande-rota-do-alva-50038202

Cunha, L., & Abrantes, A. (2013). *Introdução ao Turismo* (5ª ed.). Lidel.

ERTCP. (2008). *Guia de Percursos Pedestres*. Turismo do Centro. http://www.turismodocentro.pt/pt/produtos_.1/percursos_pedestres_.a1007.html

Estevão, C., & Costa, C. (2020). Natural disaster management in tourist destinations: A systematic literature review. *European Journal of Tourism Research, 25*, 2502. https://doi.org/10.54055/ejtr.v25i.417

Figueira, L. M. (2013). Roteirização do turismo: uma abordagem preliminar à 'apresentação-interpretação' do território. *Revista Turismo e Desenvolvimento*, (20), 121–133. https://doi.org/10.34624/rtd.v0i20.12497

Figueira, L. M., & Oosterbeek, L. (2021). *Turismo Mundial, Crise Sanitária e Futuro: novos tempos, novos ritmos*. Instituto Politécnico de Tomar. http://www.cda.ipt.pt/download/Livros/novos-tempos-novos-ritmos-ebook_compressed.pdf

Inskeep, E. (1991). *Tourism planning: An integrated and sustainable development approach*. Wiley.

Kastenholz, E., Eusébio, C., Figueiredo, E., Carneiro, M. J., & Lima, J. (Eds.). (2014). *Reinventar o turismo rural em Portugal: cocriação de experiências turísticas sustentáveis*. UA Editora.

Klimek, K. (2013). Destination management organizations and their shift to sustainable tourism development. *European Journal of Tourism, Hospitality and Recreation, 4*(2), 27–47.

Kotler, P., Bowen, J. T., Makens, J., & Baloglu, S. (2021). *Marketing for Hospitality and Tourism* (8th ed.). Pearson.

Leitão, M. L. (2004). *Análise do Mercado Turístico do Parque Natural da Serra da Estrela: Estudo de Segmentação e Implicações para o Planeamento de Marketing de Áreas-Destino*. [Master's thesis, Aveiro University]. Aveiro University Institutional Repository. https://ria.ua.pt/handle/10773/17607

Lourenço, L., Nunes, A., Gonçalves, J. (2014). Enxurradas violentas após incêndios florestais: o exemplo de praias e piscinas fluviais da bacia do Rio Alva (Portugal). In L. Lourenço (Ed.), *Multidimensão e Territórios de Risco* (pp. 691–695). Simões & Linhares. https://doi.org/10.14195/978-989-96253-3-4

Ministério da Economia e da Inovação. (2009). "Lei de Bases do Turismo - Decreto-Lei nº 191/2009". *Diário da República* 1ª Série, 158 (agosto): 5336-41. https://dre.pt/dre/detalhe/decreto-lei/191-2009-493688

Queirós, A. S. (2014). Turismo Cultural e Economia do Património. *Revista Turismo e Desenvolvimento, 21/22*(4), 107–117. https://doi.org/10.34624/rtd.v4i21/22.12293

Queirós, A. S., Veloso, A. S., & Ferreira, H. D. (2008). *Património Natural e Cultural da Serra da Estrela: Roteiro Serra da Estrela*. Liga de Amigos de Conimbriga.

Rodrigues, A., & Kastenholz, E. (2010). Sentir a Natureza – Passeios Pedestres como Elementos Centrais de uma Experiência Turística. *Revista Turismo e Desenvolvimento, 13/14*(2), 719–728. https://doi.org/10.34624/rtd.v2i13/14.12285

Rodrigues, A., Farias, K. S. D., Tavares, M. G. D., & Castro, M. D. S. (2021). Tourism planning and management during the covid-19 pandemic in the tourist regions of Belém and Campos do Marajó. *Geo Uerj*, (39), Article e61314. https://doi.org/10.12957/geouerj.2021.61314

Salgado, M. A. B. (2010). Perspetivas do turismo rural em Portugal. In F. J. J. Moreno & F. J., M. Arcos (Eds.), *Estudios de Turismo Rural y Cooperación entre Castilla y León y Portugal* (pp. 49–63). Ediciones Universidad, Salamanca.

Salgado, M. A. B., & Leitão, M. L. (2011). Estratégia de Desenvolvimento Turístico da Serra da Estrela. *Revista Turismo & Desenvolvimento*, (16), 97–113. https://doi.org/10.34624/rtd.v0i16.13373

Salgado, M. A. B., Martins, J. A. S. V., & Ramos, E. M. C. V. (2015). Desenvolvimento de "novos" produtos turísticos no vale do Alva: turismo industrial. *Inovação, Gestão e Educação em Turismo e Hotelaria: Investigação Aplicada* (Coleção Politécnico da Guarda, V. 13, pp. 83–91). Instituto Politécnico da Guarda.

Salgado, M., Oliveira, I. P., & Martins, J. A. (2019). Pedestrianismo no vale do Alva: rotas de água. *Revista Turismo & Desenvolvimento*, (32), 143–157. https://doi.org/10.34624/rtd.v0i32.20468

Saramago, T. F. A. (2021). *Paradigma de Desenvolvimento Turístico Sustentável e Pandemia da COVID-19: Espaços rurais enquanto destinos turísticos*. [Master's thesis, Guarda Polytechnic Institute]. Guarda Polytechnic Institute Institutional Repository. http://bdigital.ipg.pt/dspace/bitstream/10314/5275/1/GST%20-%20Tiago%20F%20A%20Saramago%20-%20Projeto%20Aplicado.pdf

Silva, F. (2013). Revisão Crítica ao Produto Turístico de (na) Natureza em Portugal. *Revista Da UIIPS, I*(1), 136–155.

Silva, F., & Umbelino, J. (Eds.). (2017). *Planeamento e Desenvolvimento Turístico*. Lidel.

Simões, J. M., & Ferreira, C. C. (Eds.). (2009). *Turismos de Nicho, Motivações, Produtos, Territórios*. Centro de Estudos Geográficos.

Tovar, Z. (2011). Pedestrianismo, Percursos Pedestres e Turismo de Passeio Pedestre em Portugal. *Revista Turismo & Desenvolvimento*, (16), 145–159. https://doi.org/10.34624/rtd.v0i16.13397

UNWTO. (2005). *Making Tourism More Sustainable: a Guide for Policy Makers*. UNWTO.

UNWTO. (2008). *Report Climate Change and Tourism—Responding to Global Challenges*. WTO.

UNWTO (2017). *Tourism and the Sustainable Development Goals—Journey to 2030, Highlights*. UNWTO. Retrieved October 13, 2021, from https://www.e-unwto.org/doi/book/10.18111/9789284419340

Policies and Mechanisms for Heritage Preservation

In comparison with sometimes pressing concerns like infrastructure development or the creation of jobs, the preservation of heritage may appear like a lesser priority. However, the long-term benefits of effective heritage resource conservation go beyond the preservation and protection of resources; they also assist revitalize local economies and foster a feeling of community among citizens (Srinivas, 2022), thus helping sustainable tourism development. Change is unavoidable in a historical context and to conserve anything, one must learn to deal with change in such a way that its historical values are not jeopardized (Heritage, 2007). From a global, regional, national, or sectoral perspective, there are many official policies and mechanisms for physical or digital preservation of tangible and intangible, natural and cultural, mobile and immovable, or documentary heritage (Guerzoni, 1997; Pickard, 2001; Wai-Yin, Shu-Yun, 2004). Despite all efforts, and even in cases where the heritage has already been preserved, there is a continuing threat from nature (Milošević, et al., 2017) or people that influence the heritage (Vujičić et al., 2018, Vujičić et al., 2020a). This necessitates constant innovation in the field of heritage preservation policies and procedures, in which the tourism domain plays an important role (Stankov et al., 2016; Vujičić et al., 2020b).

The third part of the *Cultural Sustainable Tourism* edition, titled *Policies and Mechanisms for Heritage Preservation* is organized into five chapters. Chapter "Municipality of Ramon, Isabela, Philippines: Sustainable Tourism and Strategic Development Plan 2021–2025", explore the case study of the municipality of Ramon in the province of Isabela, Philippines that, as an emerging tourist destination is faced with environmental, socio-cultural, and economic issues. The chapter emphasizes the importance of Ramon being fully aware of the notion of sustainable tourism and incorporating it into all of its plans, not only to conserve its natural beauty but also to improve its economy and protect its cultural legacy. The current lack of linkages between the tourism industry and the heritage preservation movement is highlighted in Chapter "Designing Conservation: From Sustainable Heritage to Sustainable Tourism (and Vice Versa)". For a long time, heritage, particularly traditional assets, have been viewed as mere tools for basic tourism services, with little regard for their cultural and educational potential. However, new narratives can be imagined as a result of new knowledge of the essential aspects of heritage. In this regard, the emphasis on circular economy and sustainability in tourism may have a significant impact on design in heritage preservation. The importance of education and professional training, which is increasingly being emphasized by key global and national tourism organizations as a vital aspect of sustainability initiatives, is discussed in Chapter "Education and Professional Training in Tourism and Hospitality: The Case of Portuguese Travel Agents". It focuses on Portuguese tourist education and training, as well as how it has been merged in response to market trends. The topic of Chapter "The Future of Historic Re-enactments in Portugal—The Case of Santa Maria da Feira", is also situated in Portugal. The authors provide an extensive literature review on the subject of reenactments, including the origins and history, the difficulty of defining the field, the dangers posed by globalization, as well as the opportunities presented by globalizing forces and tourism for historical preservation. The chapter presents the Case of Santa Maria da Feira. The final Chapter "Assessing On-Water Recreation Experiences in Midwestern U.S. Glacial Lakes: Identifying Areas of Conflict and Agreement", deals with specifics of lake-based recreation, which is often constrained by physical, social, and economic factors that can result in contradictory views and perspectives on personal experiences. The goal of this chapter was to construct a basic profile of the local recreation community by analysing current users' perspectives and levels of historical and current on-water activities on the Iowa Great Lakes. This could improve the ability to investigate broader water recreation issues like environmental impact, recreation conflict, and the development of locally led comprehensive utilization plans.

Uglješa Stankov
Department of Geography, Tourism and Hotel Management
Faculty of Sciences
University of Novi Sad
Novi Sad, Serbia
e-mail: ugljesa.stankov@dgt.uns.ac.rs

References

Guerzoni, G. (1997). Cultural heritage and preservation policies: notes on the history of the Italian case. In *Economic perspectives on cultural heritage* (pp. 107–132). Palgrave Macmillan.

Heritage, E. (2007). *Conservation principles policies and guidance: For the sustainable management of the historic environment*. English Heritage.

Milošević, D., Savić, S., Stankov, U., Žiberna, I., Pantelić, M., Dolinaj, D., & Leščešan, I. (2017). Maximum temperatures over Slovenia and their relationship with atmospheric circulation patterns. Geografie-Sbornik CGS.

Pickard, R. (2001). *Policy and law in heritage conservation* (Vol. 1). Taylor & Francis.

Srinivas, H. (2022). Heritage and Conservation Strategies: Understanding the Justifications and Implications. *GDRC Reseaarch Output-100: Global Development Research Center*. Retrieved from https://www.gdrc.org/heritage/heritage-strategies.html

Stankov, U., Klaučo, M., Dragićević, V., Vujičić, M. D., & Solarević, M. (2016). Assessing land-use changes in tourism area on the example of Čajetina municipality (Serbia). *Geographica Pannonica, 20*(2), 105–113.

Vujičić, M. D., Stamenković, I., Stankov, U., Kovačić, S., Vasiljević, Đ. A., & Popov-Locke, J. (2020a). What will prevail within citybreak travel, motivation or demotivation?: Case study of Novi Sad, *Geographica Pannonica, 24*(1), 42–55.

Vujičić, M. D., Stankov, U., Kovačić, S., Vasiljević, Đ. A., Pivac, T., Čarkadžić, J., Mujkić, D. & Cimbaljević, M. (2020b). Digital Skills and Behaviours of Youth That Are Relevant for Digital Culture: A Two-Country Self-Evaluation Perspective. In *Examining the Roles of Teachers and Students in Mastering New Technologies* (pp. 128–149). IGI Global.

Vujičić, M. D., Vasiljević, D. A., Hose, T. A., Tasić, N., Morar, C., Đurić, A., & Marković, S. B. (2018). A multi-criteria decision analysis with special reference to loess and archaeological sites in Serbia (Could geosciences and archaeology cohabitate?). *Open Geosciences, 10*(1), 333–343.

Wai-Yin, C., & Shu-Yun, M. (2004). Heritage preservation and sustainability of development. *Sustainable development, 12*(1), 15–31.

Municipality of Ramon, Isabela, Philippines: Sustainable Tourism and Strategic Development Plan 2021–2025

Antonio S. Tiong III, Miguell Jhon R. Casanova, April Dominique O. Cruz, Ma. Arna Jinia L. Sanggalang, and Kryztl Ann Felice R. Veniegas

Abstract

The municipality of Ramon, Isabela, Philippines, is an emerging tourist destination constrained by problems and impacts on the environmental, socio-cultural, and economic scale, impeding Ramon from becoming one of the leading destinations in the province of Isabela. The objectives are determined to spearhead the municipality's sustainable tourism and strategic plan. Essential principles have been incorporated, such as the seventeen (17) Sustainable Development Goals (SDGs) of the United Nations (UN) and the Destination Assessment patterned from the United Nations World Tourism Organization (UNWTO). The case study is a qualitative design that utilized document analysis and virtual interviews through purposive sampling. The key informants were the Municipal Tourism Officer and the Municipal Planning & Development Officer. Likewise, a virtual interview was conducted through random sampling of the municipality's residents aged twenty (20) to twenty-one (21) years old to gain a perspective from the younger generation. The data was analyzed using the UNWTO Destination Assessment Matrix, SWOT Analysis, and PEST Analysis. Strict enforcement of rules and regulations, standards governing tourism-related training, and formulation of the marketing strategies of the municipality as a tourist destination were proposed to mitigate the negative impacts of tourism while continuously developing its tourism sector.

Keywords

Strategic development plan • Sustainable development goals • Marketing strategies • Ramon • Philippines

A. S. Tiong III (✉) · A. D. O. Cruz · Ma. A. J. L. Sanggalang · K. A. F. R. Veniegas
University of Santo Tomas, Manila, Philippines
e-mail: antoniotiongiii@gmail.com

M. J. R. Casanova
Municipality of Ramon, Isabela, Philippines

1 Introduction

The second (2nd) class municipality of Ramon, Isabela, is located in the Northeastern Cagayan Valley Region (Bureau of Local Government Finance, 2008). It is landlocked below the Caraballo Mountain of Cordillera on the west and the Sierra Madre on the east (Perante II, 2017). It consists of thirteen thousand five hundred seventeen (13,517) meters of land area, covering one and nine hundredth percentile (1.09%) of the province of Isabela (PhilAtlas, n.d.). The municipality has numerous attractions to offer, posing great potential to become one of the leading tourism destinations in the province of Isabela. However, as time passes, negative impacts and challenges are faced—environmental, socio-cultural, and economic factors that hinder Ramon's sustainable development.

Ramon is situated at the heart of Cagayan Valley Region and is renowned for its proven strength in the business and agricultural industry (BusinessMirror, 2018). The *"Isabelinos"* livelihoods are inclined toward agriculture and fishing. The province of Isabela contributes twenty-one percentile (21%) to the annual yellow corn production of the country, in which a margin of the percentage derives from the municipality of Ramon. The province of Isabela's rice sample rate is two hundred twenty-four percentile (224%), producing more than they consume, and is exclusively responsible for providing rice necessities for some cities in the country. Animal and poultry industries, particularly milk processing, pork production, cattle breeding, and commercial poultry rearing, also flourish in the municipality. Moreover, Ramon is thriving with its aquaculture, specifically in Magat Dam, wherein there are four hundred fifty (450) hectares of artificial fish cages situated within the area (Province of Isabela, 2013).

The municipality of Ramon is accessible due to its positioning near the bustling cities of Santiago and Cauayan, the province of Ifugao, and the municipalities of San Mateo and Alicia. One can travel via air through Cauayan Domestic

A. Mandić et al. (eds.), *Cultural Sustainable Tourism*, Advances in Science, Technology & Innovation,
https://doi.org/10.1007/978-3-031-10800-6_12

Airport (Province of Isabela, 2013). Based on observation, various land transportation systems are utilized going to and from the municipality; traditional jeepneys, vans, and provincial buses traverse Ramon's central barangays to load and unload passengers. Public utility jeepneys (PUJs) were the primary transportation before the pandemic.

1.1 Statement of the Problem

The researchers have identified eight (8) problems that Ramon must recognize and resolve to sustain its tourism sector and the municipality in general. First, in the environmental context, three (3) problems were determined: (1) weather conditions in which the province of Isabela is usually hit by strong typhoons annually; (2) maintenance of attractions are not adequately sustained as claimed strengthly by Mr. Medios, a resident of Ramon during a virtual interview; and (3) landlocked municipality, making tourist choose destinations with 3S tourism that are adjacent from Ramon; Secondly, for socio-cultural there are two (2) problems specifically (1) tourism awareness as mentioned by the Tourism Officers during a virtual interview that locals are not yet adequately oriented with the benefits of tourism and (2) people's health and safety due to the continuous increase of COVID-19 transmission as of the making of this study in the locality, and a response of a resident during a virtual interview stating that the local government is not taking their duties seriously with regards in handling the pandemic. Lastly, for economic, there are three (3) problems observed: (1) lack of proper/quality training of tourism practitioners in which according to the Tourism Officer, the current training offered is inlined with dressmaking, driving, and baking; (2) having limited facilities and amenities to support the tourism development whereas based on observation and results of virtual interviews, Ramon lacks primary facilities such as toilets in the attraction; and (3) insufficient marketing strategies such as not having a dedicated website containing necessary information of the municipality's attractions, and a social media page dedicated for tourism promotions.

1.2 Objectives

The study aims to develop the following objectives to guide the sustainable tourism growth of Ramon, Isabela, to reduce the adverse effects of tourism on environmental, socio-cultural, and economic aspects.

- To propose a five (5)-year comprehensive sustainable tourism development design for Ramon, Isabela, from 2021 to 2025 to contribute to the upturn of the tourism industry from COVID-19;
- To assess the municipality's current situation regarding the three (3) sustainable tourism development goals—economic, socio-cultural, and environmental factors;
- To channel the importance of sustainability to locals of Ramon toward having an in-depth relationship with the attractions of Ramon to cultivate its presence in tourism;
- To provide strategies that will help the municipality in strengthening its tourism presence in the market as well as to mitigate the negative factors affecting tourists' choice of destination; and
- To enhance the tourism sector of Ramon by maintaining the natural features of the attractions in reference to the seventeen (17) sustainable goals proposed by the United Nations (UN).

The municipality of Ramon needs to be fully aware of the concept of sustainable tourism and incorporate it into every plan it will formulate, not just to preserve its natural beauty but also to boost its economy and protect the cultural heritage of Ramon. Carrying out sustainable tourism practices adopted from international standards will allow Ramon to fully consider its present and future state of tourism and the municipality in general.

1.3 Significance of the Study

Despite the current progress of Ramon to boost its tourism, numerous problems challenge the betterment as mentioned above, decelerating the growth of the municipality's tourism sector and in an overall sense. This study was conducted to aid the municipality significantly to acknowledge, study, and formulate solutions for the problems discussed in the preceding pages.

1.4 Literature Review

The researchers identified the related literature in three (3) categories, specifically sustainable tourism on the national, regional, and local scale. The presented literature can help Ramon provide a clear and broad perspective of sustainable tourism by identifying ideal solutions for the problem.

The sustainability of Philippine tourism industry as per the study conducted by Pilapil-Añasco and Lizada (2014) titled "*Philippine Tourism: Evolution Towards Sustainability*," there are four (4) significant stages of the Philippine sustainable tourism development based on the political spectrum of the country: Pre-Martial Law Era from 1971 and earlier, Martial Law Era from 1972 until 1986, Post-Martial Law Era from 1986 until 2000, and the twenty-first century Era from 2001 to present. Focusing on the twenty-first century, the national government officially adopted tourism as part of the country's pillar for economic growth. In line with this, the Philippine government crafted a medium-term development plan (Pilapil-Añasco & Lizada, 2014), where some chapters are patterned with the United Nations Millennium Development Goals, focusing on poverty-causing multiple deprivation effects (United Nations, 2017).

Concerning the United Nations (UN) efforts, the seventeen (17) Sustainable Development Goals (SDGs) can benefit Ramon as a guide in sustaining and developing the municipality's tourism industry. In relation, tourism has been included in Goals Eight (8th) Decent Work and Economic Growth, Twelfth (12th) Responsible Consumption and Production, and Fourteenth (14th) Life Below Water (United Nations World Tourism Organization, n.d.-b). The full context of the goals is found in the 2030 Agenda for Sustainable Development by the Department of Economic and Social Affairs of the United Nations.

To support the claim, based on various studies conducted concerning sustainable tourism, tourists expect the industry providers to initiate and undertake sustainable options for tourism. The result from the Center for Responsible Travel Study reveals that there is an increasing responsible travel choice among tourists. Supporting the assertion is the result of the 2018 study carried out by booking.com wherein over fifty-five percentile (55%) of the respondents create sustainable travel choices, while seventy-two percentile (72%) express their concerns regarding the current actions for sustainable tourism as a choice for future generations (Sustainabletourism.net, n.d.). The study of Pulido-Fernández and López-Sánchez (2016) of Universidad de Jaén, Spain, titled "*Are Tourists Really Willing to Pay More for Sustainable Destinations?*" shows that those tourists who are well-aware and knowledgeable about sustainability are willing to pay more. However, those willing demographics are relatively low, with only twenty-six and six hundredth percentile (26.6%) of the total one thousand one hundred eighteen (1118) respondents (Pulido-Fernández & López--Sánchez, 2016).

The efforts of the Department of Tourism (DOT) and its attached agencies to help the country become a world-renowned tourism destination are highly influenced and patterned to Republic Act No. 9593 or Tourism Act of 2009 signed by former President Gloria Macapagal Arroyo (Palaubsaon & Pareja, 2009). It is the national policy for tourism to become an engine of investment, employment, growth, and national development, strengthening the DOT and its attached agencies to implement policy and appropriating funds effectively and efficiently (Official Gazette of the Philippines, 2009).

Meetings, Incentives, Conferences, and Exhibitions (MICE) tourism in the Philippines has progressed since the 1970s, leading the country to be considered as Asia's leader in hosting International MICE events with its world-class Philippine International Convention Center (PICC). However, the country slipped in the rankings; that is why during the administration of the late President Benigno Aquino III, there was an attempt from the DOT to gear up in promoting the country as a MICE destination. Several promotional attempts of the Philippine Tourism Promotions Board (PTPB) were the tagline "*Business Meets Fun in the Philippines*" alongside the prominent "*It's More Fun in the Philippines*" (Oxford Business Group, 2015).

The Philippines strives to promote its local products through One Town One Product (OTOP), adopted from Japan's One Village One Product, whereas various countries have also started practicing it. OTOP aims to become more inclusive by prioritizing Micro, Small, and Medium Scale Enterprises (MSMEs) in the Philippines (Department of Trade and Industry, n.d.). Since 2002, OTOP has ventured into the country, which was further strengthened and prioritized by Executive Order No. 176, S. 2003 or the "*Isang Bayan, Isang Produkto, Isang Milyong Piso*" program (Official Gazette of the Philippines, 2003).

When it comes to agritourism, DOT Secretary, Honorable Bernadette Romulo-Puyat, stated that the province of Isabela and its municipalities are flourishing because of its natural resources; Isabela is eyed by the DOT Secretary to become an agritourism haven because of its rich and bountiful agricultural industry. She also emphasized that Isabela could become part of the world-class agri-ecotourism, whereas various experimental tour packages can be formulated, including fruit picking, fishing, rice planting, and harvesting (Campos, 2020).

In line with the 2017 long-term goals of Region 2 through "*Ambisyon Natin 2040,*" the Regional Development Plan (RDP) of Cagayan Valley for 2017–2022 was launched. It pinpoints strategies for tourism, agriculture, and environmental development. Protection of the existing forest, development of river systems as a substitute mode of transportation, and renewable energy development are also included in the RDP (National Economic and Development Authority, n.d.). It tackles the regional population distribution, access to public services, and investment guidelines for both public and private investors while sustaining both natural and artificial resources of Region 2 while also focuses on protecting the physical environment. Some of the framework topics are opportunities to grow underdeveloped resources and productive forests and grasslands (National Economic & Development Authority, 2017b).

Efforts are initiated to increase the tourism arrival in Isabela, wherein last December 3, 2019, fifteen (15) modernized jeepneys were handed over to First Isabela Bus and Van Operators Transport Cooperative by the Land Transportation Franchising and Regulatory Board (LTFRB) (Edale Jr., 2019). (1) The modernization project will help in reducing pollution by utilizing either Euro 4 engines or electricity; (2) improve public safety by providing a GPS navigation system and CCTV camera monitor; (3) increase punctuality; and (4) enhance comfort (Philippine Toyota, 2020). There will be lesser emission of air pollutants if the Philippine public transport system is improved. The data of Mariano (2019) reveals that thirty-nine percentile (39%) of greenhouse gases (GHG) will be reduced, particulate matter by ninety percentile (90%), and sixty-three percentile (63%) of nitrogen oxide (NOx) will be decreased, inclining the Philippines toward a low-carbon pathway.

In 2017, the local government of Ramon released Municipal Ordinance No. 12, which prohibits and regulates the usage of cellophane, sando bags, and foamed polystyrene products (Styrofoam) as food and beverage packaging in the municipality (Parinas Sr. & Honorio, 2017). The Office of the *Sangguniang Bayan* of Ramon released Municipal Ordinance No. 13-2017, known as the "*Integrated Solid Waste Management (ISWM) Ordinance of the Municipality of Ramon.*" Under Article 3, Section 4 states that "*The ordinance shall be used to guide, control, and regulate an orderly sanitary system, enhance the total environment by mitigating the negative impacts of solid waste by promoting, and protecting the health, safety, peace, and convenience of the municipality's inhabitants and maximizing the resource recovery and recycling.*" (Parinas Sr., 2017).

In addition, there is an initiative to rebuild the prosperous fishing industry of Ramon with the formation of the Magat Aquatic Park Project, together with the Bureau of Fisheries and Aquatic Resources (BFAR) of Region 2, SN Aboitiz, and the National Irrigation Administration (NIA); the project aims to help in attracting new investors in Ramon (Prudencio, 2018). To test the project's viability, six (6) fish cages were built and closely monitored (Towers, 2013).

2 Methodology

2.1 Research Design/Data Measure

The research is a qualitative design that utilizes the case study method through document analysis and virtual interviews, using purposive sampling. Statements were gathered and analyzed, together with the documents obtained from reliable sources such as news articles, journal articles, and government websites on national and regional scales—all accessed on the Internet. The virtual interview was divided into two (2) parts: the first part included the current tourism policies being implemented by Ramon, while the second part tackled the tourism of the municipality in general, its current landscapes, and plans for tourism recovery.

2.2 Subjects and Study Sites

The subjects were tourism stakeholders of Ramon, particularly the representatives of the local government such as the Municipal Tourism Officer and Municipal Planning & Development Officer, together with the local community of Ramon ages twenty (20) to twenty-one (21) years old residing for more than ten (10) years in Ramon. The chosen study site is the Municipality of Ramon, Isabela.

2.3 Data Gathering/Mode of Analysis

The one-on-one interview of the key informants was done in an online manner. The key informants were invited for a scheduled virtual interview via Google Meet platform. Data was analyzed using situational analyses: Strengths, Weaknesses, Opportunities, and Threats (SWOT) Analysis, together with Political, Environmental, Social, and Technological (PEST) Analysis. Moreover, the Destination Assessment Matrix from the United Nations World Tourism Organization (UNWTO) was also adopted to support the study.

3 Results and Discussion

3.1 Situational Analysis

The following situational analyses presented in Tables 12.1 and 12.2 assess both the internal and external factors affecting and determining the necessary actions and procedures that must be considered to maintain and develop Ramon's tourism industry (Tables 12.3, 12.4 and 12.5).

3.2 Destination Assessment

The destination assessment used for the study was obtained from the United Nations World Tourism Organization (UNWTO) website, in collaboration with EuropeAid through its project that focuses on enhancing sustainable tourism in developing countries (United Nations World Tourism Organization n.d.-c). The four (4) pillars specifically under Tables 12.4, 12.6, and 12.7 patterned from the UNWTO and EuropeAid were tailored for the municipality level to guide the sustainable tourism development of Ramon, Isabela. Nevertheless, Table 12.3 will be the basis for answering the assessment.

Pillar 1—Tourism Policy Governance

The Municipal Tourism Officer of Ramon, Ms. Mabilyn Tolentino, asserted during a virtual interview that tourism is

Table 12.1 Strengths, weaknesses, opportunities, and threats analysis

	Strengths	Weaknesses
Internal factors	Geography • Ramon is protected by Caraballo Mountain and Sierra Madre Mountain Ranges from strong typhoons • It is ideal for regional and provincial jamboree camping Located near bustling cities and transportation hubs of Isabela • It can be easily accessed using various land transportation systems. Public Utility Vehicles (PUVs) such as vans and jeepneys have regular trips around and within the municipality • Multiple bus companies are located in its neighboring city of Santiago, in which there are regular trips to and from various regions of Luzon • Located near the Cauayan Regional Airport Magat Dam • It sustains the irrigated water system used by locals for agriculture and aquaculture industry • Magat Dam is the only dam in the Cagayan Valley Region open for tourism Agriculture and Aquaculture • Rice Farming, Corn Farming, Tilapia Farming DOT accredited establishments • The municipality has consistently supported each establishment in obtaining a DOT accreditation	Tourism Development Plan (TDP) is not yet being developed • Formulation of Tourism Development Plan has been further delayed due to change of governance and the Novel Coronavirus 2019 Underdeveloped marketing strategy • There are scattered, unofficial, and poorly managed Facebook pages of Ramon, illustrating that there is no stable marketing strategy • When searched online, little to no information can be found regarding the municipality. Similarly, few available information is either not credible or misleading • There is no official website to gather credible sources of information about the municipality Maintenance of attractions • Attractions exist, but maintenance and sanitation are not adequately observed • Plastic bottles, disposable food containers, paper bags, and the like can be spotted in some attractions
External factors	Opportunities	Threats
	Promoting Ramon, Isabela, as an eco-tourism destination while increasing tourism arrivals • Most activities are nature-friendly, where tourists can trek, camp, and set up a picnic • Identify the different potential eco-tourism attractions • Ordinances regarding the use of plastics should be appropriately observed and implemented Creation of official website and consolidated social media accounts • Ramon can further extend its reach to potential tourists through social media platforms like Instagram, Twitter, and Facebook • A consolidated social media presence is beneficial to the municipality for information dissemination, reducing confusion when searching for credible data sources Job opportunities for locals and development of Ramon Rolling Hills • The LGU will take full charge of the site's management once the municipal ordinance for developing Ramon Rolling Hills has been approved • Jobs should be first offered to the locals, helping them further understand the benefits and impacts of tourism in their lives Creation of Tourism Development Plan (TDP) • It will help Ramon to recognize the potential attractions that can appeal to the needs and wants of the target market • Various technical reports from attractions offering comparable tourism attractions will help • It will identify the SWOT of potential site attractions Providing better health protocols and strict implementation of ordinances • General preventive measures such as disinfection of general areas • Employment of well-trained staff and medical personnel to serve and guide both locals and tourists in following various health protocols • Strict checking of tourist arrivals and departures to aid in contact tracing	The impacts of Novel Coronavirus 2019 on the municipality gradually decrease tourism • Since Camp Vizcarra Ecology Park was used as a quarantine facility, it discontinued income generation due to its public visit closure • There were no recorded tourist arrivals during the Enhanced Community Quarantine (ECQ) Competition with adjacent municipalities and provinces offering the same tourism classification • City of Ilagan (Ilagan Sanctuary) • City of Santiago (Shrine of our Lady of La Salette and Spring Garden Resort) • San Mateo, Isabela (Dagupan Resort) • Alfonso Lista, Ifugao (One thousand (1,000) Steps) • Diadi, Nueva Vizcaya (Lower Magat Eco-tourism Park) The lifespan of Magat Dam was shortened • According to Engineer Rodolfo Mejia, during his interview at one of the GMA TV shows in 2013, Magat Dam was initially planned to run for one hundred (100) years. However, due to the destructive earthquake during the 1990s, the lifespan was reduced to 2050; technically, only twenty-nine (29) years left before it became obsolete. (Philippine Book Records—GMA News TV, 2013, 03:15–05:21) • Natural calamities and illegal tree-cutting near Cagayan River had brought an irreversible effect on Magat Dam, which reduced the service availability of the dam from fifteen (15) years to twelve (12) years (Lagasca, 2005) • Prone to disasters due to natural calamities and climate change • Strong typhoons frequent the province of Isabela; according to the National Disaster Risk Reduction and Management Council (NDRRMC), last 2020, seven (7) consecutive typhoons affected Isabela and its neighboring provinces—Ofel, Pepito, Quinta, Rolly, Siony, Tonyo, and Ulysses (Nepomuceno, 2020)

Table 12.2 Political, economic, social, and technological analysis

Political factors	Economic factors
The local government of Ramon politically supports tourism • The administration of Honorable Mayor Jesus Laddaran focuses on Ramon's tourism industry development • The municipality handles two (2) attractions (Camp Vizcarra Ecology Park and Ramon Rolling Hills) • They are conducting a strict implementation and monitoring of Ramon's tourism attractions in general **The municipal government of Ramon is open to public–private partnerships (P3)** • They are currently coordinating with the National Irrigation Administration (NIA) to manage the Magat Dam Tourism Complex • Communications such as letters are being forwarded to the local government for proper action **Barangay officials are working in tourism development** • Barangay officials of Planas manage and develop tourism attractions and activities that can be offered to tourists **Misunderstanding between barangay officials** • There was a dispute between Barangays Planas and San Miguel regarding the geographical location of Ramon Rolling Hills **Municipal Ordinance No. 12–2017 (Parinas Sr. & Honorio, 2017)** • Prohibiting and regulating disposable packaging containers and polystyrene foam for beverages and food in the municipality. The ordinance encourages alternative packages such as but not limited to *"bayongs"* woven bags **Municipal Ordinance No.13–2017 (Parinas Sr., 2017)** • There are five (5) objectives identified under the ordinance: (1) ensuring cleanliness through waste management, (2) banning the usage of open garbage dumps causing diseases and infections, (3) eradicate the waste containers that are overflowing in every public space, (4) maximized and optimized sanitary recovery, and lastly (5), minimizing the pollution **Tourism is a priority of the Cagayan Valley Regional Development Plan (CVRDP) 2017–2022** (National Economic & Development Authority, 2017a) • The region will face numerous challenges that can become an opportunity to maximize prime water resources, agri-industrial hubs, and emerging tourism destinations • Lack of investment regarding the optimization of industries relating to natural resources, however not limited to tourism development, is observed based on the CVRDP 2017–2022 data	**Tourism budget and national expenditure allocation** (National Economic and Development Authoity, 2017a) • Numerous industries, including fishery and agriculture, take part in the budget distribution from the annual municipal budget • Regional economy fell continuously from 2013 to 2015 regarding national expenditure allocation • During Fiscal Year (F.Y.) 2014 and 2015, Cagayan Valley's allocation regarding the General Appropriation Act (GAA) grew. However, it is lower than other municipalities, which also recorded an increase in the national budget **Low inflation rate for Cagayan Valley Region** • Bangko Sentral ng Pilipinas (BSP) data shows that Cagayan Valley Region has a low inflation rate (National Economic & Development Authority, 2017a) **The agricultural sector is essential to Ramon's economy** • Livestock, crop production, and aquaculture support the economic growth and development of the municipality • The total share of the region's agricultural sector to the Philippines Gross Domestic Product (GDP) is eight and eighty-two hundredths' percentile (8.82%) last 2019 (O'Neill, 2021) **Magat Dam contributes to the economic growth of Ramon, Isabela** (Travel to the Philippines, 2018) • Magat Dam is the primary irrigated water used for agriculture farmlands of approximately eighty-five thousand (85,000) hectares. It is also a hydroelectric power plant generating three hundred sixty (360) megawatts of electricity under SN Aboitiz's administration **Camp Vizcarra Ecology Park is the leading tourism economic contributor in the municipality of Ramon** (Tolentino, 2021) • It usually hosts local and regional jamborees of girls and boy scouts of the Philippines • The income generated is given to the municipal treasury department allocated for social services, infrastructure, education, tourism, and more **Regional Shift to service sector from the agricultural sector** (National Economic & Development Authority, 2017a) • In 2015, the Gross Regional Domestic Product (GRDP) of Cagayan Valley showed only thirty-six and six hundredth percentile (36.6%) of agricultural shares, making the service industry more dominant • Agriculture is still perceived as the primary economic contributor next to the service sector, in which more than fifty percentile (50%) of *"Isabelinos"* heavily rely on agriculture as their livelihood
Social Factors	**Technological Factors**
The Novel Coronavirus 2019 shifted the tourism industry • Attractions are temporarily closed when there is an increase of infection in the municipality • There are only limited number of people allowed in the attractions • Minimum health and safety protocols are now observed in the attractions **Tourism-related jobs are available for locals** • The municipality of Ramon employs both male and female local tour guides, in which some are given permanent positions. In relation, some are seasonal and on-call tour guides in the municipality **The municipality is promoting gender equality and health safety to employees** • They employ anyone capable of the job, regardless of gender and sexuality. Giving equal access to opportunities, resources, and power. Most importantly, they are treated with dignity, respect, and fairness • Staff is constantly being examined to check if they are in excellent health to work and engage with others	**Launching of Cagayan Valley Integrated Agricultural Laboratory or CVIAL** • During the start of CVIAL operation in 2016, it was expected to support the agricultural sector through training, preparing, and capacitating (National Economic & Development Authority, 2017a) **Promoting Science, Technology, and Innovation (STI) stimulates economic growth through regional platforms** (National Economic & Development Authority, 2017a) • Significant science and technology program expansions are being done; these involve the upgrading and setting up of Small Enterprise Technology • It provides a timetable to boost MSME competitiveness and productivity in Cagayan Valley. Many technological involvements have been offered to LGU's MSMEs with academic endeavors in the said program, including equipment and facilities upgrade • The DOST Region 2 carries out the program on the Deployment of Early Warning Systems or DEWS to incorporate Level Monitoring Stations (LMS) and Automatic Rain Gauges (ARGs) to surveil

(continued)

Table 12.2 (continued)

Political factors	Economic factors
Programs offered by the Technical Education and Skills Development Authority (TESDA) accreditation and Department of Tourism (DOT) Region 2 tour guide training • TESDA offers various courses for free, providing certificates to individuals who finish the course. Some of the courses are (1) driving, (2) tailoring, (3) electronics, and (4) baking • The Department of Tourism (DOT) Region 2 sometimes conducts tour guiding training of Ramon's locals organized by the municipality **Education** • Numerous academic institutions offer basic and secondary education in Ramon • Universities are also present neighboring municipalities and cities—University of La Salette, Isabela State University, Northeastern College, and Ifugao State University • There is no standalone library that can be accessed	parameters such as rainfall, water level, and air pressure along headwaters of the Cagayan River basin as well as those small river basins located in the region **Limited budget for research and development (R&D) spending is one of the region's main challenges** (National Economic & Development Authority, 2017a) • Cagayan Valley is one of the regions behind research and development • Much of the research and development undertakings are focused on several regions such as Central Luzon, CALABARZON, and NCR • The data of the DOST Survey on R&D human resources and disbursement in 2013, as indicated in CVRDP 2017–2022, Cagayan Valley ranked seventh (7th) and fifth (5th) with the lowest number of researchers and R&D spending, respectively

Table 12.3 Destination assessment criterion (United Nations World Tourism Organization, n.d.-c)

1	The area with the highest priority for intervention—opportunity	High priority for action Low priority for action
2	The area should be considered for intervention/support—needs improvement	
3	The area may benefit from some improvement—less need for intervention	
4	The area to be maintained—satisfactory actions were already made	
5	The area may provide as an example to others—considered as the strength of the area	
x	There is no information available for answering the question	

indeed one of the priorities of the local government at present. It is one of the three (3) areas of focus of the Cagayan Valley Regional Development Plan (CVRDP) 2017–2022. However, she stated that the office for tourism of the municipality is not yet standalone, meaning it is still under the Mayor's Office. Other industries such as agriculture and aquaculture are still the initial priorities of the municipality. Hence, the degree of priority of tourism development strategies is at a midpoint in focus for action and must be improved but does not necessarily need intervention.

There are notable efforts at an international level, such as the ReBUILD Project with United Nations Development Plan (UNDP), New Zealand Aid Program (NZAP), and Climate Change Commission (CCC), to help Cagayan Valley Region in mitigating the impacts of climate change. Despite this, the 2020 typhoon Ulysses has brought about significant flooding in Cagayan Valley, which led to massive destruction of agriculture and infrastructure, thus becoming a national concern.

As for the tourism policy and regulatory framework, Ramon currently does not have a Tourism Development Plan (TDP), a significant concern for the municipality in developing its tourism industry. The TDP will directly create the tourism policy for Ramon and ensure sustainable tourism. Concerning the second (2nd) issue in pillar 1, it needs improvement since ordinances in the municipality must also consider tourism, and the locals are not yet fully involved because of insufficient awareness.

Regarding regulations and ordinances affecting tourism, Ramon only needs to maintain its performance. However, it still needs improvement in the local sector regarding sustainable tourism legislation. Regarding the tourism governance and institutional setup, there is somehow support for Ramon's tourism concerning environmental management, which provided broad ordinances. Tolentino mentioned that Ramon is open to private and public partnerships (P3), but structures and invitations to the private sector can still be enhanced. The same shall apply to the efficiency of the tourism governance of Ramon.

Pillar 2—Economic Competitiveness, Performance, and Investment

The quality of Ramon's data collection and analysis can be enhanced through consistent and stringent documentation of visitor arrivals, profiles, and activities annually, crucial for future research concerning but not limited to tourism in general. In contrast, one factor that impedes an effective data

Table 12.4 Pillar 1—tourism policy governance (United Nations World Tourism Organization, n.d.-c)

Tourism Governance and Institutional Setup in the Municipal Level								
	Question	Comment	Priority for action					
			High			Low		
			1	2	3	4	5	X
Issue 1: Visibility of Public–Private Partnership (P3) and Private Organizations Involvement to Development								
1	Are there various forms of business model presented, and is structure development in tourism encouraged?	There is no municipal TDP right now	*					
2	In tourism development and management, is P3 partnership being utilized frequently?	There is no municipal TDP right now	*					
Issue 2: The Local Tourism Governance Effectiveness in the Municipality								
1	Is the local government presence effective in the process of tourism development?	The current farm supervisor of Ramon, Isabela, is also appointed as the Tourism Officer			*			
2	Do the national, regional, and local tourism governance effectively coordinate with each other?				*			
3	Is there any involvement of private sectors or other tourism-related establishments present in tourism governance and delivery?	Magat Dam (NIA); hydro power plant (SN Aboitiz)					*	
4	Do the local communities in the municipality get involved in tourism governance?				*			
5	Are the civil service bodies and Non-Government Organizations involved in the area's tourism governance?							*
6	Is there an effect of local-level structures working with tourism development and does it have a voice in raising concerns?			*				
7	Is there sufficient training for skill development regarding tourism sustainability in the local governance? Do they have essential experience in the field?					*		

collection system is the change of leadership/governance. Still, it can be acknowledged that Ramon covers both international and domestic visitors based on the Tourism Officer's three (3) years of data.

The municipality allots a budget for tourism-related infrastructure investments regarding the extent of tourism and business environment's conduciveness. As observed, food and beverage establishments are increasing in number. However, the municipality needs to boost its marketing strategies to encourage investments in tourism awareness in general. The issue of strategic marketing plans' unavailability will affect stakeholders' engagement, which is a barrier for the municipality's brand, marketing, and product positioning in terms of tourism.

Regarding Ramon's competitiveness, the diversity and quality of the products it offers could benefit more from improvement and ensuring quality standards. For risk and crisis management, ReBUILD Project directly targets the said issue. However, it is not directly related to tourism. Also, the municipality needs proper monitoring and implementation in terms of the security and safety of the tourists.

Pillar 3—Livelihood, Decent Work Condition, and Human Resources

Other municipalities in Region 2 can learn from Ramon regarding employment and a decent work environment. They promote gender equality in their workforce, allowing anyone to be employed if they can prove that they are fit and can perform the job successfully, as stated by Tolentino during a virtual interview.

Consultation and dialog in human resources are quite observed in the municipality where employees hold training, although the occupations are not directly associated with tourism. During a virtual interview, Tolentino briefly discussed that the employees' safety is a priority, in line with the workers' rights and decent working conditions. The municipality also provides job opportunities for locals, especially those residing in nearby attractions.

In addressing training needs and skill gaps, the Department of Tourism (DOT) Region 2 facilitates frequent training for tourism-related occupations in the municipality. About the interview regarding deficiency in labor

Table 12.5 Pillar 2—Economic competitiveness, performance, and investment (United Nations World Tourism Organization, n.d.-c)

Measuring Tourism Contribution to the Economy of Municipality								
	Question	Comment	Priority for Action					
			High			Low		
			1	2	3	4	5	X
Issue 1: The Data Collection and Analysis's Value to Growth								
1	Is the collection of tourist arrivals, activities, and profiles frequent and detailed?	Data from the Municipal Tourism Officer is lacking and quite challenging to interpret		*				
2	Are both international and local visitors sufficiently covered in the data collection? What is their purpose of visit?					*		
3	Is there collected information regarding the performance of the tourism sector and its employees in the area?					*		
4	What are the hindering factors that may be affecting the effectiveness of data collection?	Change of governance		*				
5	What are the analyses currently used to estimate the contribution of tourism? Does it include TSA?							*
6	Are the TSA's requirements and needs being understood?							*
7	What are the factors preventing the development of a TSA in the locality?							*
8	Can the present tourism data be sufficiently separated to analyze the current issues in sustainability?	Lacking credible sources		*				
9	Is there robust data collection, and has it been prioritized for external validations?			*				
The Investment, Trade, and Business Environment in the Municipality								
Issue 1: Recognition and Treatment of Tourism in Investment and Trade Policies and Commitments								
1	Are the tourism policies concerning trade and investment promotion reflected comprehensively in the national policies?	One Town One Product (OTOP) center is being observed			*			
2	Do the policies regarding trade and investment of the Philippines incorporated in the tourism policy?	Executive Order No. 175, S. 2003 (SMEs)			*			
3	Has the municipality made international trade commitments concerning tourism development?							*
4	Do trade duties value sustainable tourism in terms of impacts on the environment, society, and local economy?							*
5	Regarding the principles of sustainable tourism, do FDI policies encourage investment?							*
6	In encouraging tourism arrivals, do the current visa policies and procedures sufficient?							*
Issue 2: Local Tourism Investment and Business Environment's Conduciveness in the Municipality								
1	From what sources have the municipality recently gotten its investments for significant tourism establishment development?	Municipal budget			*			
2	In recent years, have there been significant tourism facilities/business investments in the municipality? From what sources?	More on food and beverage establishments are observed			*			
3	Were there actions in place to support and promote the growing tourism investments in the municipality?	A comprehensive marketing strategy is not observed	*					
4	What are the primary challenges seen in the tourism development and operation of tourism-related establishments?	A comprehensive marketing strategy is not observed	*					
5	Is financing accessible for tourism businesses' operations, development, and promotions?				*			

(continued)

Table 12.5 (continued)

Measuring Tourism Contribution to the Economy of Municipality								
	Question	Comment	Priority for Action					
			High			Low		
			1	2	3	4	5	X
6	Conduciveness of doing tourism-related business in general; where can most of the challenges be seen?				*			
7	Are there actions being observed toward strengthening the business linkages of tourism to other significant industries?				*			
Product Branding, Positioning, and Marketing in the Municipality								
Issue 1: The Brand and Marketing Plan's Target and Scope								
1	Does the municipality have a recognized brand identity? Is it appropriately maintained?	There is no established branding observed	*					
2	Does the municipality have a marketing plan to identify the target market to plan a promotional framework?	No strategic marketing plan for tourism	*					
3	Does the major private and public sector's participation in implementing the marketing plan observe?	No strategic marketing plan for tourism	*					
4	Is there a full implementation of the marketing plan in the municipality? What are the challenges observed?	No strategic marketing plan for tourism	*					
5	Are social media, e-marketing, and other technology utilized in the current marketing activity?	Marketing is being observed, but it must be further enhanced		*				
Issue 2: Diversity and Quality of Municipality's Product Offerings								
1	Is the consistency of the tourism product offering and services offered in the area sufficient and improving?				*			
2	Are there effective methods currently used regarding inspection, setting, and reporting standards?				*			
3	Are there efforts being made to identify product gaps and improve its diversification?	There is no municipal TDP right now		*				
Risk Management, Security, and Resiliency of the Municipality								
Issue 1: Crisis and Risk Management's Level of Awareness								
1	Are risk and crisis management plans observed in the tourism development strategies and planning?	There are general plans, but not directly aligned for tourism (e.g., ReBUILD)		*				
2	What are concrete measures being taken to provide a secure environment for tourists' well-being?	Needs proper implementation and monitoring		*				
3	Are there recorded accounts regarding risk spreading and dependency avoidance in selecting products and markets?							*
4	Are there concrete plans developed to handle such situations during emergencies and crises?		*					

availability in the tourism sector, Tolentino indicated in a virtual interview that there are permanent, freelance, and on-call tour guides in Ramon. Training conducted in the municipality is usually in collaboration with TESDA and its offered courses. Other than that, no other private sector was mentioned.

Pillar 4—Poverty Diminution and Social Involvement

The involvement of poor communities in tourism is somehow evident since Tolentino noted that everyone could be employed for tourism-related occupations if they willingly do so. Continuing, the connection between tourism development relating to agriculture is observed since the province of Isabela, where Ramon belongs, is dubbed "*Rice Granary of Luzon.*" With this note, seldomly, informal trading is observed in Ramon through unidentified underaged tour guides based on observation. The engagement of poor communities in providing tourism facilities is nonexistent, so it must be highly prioritized to strengthen pro-poor initiatives. Fundings for MSMEs who qualify under Section 2 of Executive Order No. 175, S. 2003 are and can be further supported by Ramon (Official Gazette of the Philippines, 2003).

Table 12.6 Pillar 3—livelihood, decent work condition, and human resources (United Nations World Tourism Organization, n.d.-c)

Working Conditions and Human Resources Planning								
	Question	Comment	Priority for Action					
			High			Low		
			1	2	3	4	5	X
Issue 1: The Existing Human Resource Plans, Policies, and Actions in Collaboration with the Private Sector and Relevant Stakeholders								
1	Among all plans and policies in human resources, is tourism exhibited?	Tolentino (2021), the municipality is promoting gender equality				*		
2	Is there a clear plan and policy in human resources which can be connected to a Development Plan/Tourism Policy?	There is no municipal TDP available as of the making of this study	*					
3	Are there any existing organizations representing workers or trade unions in the tourism sector?							*
4	Do parties that stand for the private sector's tourism enterprises handle issues in human resources?							*
5	Do discussions in HR Planning take place with stakeholder spokespersons?	There were trainings conducted but not directly related to tourism		*				
6	Are there any strategies to foster employment, including specific groups? (e.g., incentives and initiatives)				*			
Issue 2: Safeguarding the Rights and Conditions of Workers in the Municipality								
1	Does the tourism sector comply with decent work conditions and value workers' rights?	Tolentino (2021), employees' safety is a priority				*		
2	Does legislation serve its purpose in necessitating and imposing decent work in the tourism sector?	Tolentino (2021), tourism-related jobs are available for locals				*		
3	Are relevant labor laws and regulations effectively enforced in the tourism sector?	As observed, unaccounted minor tour guides seldom approach tourists			*			
Skills Assessment and the Provision of Training								
Issue 1: The Degree of Understanding of Training Needs and Skills Gaps								
1	Is there a skill gap assessment and training plan for the current industry and future deployment?	DOT Region 2 often conducts trainings for tourism jobs in Ramon			*			
2	Is there a shortage of competent tourism employees present, and if so, in what areas?	Marketing and promoting the area		*				
3	Is there a shortage of available workers with the appropriate abilities, and if so, in what areas?	There are permanent and on-call tour guides				*		
4	Is there a perceived deficiency in sustainable tourism knowledge and corresponding practical skills in the municipality?			*				
Issue 2: The Readiness of Training and Capacity-Building Programs, Standards, and Institutions in the Municipality								
1	Is there a comprehensive and methodical evaluation of tourist training available?	Currently, it is more inclined with TESDA courses			*			
2	Is it potential to get quality accreditation for tourist training institutes and courses?	Currently, it is more inclined with TESDA courses			*			
3	Is there enough competence and availability of training institutions, bodies, and teachers to meet the sector's needs?	More trainings directly related to tourism		*				
4	Do training institutes have enough resources such as skilled personnel and equipment?	The current Tourism Officer of Ramon is initially based on the agriculture sector			*			
5	Is the range of administrative and technical courses provided at various levels sufficient for the sector's demands?	Currently, it is more inclined with TESDA courses		*				
6	Is there a standard skill requirement and certification appropriate to the needs of the tourism industry?				*			

(continued)

Table 12.6 (continued)

Working Conditions and Human Resources Planning								
	Question	Comment	Priority for Action					
			High			Low		
			1	2	3	4	5	X
7	Are the youth educated in tourism training and opportunities as a future career?		*					
8	Is proper training in tourist sustainability issues and related skills accessible and provided?	DOT Region 2 often conducts trainings for tourism jobs in Ramon		*				
Issue 3: The Private Sector's Level of Commitment to Capacity Development and Training								
1	Has the private sector been involved in designing and delivering training courses and evaluating skills?				*			
2	Is there adequate career development and training for employees and the local community in the private sector?				*			

Poor communities are aided by taxes and charges on tourism businesses, as the budget is being allocated for Ramon's neighborhoods, asserted by Tolentino during a virtual interview. However, initiatives must be improved regarding the disadvantaged tourism groups (young generations, the elderly, and the disabled). In contrast, gender equality was emphasized during the topic of inclusion in the tourism sector. Local communities are informed when developments concerning the tourism sector occur in their areas, since barangay officials also take part in the said operations. Nevertheless, there should still be an increased effort to enhance the knowledge and understanding of locals regarding the benefits of tourism.

Lastly, when it comes to the impacts on natural resources, Tolentino reassured during a virtual interview that the attractions are monitored, and there are no reports involving crime in tourism operations, as mentioned. However, there are no credible online articles to strengthen the claims.

4 Recommendations

The thirteen (13) independent recommendations identified by the researchers are presented, patterned with various analyses, specifically SWOT Analysis, PEST Analysis, and UNWTO Destination Assessment Matrix that will aid Ramon in developing its tourism sector encompassing economic, social, and environmental factors.

1. **Develop a 5-Year Tourism Development Plan (TDP) for Ramon, Isabela, Highlighting Sustainable Tourism Development**
   The comprehensive TDP will enormously aid the municipality in developing its tourism sector, creating its tourism policy, and ensuring sustainable development. The researchers recommend the following scopes: (1) assessment of tourism supply and demand, (2) the setting of goals and objectives, (3) zoning, (4) essential infrastructure, financial, and human resource development, and (5) marketing and promotion. The main objectives are to organize and develop Ramon aligned with the Cagayan Valley Regional Development Plan (CVRDP) for the prosperity and sustainability of the municipality's tourism. In addition, it is to ensure the market viability of the attractions to be promoted and developed by Ramon.
   The TDP will include sustainable tourism and advancement concerning progressive practices to ensure that every task and project are inclined with the seventeen (17) Sustainable Development Goals (SDGs). In addition, the researchers recommend that the local tourism office emphasizes the third (3rd) Good Health and Well-being and fourth (4th) Quality Education SDGs. Moreover, the planners must conduct extensive research that the Local Tourism Officer will spearhead, and the researchers recommend that they study and evaluate other destinations in the region through benchmarking.
2. **Establishment of a Standalone Tourism Office for the Municipality of Ramon**
   For Ramon to be a competitive tourist destination in Isabela, the researchers recommend that the municipality must establish its local tourism office and appoint a permanent Municipal Tourism Officer who is capable, competent, and qualified in pursual with Section 42 of Republic Act No. 9593 or Tourism Act of 2009 (Official Gazette of the Philippines, 2009). The appointed Municipal Tourism Officer must distinguish Ramon as a tourist destination from its neighboring cities/municipalities that possess similar geographical features and site attractions. In connection with the previous recommendation, he/she must be knowledgeable about sustainable tourism or utilize this study to better aid the essential authors, including him/her, in formulating the TDP for Ramon. The tourism sector of the municipality

Table 12.7 Pillar 4—poverty diminution and social involvement (United Nations World Tourism Organization, n.d.-c)

A Comprehensive Strategy for Alleviating Poverty Through Tourism								
	Question	Comment	Priority for Action					
			High			Low		
			1	2	3	4	5	X
Issue 1: The Degree of Involvement to Pro-poor Tourism								
1	Is tourism recognized as a platform for alleviating poverty in National Strategy Papers or Poverty Reduction Strategy Papers (PRSP)?							*
2	Is poverty alleviation a goal of tourism? Does it involve actions and produce quantifiable results?	There is no municipal TDP right now						*
3	How well-informed is the tourism industry about pro-poor concerns and approaches?	There is no municipal TDP right now						*
4	How committed is the private sector to working with local citizens and promoting pro-poor tourism?	There is no municipal TDP right now						*
Issue 2: Instigating a National and Destination-Level Approach for Easing Poverty Through Tourism								
1	Has any national action been taken to enhance the pro-poor tourism approach?							*
2	Is there an approach for developing rural tourism or some types of destinations focusing on pro-poor?							*
3	Is there any deliberation given to destinations identified for pro-poor tourism initiatives?							*
4	Is there any specific attraction implementing pro-poor tourism approaches and/or activities?							*
5	Has a study been conducted to determine the percentage of tourist expenditure that reaches the marginalized and under what aspects?							*
Initiatives to Strengthen Pro-poor Tourism								
Issue 1: Increasing the Amount of Financial Support to Disadvantaged Communities Through Trading and Tourism Businesses								
1	Have efforts been made to increase tourism employment in disadvantaged communities?	Tolentino (2021) stated that they are employing all capable individuals for tourism jobs			*			
2	To what extent is the connection between agriculture and aquaculture to the tourism industry?						*	
3	When it comes to tourism supply chains with poor communities, how well developed are these linkages at a local level?			*				
4	Is there extensive informal trading by the poor in a tourist destination, and has it been managed?	As observed, unaccounted minor tour guides seldom approach tourists			*			
5	Are poor communities involved in providing tourism facilities and visitor arrivals, and is there a possibility of improving or increasing this engagement?		*					
6	Are there programs to educate and support poor communities in developing, operating, and marketing tourism businesses?							*
7	Is financial assistance available MSMEs through microfinance or other programs?	E.O. No. 175, S. 2003 Sect. 2			*			
Issue 2: Ensuring That Tourism Favors the Wider Population								
1	Are taxation and charges on tourism businesses and visitors assisting marginalized communities, and may this be expanded?				*			
2	Are there any outreach donations present within the tourism sector to help marginalized communities?				*			
3	Is the impact of modern tourism infrastructure on marginalized communities considered in its planning?				*			

(continued)

Table 12.7 (continued)

A Comprehensive Strategy for Alleviating Poverty Through Tourism								
	Question	Comment	Priority for Action					
			High			Low		
			1	2	3	4	5	X
Inclusion of Marginalized Communities in the Tourism Industry								
Issue 1: An Attempt to Involve Marginalized Communities in Tourism								
1	Are the needs of the marginalized communities considered in tourism strategies and policies?				*			
2	Are legislations in place to protect the rights and needs of the marginalized groups?	There is no municipal TDP right now						*
3	How successful are women inclusion into the tourism industry, and what initiatives have been taken to improve their status and opportunities?	Tolentino (2021) stated that they are practicing gender equality in the workforce					*	
4	How successful are youth inclusion into the tourism industry, and what initiatives have been taken to improve their status and opportunities?	Currently, it is more inclined with TESDA courses			*			
5	How successful are elderly inclusion into the tourism industry, and what initiatives have been taken to improve their status and opportunities?				*			
6	How successful are disabled person inclusion into the tourism industry, and what initiatives have been taken to improve their status and opportunities?				*			
7	How successful are ethnic minority inclusion into the tourism industry, and what initiatives have been taken to improve their status and opportunities?				*			
Preventing Adverse Social Outcomes								
Issue 1: Efforts to Recognize and Mitigate Harmful Societal Consequences								
1	Are residents consulted about the operation and development of tourism in their community?					*		
2	Are residents' views and opinions included and empowered about tourism operation and development in their communities?	Barangay officials are working in tourism development				*		
3	Are residents' perspectives on the impact of tourism on their livelihoods addressed and monitored?	Thoughts			*			
4	Is the access to resources particular to land considered for local populations in planning and managing?	Attractions are monitored			*			
5	Concerning the issues of detrimental impacts on resources, to what extent are residents considered in the tourism development?				*			
6	Is there proof that tourism is causing an increase in crime and other social problems, such as sexual exploitation, and what steps are being taken?						*	
7	How knowledgeable and attentive is the private sector about tourism's negative social consequences?				*			

will become independent in enhancing its plans for the development of the attractions, in which approvals and implementations will become more efficient.

3. **Publication of Tourism Journals of Tourism Attractions and Activities for the Database of Ramon, Isabela**
   The researchers recommend that this sustainable tourism study by the students from the College of Tourism and Hospitality Management of the University of Santo Tomas—Manila be included in the suggested database. In addition, planners must create a comprehensive inventory sheet containing all but not limited to the potential attractions for development. The data collected should be accurate and well-researched to avoid misinformation. Activities must be thoroughly designed and deliberated to avoid unprecedented accessibility and safety issues. In this manner, it will assist the municipality in identifying which interests should be

given greater attention. In case of a change in governance, the inventory sheet should be handed over to the relevant manager or developers, such as the Assistant Tourism Officer or the newly appointed Municipal Tourism Officer. Such essential documents must be stored or appropriately kept, avoiding mishandling.

4. **Strengthening Agriculture and Aquaculture Through the Establishment of a Cooperative in Line with the TDP of Ramon While Promoting Sustainable Tourism Practices**
   Being part of the major rice and corn contributors in Region 2, Ramon can align its development of tourism attractions with the agricultural sector. Aside from agriculture, aquaculture can also be improved, specifically near Magat Dam Tourism Complex and Ramon Rolling Hills. There should be an established cooperative in partnership with the local tourism office to integrate fishponds in the nearby attractions in which, based on researchers' observation, plenty of fishponds are present in the municipality, especially in Barangays of Planas and General Aguinaldo where Ramon Rolling Hills and Magat Dam Tourism Complex are situated. With its integration, tourists will have an opportunity to get their food (live fish) cooked in a designated area approved and accredited by the proper local authorities. A new proposed tourism activity, through the integration of rice, harvesting especially during peak season (month of December) (International Rice Research Institute, 2013) can help the municipality on a social scale as the abovementioned could encourage more people to visit Ramon especially those who have not experienced harvesting yet. However, only designated rice fields will be used for the proposed activity to avoid disrupting the farmers' operations. The local tourism planners must ascertain the activity's viability. With all this in mind, the researchers recommend that preemptive measures such as limitation of daily tourist arrivals in the rice and corn fields must strictly be observed, especially during the high months of harvest. Also, there should be an environmental fee that must be collected and an initial activity fee to have more funding in preserving and maintaining the farm attraction.
5. **Rethinking Transportation Within the Neighboring Attractions of Ramon, Isabela**
   The researchers recommend using bicycles as alternative transportation and as an engaging tourism activity in Ramon, specifically in the attractions adjacent to each other, and protecting the environment and people's health. Thus, the local tourism office and the local government unit must work together to promote cycling when moving from one place to another. Both parties must emphasize the environmental benefits of riding a bike, such as reducing air and noise pollution, minimizing petroleum consumption, and generally lowering one's ecological footprint (Green, 2020).
   To make this project possible, the local tourism office must collaborate with the local government to construct trails that are safe and accessible for bikers. There should also be personnel checking both points to monitor if the bikes are well-kept and maintained, and a payment matrix should be developed to produce strategic and competitive pricing, encouraging tourists to try cycling as substitute transportation around the attraction.
6. **Standardized Training of Tourism Practitioners**
   Ramon's tourism practitioners should have a standardized service quality in line with Republic Act No. 9593 or Tourism Act of 2009. The study of Rocamora & Aguiling (2019) titled *"Competencies of Tourism Officers in the Philippines: Toward a Development of Competency Model for Managing Destinations in Public Governance"* has recommended various capacity-building activities as part of programs intended to improve their competencies. Concerning the abovementioned study, Ramon should have training for all the tourism employees that concerns adequate knowledge and skills to direct the service offered to the tourists, attributable to the tourism practitioners' confidence, competence, and efficiency.
   The following subjects were observed and stated on the list of Rocamora & Aguiling (2019) as it is deemed relevant: "*Strategic planning and tourism; Tourism statistics training; Disaster risk reduction and management; Marketing and product; Public–private partnerships and Relationship management; Cultural tourism including history; Environmental impact; Management of parks and recreation; and Personality development."*
   In line with the mentioned factors, the Tourism Officer's conviction in the municipality's offering will be reflected in their interactions with tourists, leading to a greater probability of tourism growth in the area and tourist satisfaction. These skills and knowledge positively reveal Ramon, allowing private sectors to invest in the area.
   The Department of Tourism (DOT) launched a virtual educational course for the tourism stakeholders of the Philippines, to aid in staying significant, to understand how to manage the challenges posed by COVID-19, and to promote the transition toward recovery (Department of Tourism, 2020). This training course was envisioned as a counterpart to the Office of Industry Manpower Development's regular training courses. Thus, Ramon must require all tourism professionals to complete the DOT "*Enhanced Opportunity*" training. The said training would assist Ramon in adapting to the tourism industry's new quality of service toward their community, catering to the needs of potential tourists.

In line with the abovementioned, participants or tourism practitioners should complete two (2) training programs: first, Tourism Enterprise, which encompasses monitoring, evaluating, and learning from past and present occurrences to prepare tourism practitioners for a better future (Department of Tourism, 2020). The second is Filipino Brand of Service, which integrates the seven Filipino values, commonly referred to as the "*7Ms*," defining that the municipality should incorporate Filipino hospitality in their value as tourism practitioners, namely (1) *May-Maylikha* (Respect for the Divine Providence), (2) *Makatao* (People-oriented), (3) *Maka-Kalikasan* (Steward of the Environment), (4) *Makabansa* (Patriotism), (5) *Masayahin* (Fun-loving), (6) *May-Bayanihan* (Collaborative), and (7) *May-Pag-Asa* (Resilient) (Callar, 2020).

7. **Developing Rice Farms as Part of the Community-Based Tourism of Barangay Planas**
   The Philippines recognized the potential of farm tourism (Philippine News Agency, 2020). Hence, one of the best regional candidates for this sector is provinces in the northern part of Luzon. Farm tourism is strengthened in the Republic Act No. 10816 or Farm Tourism Act of 2016, corresponding to developing and creating farm tourism in the country. Isabela, being part of nature-based tourism centers around low-sway, nature-based, and local area-based (community-based) exercises, including the local people in manners of their socially, culturally, and monetarily developing allows it to take part in the said legislation (Official Gazette of the Philippines, 2016).
   The researchers recommend Barangay Planas to develop its farm tourism as a bounce-back strategy. In line with the mentioned sector, such benefits for Ramon are not limited to income generation since farming contributes to ten and two hundredth percentile (10.25%) of the country's Gross Tourism Product (GTP) (Statista, 2021). Nevertheless, farm tourism and its community-based tourism concept can sustain the social and economic aspects of the area as locals gravitate toward agriculture.
8. **Strengthening Private–Public Partnership (P3)**
   The researchers highly recommend that the municipality of Ramon must strengthen its partnerships. The local tourism office must recognize the benefits of having a private–public partnership (P3). The partnership may include financial needs, human resources, and marketing, allowing the municipality to be noticed by the public in a broader scope (Sa, 2017). For instance, building facilities and amenities is a product of such a partnership. In line with strengthening the agriculture and aquaculture of Ramon, the researchers recommend that the cooperative (mentioned in the 3rd recommendation) will partner with the local government in which they can create a tourism map of Ramon, consolidating all the attractions near fishponds and crop sites that can be offered to tourists as a package.
9. **Establishing an OTOP Facility in the Municipal Proper**
   The researchers recommend putting up a supplementary *pasalubong* center (souvenir shop) in Bugallon Proper to increase the satisfaction of potential tourists. The *pasalubong* center will house the best of the province's delicacies such as *moriecos* (rice cake), *inatata, bibingkang kanin* (sticky rice cake), and *pinataro*. Popular corn products such as corn coffee and corn pastillas should also be sold to showcase the local products, which will help circulate money for the local economy. Aside from creating different delicacies, corn or its husk would make a good resource material in creating a variety of souvenir products. Corn husks are versatile, making them an interesting material to create handmade products like dolls, garlands, ribbons, and other souvenir items that locals from the province can produce. It can also be woven into dish scrubbers and be used as wrappers for food like pudding and rice-made delicacies and pastries.
10. **Reduce the Transmission of COVID-19 in the Municipality and Prioritizing Vaccination While Eliminating Fake News**
    The researchers recommend creating a long-term strategic plan of action focusing on the challenges and the effects of COVID-19 on Ramon. The tourism planners must be robust in coordinating with the Rural Health Unit (RHU), Local Disaster Risk Reduction Management Council (LDRRMC), and the Local Police Office (LPO) of Ramon.
    New guidelines must be published and implemented by the municipality, including a designated fine for a specific quarantine violation. In relation, the municipal government must emphasize information dissemination regarding the effects of COVID-19 and set an excellent example for locals to abide by the guidelines. It is highly recommended that multiple publication materials be more readable, appealing, and informative for an easier understanding of the changes that will be implemented. Publication materials must be published online and printed so that a greater pool of people can read and comprehend the material. Such materials are signages that remind tourists to follow social distancing and proper wearing of accredited personal protective gears such as the N95 mask recommended by the Department of Health (DOH) and the World Health Organization (WHO). Similarly, an article published by Harvard Medical School highlighted that to help contain the virus, the foot traffic should be managed and well-maintained (Harvard Health Publishing, 2021).

A preemptive solution for Ramon to combat COVID-19 is mass vaccination, which the local government must prioritize to attain herd immunity, gradually reducing COVID-19 cases. Listing and identifying priority groups to be vaccinated are must; however, locals must be educated about the purpose of vaccines. In contrast, adult Filipinos topped the result of a survey conducted by OCTA about vaccination in which forty-seven percentile (47%) expressed that they would not get vaccinated and nine percentile (9%) of the latter conveyed that they do not believe in vaccines (Lalu, 2021). The researchers recommend that incentives such as gift certificates and/or a small amount of cash be given to people willing to get vaccinated.

11. **Temporary Closure of All Tourism Attractions to Ensure Proper Hygiene Protocol and Safety Standard are in Place while Rehabilitating its Tourism Attractions**
The researchers highly recommend the temporary closure of Camp Vizcarra Ecology Park since it had been previously used as a quarantine facility. Aside from Camp Vizcarra, other tourist attractions are also recommended to be closed to rethink ways on how to promote health and safety protocols. Moreover, they must further enhance its facilities' sanitation and hygiene protocols to ensure the absence of COVID-19 within the vicinity.
Rehabilitation while on temporary closure will enable initiatives for sustainable tourism practices to be in place when attractions re-open. Together with the improvement of sanitation, it can be made feasible by having a concrete procedure on the systematic plan regarding the cleaning of attraction's surfaces, such as but not limited to door handles, light switches, tables, benches, and other facilities of the attraction. To further maintain safety, the researchers recommend that when the attraction is open to the public, a specific area for conducting the disinfection procedures should be monitored by facilitators. Moreover, to continuously sustain the protection from COVID-19 and safety of all individuals within the vicinity of the attractions, publication materials comprising proper hand-washing procedure, correct usage of safety equipment inside the facility, and the set of rules that are implemented should be published clearly and appealingly.
12. **Cohesive Marketing Strategy of All Attractions**
It is recommended to have a tourism website updated regularly, whereas information on attractions must be found on the website. Furthermore, among the platforms that can be utilized is the formulation of Ramon's local tourism office's official Facebook page, which must contain relevant information. Hyperlink should be included within the official website of Ramon. There must be an appointed social media manager who has essential knowledge and qualification in operating the website to have a comprehensive and appealing online presence. A team must be put together composed of but not limited to researchers, influencers, creatives, and technical teams to work collaboratively and come up with a contemporary and well-thought-out marketing strategy. In line with the pandemic, the researchers highly recommend using a Quick Response code (QR code) as part of the registration procedure when going to and from a particular attraction. By simply scanning the printed code found in the attractions, tourists can easily be tracked and checked if an unprecedented outbreak in one of the attractions occurs. The municipality can pattern its QR code with the guidelines and checklist imposed by the Inter-Agency Task Force for the Management of Emerging Infectious Diseases (IATF) and *Napanam* QR pass of La Union (Province of La Union, n.d.). Everyone logged in with a QR code is protected under the Republic Act No. 10173 or Data Privacy Act of 2012 (Official Gazette of the Philippines, 2012).
13. **Monitoring and Evaluation Procedures**
The researchers highly recommend the implementation of the Kaizen method alongside the Plan, Do, Check, Act (PDCA) cycle of William Edward Deming for frequent improvement of existing processes through a Specific, Measurable, Attainable, Relevant, and Time-bound (SMART) approach (Corrigan, 2018). The six (6)-step method includes (1) identifying potential improvement; (2) evaluating the existing approach; (3) producing original concepts; (4) enhancement of deliberated plan; (5) executing the plan; and lastly, (6) evaluate the method.

Aside from the PDCA cycle, to monitor tourist satisfaction, the researchers recommend that a questionnaire be given to tourists, which can be answered through Google Forms, printed materials, and other relevant tools for surveys. The use of the Likert Scale Questionnaire, specifically adding bipolar and unipolar questions, is encouraged. Both bipolar and unipolar must be considered when crafting survey questionnaires, wherein unipolar type questions have one pole that focuses on extraordinarily negative and positive resulting in a more neutral answer. In contrast, the bipolar type has two poles that focus on the extreme negative and positive results (Fleetwood, 2020).

5 Conclusion

As tourism becomes a priority for the continuous progress of Ramon, environmental, socio-cultural, and economic problems hinder the municipality to immensely grow.

The evidence presented in the study will futureproof Ramon in implementing its plans for its tourism development. The thirteen (13) cited recommendations of the researchers such as the Plan, Do, Check, Act (PDCA) cycle, Kaizen method for continuous improvement, and the strengthening of agriculture and aquaculture tackle the abovementioned problems extensively while being patterned with the United Nations World Tourism Organization (UNWTO) standard assessment. While two analyses were utilized specifically (1) Strengths, Weaknesses, Opportunities, and Threats (SWOT) Analysis together with (2) Political, Environmental, Social, and Technological (PEST) Analysis to determine internal and external factors affecting the tourism of Ramon, Isabela.

Therefore, if approached correctly by the local tourism office with the support of the local government unit of Ramon, Isabela, it will have a more straightforward guide in planning and implementation procedures for sustainable tourism develo, viewed through the international, national, and local perspectives. Nevertheless, the municipality must be studied further to develop more innovative ideas, widening the scope, and supporting this study's development objectives.

Acknowledgements The researchers extend their profound appreciation and gratitude to the following individuals and institutions that provided their resources in the conduct of conceptualizing and finalizing the Municipality of Ramon, Isabela's Sustainable Tourism and Strategic Development Plan 2021-2025:**Municipal Government of Ramon, Isabela**, headed by **Honorable Mayor Jesus D. Laddaran;Ms. Mabilyn Viernes B. Tolentino**, Tourism Officer, Municipal Government of Ramon, Isabela;**Associate Professor Attorney Gezzez Giezi G. Granado, DCL, CHE,** Dean, College of Tourism and Hospitality Management, University of Santo Tomas, Manila and; **Reverend Father Roland D. Mactal, OP, SThD**, Regent, College of Tourism and Hospitality Management, University of Santo Tomas, Manila.

Appendix: Product Development Strategies

The municipality of Ramon promotes eco-tourism and MICE tourism, as confirmed by Ms. Mabilyn Tolentino of the Local Tourism Office of Ramon, Isabela. There are established and emerging destinations in the municipality in which the attractions contribute to the municipality's annual income used for the development of Ramon in general. However, Ramon is currently experiencing a decline in tourism arrivals because of the Novel Coronavirus 2019 and problems linked to the environment, socio-cultural, and economy.

Table 12.8 is presented to sustain its tourism industry. Data was gathered through a virtual interview with Ms. Mabilyn Tolentino, the incumbent Tourism Officer, and further research.

Table 12.8 Product Development Strategies of Municipality of Ramon, Isabela (Tolentino, 2021)

Problems/ Negative impacts of tourism	Solutions	Current Activities/ Methodologies	Output	Economic, Environment, and Socio-cultural Benefits
Environmental				
Extreme weather conditions brought by typhoons and global warming	Rehabilitation of Camp Vizcarra Ecology Park	Community clean-up drive spearheaded by the Local Government Unit Temporary closure of Camp Vizcarra Ecology Park to check for damages brought by typhoons	Seek assistance from the PDRRMC Isabela Strengthening the Republic Act No. 10121 (The LAWPhil, 2010) Participation of youth in the seminars	Uninterrupted water flows in Camp Vizcarra Ecology Park Residents will be given employment Local youths will become more knowledgeable about the effects of natural disasters
Attractions are not well-maintained	Creating municipal ordinances Installation of water filtration system Development of solid waste management program	Implementation of Municipal Ordinances No.12 and 13 of 2017 Installation of a sound water system Inter-barangay cleanliness competition	A task force from Ramon's LGU monitors prohibited food containers Violators will be penalized Free elf trucks are given to each barangay	It will incredibly reduce the waste in the landfill of Ramon Increased tourist arrivals of the municipality Perception toward cleanliness will improve
Landlocked municipality	Development of tourism attractions	The tourism office highlights eco-tourism attractions and MICE to suffice the lack of 3S tourism Development of Magat Dam Tourism Complex	Ramon is currently developing the Rolling Hills Camp Vizcarra Ecology Park and Water World Grand Resort are the MICE venues of the municipality More greens will be planted like trees and plants	The municipality will increase the number of trees in Magat Dam by replanting it, deriving from other areas Tourism-related jobs will be offered to locals People will become more appreciative of what the municipality can offer

(continued)

Table 12.8 (continued)

Problems/ Negative impacts of tourism	Solutions	Current Activities/ Methodologies	Output	Economic, Environment, and Socio-cultural Benefits
Socio-cultural				
Lack of tourism awareness	Emphasizing the benefits of tourism to locals Formulation of TDP	Encouraging different barangays to identify attractions The local tourism office has identified the planning team for the TDP	Focusing on the eco-tourism development Emphasizing sustainable development Development of TDP	New attractions and facilities will be developed in the area More investors will be encouraged to take part in the development of Ramon, Isabela
Insufficient health and safety protocols	Creation of Local Executive Orders Reducing tourism-related injuries, especially in Camp Vizcarra Ecology Park	Implementation of Executive Order No. 11, S., 2021 Implementation of guidelines for the tourism attractions	Travel restrictions have been imposed Prohibition of swimming in Camp Vizcarra Ecology Park since 2014	Fewer tourists will be flocking to a single tourism attraction Reevaluate the ordinances and guidelines for tourist arrivals It will aid in avoiding accidents relating to aquatic activities
Economic				
Insufficient marketing strategy	Social media and word-of-mouth marketing	Multiple Facebook accounts and pages are present for viewing	Posting of the different attractions in Ramon The latest news and information about the municipality are seldom posted	Easier delivery of information toward the public Engagement in social media posting will increase, reducing the cost of advertisement Locals will become more cooperative in promoting the attractions
Lack of proper/quality training	The assistance of the Regional Tourism Office of Isabela	There are some instances that the Regional Tourism Office is conducting hospitality training for local tour guides Courses such as proper tour guiding, speech, posture, and the like constitute the training	Standardize training procedures for the locals of Ramon Formulating a checklist as a monitor for proper training and evaluation of tourism employees Employees will become more knowledgeable and skilled	More awareness of how to properly maintain and develop the environment while conducting tours Tourists will be satisfied More locals will have the opportunity to work Respect toward the culture of other people visiting the attractions will enhance
Limited facilities and activities	Development of facilities, infrastructure, and tourist activities through strategic planning	The local government is supportive of the development of the attractions Road rehabilitation and construction are primarily carried out in Rolling Hills The local tourism office is formulating water activities	Development of TDP Partnership for the development of the Magat Dam Tourism Complex Positioning Ramon as a family-friendly tourism destination Construction of gazebos and function halls for significant events	Environmentally sound and acceptable practices will reduce the degradation of tourism attractions Tourists will become more eager to visit the municipality More investors will become interested To improve the lifestyle of locals More activities can be done in the attractions

References

BusinessMirror. (2018). *Isabela's Bountiful farm-tourism destinations rediscovered*. https://businessmirror.com.ph/2018/07/01/isabelas-bountiful-farm-tourism-destinations-rediscovered/

Callar RR. (2020). *Filipino Brand of Service: 7 Filipino Values During COVID 19 Crisis*. Lyceum of the Philippines University Cavite Campus. https://cavite.lpu.edu.ph/index.php/en/cithm-news-events/677-filipino-brand-of-service-7-filipino-values-during-covid-19-crisis

Campos V. (2020). *Puyat pushes development of Isabela tourism, eyes more festivities*. Manila Standard. https://manilastandard.net/business/biz-plus/316220/puyat-pushes-development-of-isabela-tourism-eyes-more-festivities.html

Corrigan, L. (2018). *How to use Kaizen Methodology to improve business processes*. Lucidchart Blog. https://www.lucidchart.com/blog/kaizen-methodology

Department of Tourism. (2020). *DOT offers "enhanced opportunity" training for tourism stakeholders*. http://tourism.gov.ph/news_features/enhancedopportunitytrainingfortourismstakeholders.aspx

Edale Jr., M. (2019, December 3). *Isabela transport coop get 15 modernized jeepneys*. Philippine Information Agency. https://pia.gov.ph/news/articles/1030922

Fleetwood D. (2020). Unipolar vs Bipolar Likert Scale questions. QuestionPro. https://www.questionpro.com/blog/unipolar-likert-scale/

Green, K. (2020). *5 health and environmental benefits of riding your bike*. Biofriendly Planet|For a Cooler Environment. https://biofriendlyplanet.com/green-alternatives/5-health-and-environmental-benefits-of-riding-your-bike/

Harvard Health Publishing. (2021). *Preventing the spread of the coronavirus*. Harvard Health. https://www.health.harvard.edu/diseases-and-conditions/preventing-the-spread-of-the-coronavirus

International Rice Research Institute. (2013). *Rice harvesting calendar —Philippines*. Flickr. https://www.flickr.com/photos/ricephotos/10924022186/

Lagasca, C. (2005). *Siltation, illegal tree-cutting imperil Isabela's Magat Dam*. Philstar.Com. https://www.philstar.com/nation/2005/10/15/301856/siltation-illegal-tree-cutting-imperil-isabela146s-magat-dam

Lalu, G. P. (2021). *46% of adult Filipinos unwilling to get COVID-19 vaccine; only 19% willing*. INQUIRER.Net. https://newsinfo.inquirer.net/1399844/only-19-of-adult-pinoys-ok-with-getting-covid-19-vaccines-46-not-willing

Mariano, P. M. (2019). *Modernizing public transport in the Philippines*. Changing Transport. https://www.changing-transport.org/modernizing-public-transport-in-the-philippines/

National Economic and Development Authority. (n.d.). *Regional physical framework plan 2001–2030*. http://neda.rdc2.gov.ph/?page_id=394

National Economic and Development Authority. (2017a). *Cagayan Valley Regional Development Plan 2017a—2022*. http://www.neda.gov.ph/wp-content/uploads/2018/02/2-Cagayan-Valley-RDP-2017a-2022-19Jan18.pdf

National Economic and Development Authority. (2017b). *Cagayan Valley Development Plan to focus on environment, agri-tourism*. http://www.neda.gov.ph/cagayan-valley-development-plan-to-focus-on-environment-agri-tourism/

Nepomuceno, P. (2020). *7 consecutive typhoons caused Cagayan flooding*. Philippine News Agency. https://www.pna.gov.ph/articles/1121856

Official Gazette of the Philippines. (2003). *Executive Order No. 176, s. 2003*. GOVPH. https://www.officialgazette.gov.ph/2003/02/11/executive-order-no-176-s-2003/

Official Gazette of the Philippines. (2009). *Republic Act No. 9593*. GOVPH. https://www.officialgazette.gov.ph/2009/05/12/republic-act-no-9593-s-2009/

Official Gazette of the Philippines. (2012). *Republic Act No. 10173*. GOVPH. https://www.officialgazette.gov.ph/2012/08/15/republic-act-no-10173

Official Gazette of the Philippines. (2016). *Republic Act No. 10816*. GOVPH. https://www.officialgazette.gov.ph/2016/05/16/republic-act-no-10816/

O'Neill, A. (2021). *Share of economic sectors in the GDP in Philippines 2019*. Statista. https://www.statista.com/statistics/578787/share-of-economic-sectors-in-the-gdp-in-philippines

Oxford Business Group. (2015). *The Philippines has an attractive MICE segment*. https://oxfordbusinessgroup.com/analysis/let%e2%80%99s-meet-attractive-mice-segment

Palaubsaon M., & Pareja, M. (2009). *Big boost to industry: Arroyo signs tourism act*. Philstar.Com. https://www.philstar.com/cebu-news/2009/05/13/466766/big-boost-industry-arroyo-signs-tourism-act

Parinas Sr., J, (2017). *Municipal Ordinance No.13–2017*.

Parinas Sr., J., & Honorio, G. (2017). *Municipal Ordinance No.12—*2017.

Peña, D. (2020). *As cycling booms during pandemic, advocates pedal toward sustainable transport*. Rappler. https://www.rappler.com/moveph/cycling-advocates-pedal-sustainable-transport-forward

Perante II L. (2017). *Cagayan Valley: Grains granary of the Philippines*. BusinessMirror. https://businessmirror.com.ph/2017/10/01/cagayan-valley-grains-granary-of-the-philippines/

PhilAtlas. (n.d.). *Ramon, Isabela Profile*. PhilAtlas. https://www.philatlas.com/luzon/r02/isabela/ramon.html

Philippine Book Records—GMA News TV. (2013). *1st largest multi-purpose dam in Southeast Asia*. Philippine Book of Records [Video]. YouTube. https://www.youtube.com/watch?v=b9p997j7mXU

Philippine News Agency. (2020). *DOT, FAO team up for sustainable farm tourism in PH*. https://www.pna.gov.ph/articles/1109133

Philippine Toyota. (2020). *Jeepney modernization debate explained: Pros & Cons & latest updates*. Philtoyota.Com. https://philtoyota.com/stories/jeepney-modernization-debate-explained-pros-cons-latest-updates-str56

Pilapil-Añasco, C., & Lizada, J. (2014). Philippine tourism: Evolution towards Sustainability. In *SHS Web of Conferences*. https://www.shs-conferences.org/articles/shsconf/abs/2014/09/shsconf_4ictr2014_01032/shsconf_4ictr2014_01032.html

Province of Isabela. (2013). *Official website of the Province of Isabela —Economic Profile*. https://provinceofisabela.ph/index.php/doing-business-in-isabela/2013-07-09-01-13-03

Province of La Union. (n.d.). *Napanam QR Pass*. Launion.Gov.Ph. https://napanam.launion.gov.ph/

Prudencio, M. (2018). *Fish caging flourishes anew in Magat Dam*. Philippine News Agency. https://www.pna.gov.ph/articles/1047692

Pulido-Fernández, & J., López-Sánchez, Y. (2016). *Are tourists really willing to pay more for sustainable destinations?* ResearchGate. https://www.researchgate.net/publication/311248153

Rocamora, J., Aguiling, H. (2020). Competencies of tourism officers in the Philippines: Toward a development of competency model for managing destinations in public governance. *Review of Integrative Business and Economics Research, 9*(3), 20–66. http://www.buscompress.com/uploads/3/4/9/8/34980536/riber_9-s3_04_h19-111_20-66.pdf

Sa, I. (2017). *How do you build effective public-private partnerships?* Yale Insights. https://insights.som.yale.edu/insights/how-do-you-build-effective-public-private-partnerships

Statista. (2021). *Agriculture in the Philippines—Statistics & facts*. https://www.statista.com/topics/5744/agriculture-industry-in-the-philippines/

Sustainabletourism.net. (n.d.). *Sustainable tourism*. https://sustainabletourism.net/sustainable-tourism/

Taga Ramon Nak. (2021). *Latest update on COVID cases* [Update]. Facebook. https://www.facebook.com/unsupportedbrowser

Towers, L. (2013). *How to become data-driven in aquaculture*. The Fish Site. https://thefishsite.com/articles/new-fish-cage-operations-technique-in-proposed-in-magat

Tolentino (2021). *Product development strategies of Municipality of Ramon, Isabela* [Unpublished].

Travel to the Philippines. (2018). *The powerful Magat Dam in Isabela*. http://www.traveltothephilippines.info/2012/11/22/the-powerful-magat-dam-in-isabela/

United Nations. (2017). *Millennium development Goals—MDGs*. Millennium Development Goals. https://www.mdgmonitor.org/millennium-development-goals/

United Nations World Tourism Organization. (n.d.-c). *EU guidebook on sustainable tourism for development*. UNWTO. https://www.unwto.org/EU-guidebook-on-sustainable-tourism-for-development

Designing Conservation: From Sustainable Heritage to Sustainable Tourism (and Vice Versa)

Marco Acri

Abstract

Today, cultural heritage preservation benefits from the growing international interests in the green and circular economy. Such benefit depends mostly on the very nature of our cultural heritage that has been evolving by relying on traditional techniques, regional materials, and local expertise and knowledge. For such, traditional heritage has been always more sustainable in terms of resources consumption and preservation actions needed, which were in most cases based on periodic maintenance and easy replacement of parts, with relatively low impact on the environment. In this light, cultural tourism is more and more contributing to or taking advantage of, an increasing aesthetic, heritage community that is aware of the importance of sustainable consumption as well as sustainable accommodation. To enforce such mental and behavioural change, beyond a solid institutional and legal framework for heritage protection aligned with the international standards and with the doctrine, the role of the architect is essential. The project is indeed the form by which the intrinsic qualities of the built environment and cultural heritage in general, including movable objects from crafts and arts, may enhance the true experience, as well as entrust the *genius loci*, inspiring so multiple other services that favour a sustainable attitude of users. A good project is indeed capable to reinterpret the tradition without mystifying it and displaying how new "residential" models are possible, with a complete, true experience of the site. By using theoretical achievements from recent research on nature-based solutions and circular economy on cultural heritage and urban regeneration, as well as examples from the practice, the paper will highlight some of the principles behind the relationship among heritage reuse, tourism, and sustainability.

Keywords

Adequate conservation • Circular economy • Conservation design • Dense cultural experience • Integration of primary and secondary cultural tourism products

1 Introduction

The theorization of the tourism phenomenon, lately leading to the specific sphere of cultural tourism, has always stressed the existence of one generating condition: the attraction, namely the reason to move. Such attraction has been always considered unique and, as such, scarce, irreplaceable, fragile. What is indeed called the primary touristic product (Burdet, 2018; Russo & Van Der Borg, 2002). Primary touristic products are characterized by uniqueness, scarcity, and tendency to depletion by (over)use, with the consequent need to put in place either conservation actions or replication or substitutions in the long run. If we look back to the history at the early stages of "cultural" tourism, witnessed by the eighteenth-century Grand Tour with its archaeological attractions (Bertrand, 2008; DeSeta, 1982) or by the diffusion of the Palace Hotels between the nineteenth and the twentieth century even in non-cultural destinations, like lakes, mountains, seaside, generating the initial second housing effects by the construction of villas, as in the Venetian lido or the Croatian Opatija during the Austro-Hungarian period. In most cases, the primary attraction was an asset or a homogeneous set of assets that were characterized by uniqueness and cultural or natural richness to motivate a journey. The theory of tourism also underlines the importance of the second segment of tourism products, called "secondary tourism products" that are those enabling the enjoyment of the primary ones, and include the hotels, the restaurants, the cafes, the shops, that highly contribute to the full experience of the attraction. We too often underestimate the impact of these products on the

M. Acri (✉)
University of Nova Gorica, Nova Gorica, Slovenia
e-mail: marco.acri@ung.si

A. Mandić et al. (eds.), *Cultural Sustainable Tourism*, Advances in Science, Technology & Innovation,
https://doi.org/10.1007/978-3-031-10800-6_13

overall quality of the experience, but mostly we too often forget that these products are becoming more and more integrated into the primary products, if not fully part of it. Recalling the origins of the tourism industry, all the hotels and infrastructures that were originally built to permit the experience of the attraction became attractions themselves, as the Austro-Hungarian hotels, villas, and promenades in the Adriatic coast, as much as the grand hotels in Rome, Athens, Venice, and Naples that hosted stories, narratives, important cultural figures of the past. In cultural tourism, considering the extension of the concept of heritage from single objects, as monuments and archaeological sites initially, to cultural landscapes, historic towns, and now historic urban landscape later (Bandarin & Oers, 2012; CoE, 2000; Dobričić & Acri, 2018; UNESCO, 1976), we should have enlarged the number of factors that influence the cultural experience, among which the secondary and ancillary tourism products. In this respect, regardless of the very definition of culture as given for example by UNESCO and of the consolidated fact that in a global society individuals are tourists longer than in their travel (Urry, 1990), secondary and ancillary tourism products are less investigated as contributors to the overall cultural experience. On one side, we assist to the increase will of destinations to communicate their specificity and uniqueness, which can be addressed by the concepts of *genius loci* (Markeviciene, 2008; Norberg-Schulz, 1992) and the territorial identity (which does not correspond at all to the community identity), and on the other side, tourists are looking for more and more authentic experiences on sites, by planning their travels through blogs, social media, and less formal travel guidance sources. Beyond the official narrative on the cultural attractiveness of the site, much is still relying on the willingness to learn of the traveller, the tourist, on his capability to grasp the details that integrate the official narrative itself and this is of course depending on the capacity of the site itself to make all these authentic details available to be grasped. In this perspective, cultural tourism should offer help to the conservation movement in optimal preservation of cultural sites in all their complexity, and not be the instrument, as witnessed so far, for drastic changes moved by quick and short-term investments. In this perspective, this paper is meant to stress the importance to consider secondary tourism products, in specific the accommodation facilities, as essential contributors to the tourism authentic experience, and as such, they need to be carefully considered under the heritage protection perspective, with adequate conservation measures and/or proper compatible design solutions. The new wave of the circular economy helps considerably heritage preservationists and designers in this difficult challenge, which hopefully encounters favourable demand by the customers, as seen in the coming methodologies containing questionnaire shreds of evidence.

2 Conceptual and Methodological Background: The Need for New Heritage Narratives in Tourism Through Circular Economy

In the last 150 years, the movement for heritage conservation had been perfecting by maturing different approaches with a unique purpose: the extension of the life of historic witnesses. Archaeological artefacts, movable and immobile assets, are the tangible permanence of humans after their death and disappearance (Leroi-Gourhan, 1943). History cannot be acknowledged only through written texts, but also by interpreting, understanding the unwritten that is held, contained, in objects. Such understanding had been permeating the heritage conservation doctrine, being applied also to traditional and vernacular heritage, referring to the following fundamentals:

- The *genius loci*, often described in the built environment (Markeviciene, 2008; Norberg-Schulz, 1992), mainly refers to the manifestation of the human adaptation to the context. Although the *genius loci* are a concept referring to the intangible sphere, its perception, and understanding, locally, is today left to the tangible assets, movable or immovable, they may be. Examples that we often face in this respect are walls and ceilings structures and textures, windows frames and shutters, plasters and their composition, but it may extend to the relationship with nature as well explained by Pietro Laureano in his work at ITKI on traditional knowledge (Laureano, 2001).
- The authenticity and the integrity, concepts heavily treated by the literature in numerous international contexts as the Nara Conference of 1994 (Acri et al., 2019; Hafstein, 2004; ICOMOS, 1994; Jokilehto, 2006; Stovel, 2007). These terms refer to different aspects of heritage assets, and they are both equally important in transmitting its qualities. Authenticity addresses the truthfulness of the source of information about a resource, namely stressing the importance to have "original" witnesses of the different changes in time. Integrity instead addresses the "wholeness" of a resource, which may be structural, visual, material. As an example, a traditional house of the eighteenth century holds a full authenticity and integrity if less is changed and adaptations are clear by permitting the understanding of the original concept, while loses authenticity and integrity the more interventions are or have been drastic. Being located in a historic environment is not sufficient to be understood as authentic and integral. Renovation and repair work especially in the tourism industry, unfortunately, is most of the time driven by the market logic, adopting renovation solutions that erase the historic layers, including the substitution of most components with globalized finishing

(PVC windows, contemporary low-cost furniture, ceramic tiles, MDF doors, and sockets, etc.).

- The *Patina* integrates the authenticity and integrity concepts but applies equally to movable and immovable cultural and traditional heritage. Patina addresses the "physical" layers of history on assets. It may tell about years of exposure to weather and use. Its preservation permits an immediate communication of the historicity of a resource, holding the additional cultural dimension of the tangible traces of time. The issue of patina has been impregnating the debates on heritage and art conservation (Brandi, 2000), confirming the equally important scientific and humanistic dimensions of the doctrine.
- Significance has been introduced, together with authenticity and integrity, as a determinant concept in WH sites nomination (Marconi, 1999). It refers to the set of justifications why an asset is worthy to be preserved. In a global context, the significance must narrate the uniqueness on a global scale, but at a local level, this concept may be used to contribute to the local specificity, toward a territorial identity that can be easily lost in absence of monumental presences.

As said, the conservation doctrine had been inspired by and developed for outstanding pieces of art and culture (as a reaction to the World Wars destructions, to the creation of the national identities, to preserve masterpieces, etc.), but had been expanding more and more its umbrella, recognising the essential role of traditional knowledge and its expression in architecture, historic cities, daily objects. Nevertheless, the conservation movement had been facing an enormous obstacle for its application *tout-court*, the market logic in a globalization era, that over-emphasized the paradoxes of conservation (Mossetto & ICARE, 2001).

In this light, a more landscape-oriented approach in heritage preservation, that started in the 60 s already with the typological conservation (Cervellati, 1991; Cristinelli, 2013; Muratori, 1967), finding in the international organizations, as ICOMOS, CoE, and UNESCO, following documents a valid backing, leading to the present historic urban landscape approach (Bandarin & Oers, 2012; CoE, 2000; Greffe, 2012; Bandarin & Pereira Roders, 2019; UNESCO, 1976), where tangible and intangible social and individual components are equally weighted. If tangible and intangible are inseparable, and mostly there is no tangible without intangible, any form of cancellation due to unwise change is questionable. On the social dimension of the historic landscape, being it rural or urban, important contributions were already made addressing the social space (Bourdieu, 1985; Lefebvre, 1974) or the sense of place in the difference between space and place (Tuan, 2011), highlighting the fundamental difference given by the appropriateness of use and inhabiting.

A very positive impulse toward the conservation of traditional and vernacular heritage comes from the circular economy. The rise of the practice and theorization of circular economy as a response to the human impact on climate change had positively included the built environment and cultural heritage (ARUP, 2016; Foster, 2020; Fusco Girard, 2019; MacArthur Foundation, 2021) and research carried out in the CLIC project[1] have highlighted the strict connection between traditional techniques of production and maintenance and circularity (Acri et al., 2019; Dobričić et al., 2019), resilience and sustainability (Foster, 2020; Foster & Saleh, 2021; Gravagnuolo et al., 2019). The effort of importing today the principles of circularity is reproducing, in a globalized economy, the basic guiding principles of *OIKOS,* home management, that had led for centuries the making of our traditional heritage (Dobričić et al., 2019). "Importing" the principles of the circular economy in heritage preservation and tourism means talking more about destination re-greening, about 0 km food and services, about traditional techniques and products preservation, about local savoir-faire and businesses, and, last but not least, about the reduction of the use of resources and consequent waste, with preference given to sustainable materials and technologies, maintenance, and conservation. Dealing with the circular economy approach to the built environment, ARUP addresses as key principles the design out of waste and pollution, the regeneration of natural systems, and the maintenance of products and materials in use as long as possible (ARUP, 2016). Tourism itself looks more and more to a circular economy for aligning responsible markets to responsible users: This became evident in the transportation industry, with increasing groups of individuals choosing sustainable transportation means, in the food industry, with more and more attention to local and sustainable foods, as well as in the accommodation, with enhancing inclination toward impact hotels and BnBs. This signifies that circular economy is not only a means to support the long-lasting efforts of heritage "conservationists", but is also the ground for the development of a larger set of narratives that enforce local specificities and traditions.

If there are no doubts that the tourism industry is committed to re-greening and waste reduction, the adoption of positive attitudes toward the materials and product preservation is less evident, especially with regard to traditional movable and immovable heritage. This trend is even more evident in historic cities by looking at the construction sites intensity, given that there is a continuous real estate market devoted to tourism that pushes renovation projects. As said,

[1] CLIC, Circular models Leveraging Investments in Cultural heritage adaptive reuse, HORIZON2020 funded, research and innovation program under Grant Agreement No. 776758, www.clicproject.eu.

conservation works are often driven by motivation that does contrast with the conservation doctrine and in general with the sustainability principles behind the circular economy approach.

The connection between cultural tourism experience and cultural heritage preservation and design may be measured in different forms. The most common one relies on the cultural tourism offers of tourism destinations and is expressed usually in events, cultural heritage assets, museums, local specialties, and recipes as proposed by local bars and restaurants. In this research, we have tried to highlight the hidden, tacit dimension (Polanyi, 1966) of heritage, the one that is grasped by visitors in their capability to tell the local history and uniqueness. Visitors are often explicitly guided in their cultural offer, being told about all possible inputs of the destination, but they are in contact with the urban or rural environment and the way it is used and preserved by the locals. Accustomed, skilled visitors, as architects, artists, craftsmen may have an introspective, critical, observation of the environment, noticing details that lead to or off the local specificity, the local genius loci, the local significance, regardless of the stories told in the booklets and brochures. In the case of these visitors, even the selection of the services may be dictated by their professional knowledge, thus choosing hotels *de Charme*, historic shops and restaurants and visiting second-hand or antique shops or junk dealers. For these visitors, the history of the place and all its tangible manifestations are important and their visit includes such content. Other visitors, less skilled or habituated to heritage and its forms, may unconsciously rely on the usual cultural tourism narrative that selects the set of marketable cultural information. These visitors do miss the advantage of having a more realistic experience of the site, which may even lead to a higher or lower final evaluation, depending on the truthfulness of the historic environment, on its authenticity and integrity, its display of the genius loci, its sense of place. Our perspective as conservationists dealing with governance by including tourism puts the second set of visitors as worthy to reach a positive tacit knowledge by an educational experience of the historic environment they visit and use. Reaching the local capacity to draw attention to the untold is the real challenge to a full cultural tourism experience. It is by offering a real, truthful, authentic, and integral environment that the cultural visit (and the cultural use) of the site is achieved, and this can be obtained solely by designing the conservation.

3 Methodology

As a confirmation of the above, I have assessed the expected satisfaction of users in tourism accommodation in different heritage contexts and have lately discussed two successful stories, from an architect conservation specialist perspective. The investigation considers the present relevant criteria for tourism accommodations quality, putting in evidence the complete absence of "heritage quality" criteria, which demonstrates that being alienated from the primary product, the reason for the visit is considered a normal condition. The investigation stresses at a later stage that niche or club forms of accommodation are focusing on the heritage perspective, evidencing even more that there is no achieved global integration between the primary and the secondary tourism products. Parallel work has been done by questioning users about their perception of accommodations qualities in more and less dense heritage destinations, showing that there is an actual demand for integration between primary and secondary tourism products, highlighting the assumption that the heritage experience should start inside the accommodation.

Both the investigation and the questionnaire put important stress on the need to work more on good architectural and object preservation practices or better compatible design of secondary tourism products, especially accommodation facilities. As a response to this, I have shown two case studies, extracted from the questionnaire and researched personally on-site that demonstrate how good adequate, conservation, as well as good compatible design, both performed through the lens of circular economy boost the heritage experience and enhance the socio-cultural level of the visitors.

4 The International Tourism Accommodation Criteria in a Heritage Perspective

A preliminary analysis, here, presented shortly to leave room for other complementary considerations, was carried out by looking at the recognition of the importance of heritage tacit knowledge in heritage accommodation by the tourism organizations in the international and some national contexts. In particular, we have looked at the categories of heritage and the criteria for hotel classifications, using the reference of UNWTO, the World Tourism Organization (UNWTO, 2015), that had already summarized and compared the different classification systems in different countries. In addition, we have selected two examples of tourism associations and hotel searches that put the heritage and sustainability issue forward.

The UNWTO is recognising in advance different tourism niches or categories, like cultural heritage tourism, rural tourism, gastronomy and wine tourism, mountain tourism, urban tourism, sport tourism, shopping tourism, to which we can add many more such as casino tourism, sex tourism, dark tourism, war tourism, etc. Such distinction is particularly advantageous for statistical data and to assess the needs of travellers and visitors, their interests, and react within the policy and market dynamics. Nevertheless, these categorizations have an impact in the local sustainable trends, unofficially assigning an acceptable standard to the local services according to the typology of tourism that is served. In itself, categorization stands in opposition to the same definition of cultural tourism product as defined by the organization itself, as

> a combination of tangible and intangible elements, such as natural, cultural and man-made resources, attractions, facilities, services and activities around a specific centre of interest which represents the core of the destination marketing mix and creates an overall visitor experience including emotional aspects for the potential customers. A tourism product is priced and sold through distribution channels and it has a life-cycle.

That clearly addresses a standard cultural basis whatever the tourism category is. For our topic of interest, what is in common to all these categories are the second tourism products, mainly accommodations, which are used by all "categories of tourists" and thus have a primary role in telling, even if tacitly, the sense of place, the cultural history, the *genius loci*.

UNWTO assessed the different services and facilities required to be classified in the different grades, for 4 and 5 stars' hotels, comparing more than 30 different global countries. The advantage of classification is in the predictive set of options of the traveller in terms of expectation of what could be found at destination.

These criteria are rated for their impact in percentage and divided per room facilities, bathroom facilities, front desk facilities, food and beverages, and available additional services such as business centre, laundry service, ticketing, and transport services always comparing the European and the global impact percentages. As an example, we have reported in Table 1 the room facilities, but the entire set of criteria never list the sustainability of the hotel in terms of natural and cultural impact.

This lack is also evident in the standards of rooms (as an example in Table 2) both in the global operators and in Europe, where the "heritage" dimension is always present due to the heavy sedimentation of centuries of cultural and historic layers all over the continent.

It is worthy to be noticed that this comparison does not take into account niche hotel chains that specifically address the "heritage" dimension but which, as previously noted, extract a "monumental to elitist" value which transform the tacit experience into an explicit one, thus being beyond our focus.

However, there are examples out of the niches that take sustainability as a criterion. In the HotelTalk[2] list for 4 and 5 stars hotel among the other facilities and requirements, the following infrastructural solutions are measured:

- Rainwater harvesting.
- Waste management.
- Pollution control methods for air, water, and light.
- Introduction of non-CFC equipment for refrigeration and air conditioning and other eco-friendly measures and initiatives.
- Solar power panels.

The HMHub, ultimate place for hospitability students, recognizes the important of the heritage dimension, including in the heritage hotels[3] any building made prior the 50s', although this may be questionable when dealing with modernist, socialist, and contemporary architectures and furniture, addressing

> The facade, architectural features and general construction should have the distinctive qualities and ambience in keeping with the traditional way of life of the area. The architecture of the property to be considered for this category should not normally be interfered with. Any extension, improvement, renovation, change in the existing structures should be in keeping with the traditional architectural styles and constructional techniques harmonising the new with the old.

A potentially complementary initiative is the so-called Ecobnb[4] hotel search engine that, regardless of the heritage dimension, had elaborated ten criteria to be considered an eco-accommodation, namely:

- Organic food provision;
- Green building;
- 100% renewable energy;
- Solar thermal panels;
- Ecological cleaning products;
- More than 80% waste recycling;
- Car-free accessibility;
- Energy saving lights;
- Water flow reducers;
- Recovery and reuse of rainwater.

The integration of such criteria with others, recognising the full circular economy process that starts from appropriate conservation of the buildings and the related objects as furniture and tools, would complete the evaluation for the completion of the ECO target.

It seems obvious that there are structured connections between the tourism market and the heritage preservation policies, with a situation that is presently leaving to the

[2] https://hoteltalk.app/t/4-star-hotel-classification-list-of-required-facilities-and-services/2411.

[3] https://hmhub.me/criteria-star-classification-hotel-five-four-three-two-one-heritage/.

[4] https://ecobnb.com/.

Table 1 Most recurring criteria in 4 and 5 star hotels across European and global groups

Most recurring criteria in 4 and 5 star hotels across European and global groups		Percentage recurrence				
	Criteria	4 stars		5 stars		Average (%)
		Europe (%)	Global (%)	Europe (%)	Global (%)	
Room	Telephone with external line	97	83	97	83	90
	Desk, work table, chair	76	100	79	100	89
	1 seat/chair per bed	90	83	86	83	85
	Wardrobe or clothes niche	86	83	90	83	85
	Reading light by each bed	86	83	86	83	84
	Minibar	81	67	88	83	80
	Adequate number of hangers	79	83	72	83	79
	Bedside table/tray	76	83	76	83	79
	Internet	66	83	72	83	76
	Wardrobe or clothes hangers	69	83	69	83	76
	Safe in room	59	67	93	83	76
	Power socket	67	83	69	83	76
	Full length mirror	79	67	79	75	75
	Luggage rack	76	67	76	67	71
	Waste basket	76	67	76	67	71
	Quality requirements mattress/bed	59	83	59	83	71
	Appropriate room lighting	55	83	55	83	69
	Bed size specifications single and double	66	67	66	67	66
	Radio (any device to listen to radio channels	76	50	79	50	64
	Writing utensils and note pad	74	33	78	50	64
	Blinds to fully darken room	41	50	76	67	58
	Curtains or similar	31	67	45	67	52
	2 Pillows per person	36	33	78	50	49
	Bathrobes and slippers in room	31		90	67	47
	Correspondence folder	53	33	60	33	45
	Pay TV or similar	3	67	20	67	39
	Spy hole/peek hole in door	3	67	10	67	37
	Suite requirements	1	17	79	50	37

Credits UNWTO (2015)

owners and their planning specialists the ethical decision to preserve heritage or not according to the doctrine and the logic. Mostly, it is clear that the built environment is considered a simple mean to host tourism services, not a mean to educate on and communicate intrinsic culture contents during the tourism experience.

5 The Tourism Experience in Heritage Perspective

A parallel research based on a questionnaire was carried out to measure the selection attitudes of potential visitors when planning a cultural experience, both in historic cities and in rural areas of Europe. The questionnaire is a preliminary attempt to grasp the potential tacit knowledge of users to select accommodation facilities that may increase their cultural experience by simply displaying the genius loci, the authenticity, the integrity the sense of place. The questionnaire, rapidly spread online by means of "Google Form" and through social media and emailing non-selected users have been asked to select options of accommodation from different destinations and using as a reference only two pictures taken from similar hotels/accommodation facilities, but with different interior designs. It was anticipated by the following text:

> Dear reader, this questionnaire is meant to understand your expectations in a tourism experience, with a specific focus on your accommodation. We are interested in understanding what

Table 2 Measurable criteria in Europe

Criteria	4 stars	5 stars
No. of bathrooms amenities	3.68 units	5.21 units
Percentage of in suite bathrooms	100%	100%
No. of floor before lift is required	1.9 floors	1.4 floors
No. of towels per person	2.3 towels	2.3 towels
Room service operation	16 h	23.5 h
Front office operation	21 h	23.5 h
Room size single room	12.5 m^2	14.8 m^2
Room size double room	18 m^2	20.9 m^2
Language skills front of house	2 languages	2.5 languages
Bed size single bed	92 cm * 200 cm	92 cm * 200 cm
Bed size double bed	163 cm * 200 cm	177 cm * 200 cm
No. of power socket in the room	1.8	2

Credits UNWTO (2015)

> is attracting you in the building you select for your holidays and if you notice differences in its qualities.
>
> In this respect, this questionnaire is simply proposing you, in different destinations, 2 accommodation options that look differently but are similar in standards (we chose randomly from Booking on a base of a high standards hotel, double room). Rooms are in historic buildings adaptations.
>
> What we ask you is to choose one for each location and select, even in multiple choice, the determining conditions for your choice. Pictures do only display the rooms not to influence your impression. At the end of the questionnaire, we also propose 3 different historic environments differently preserved and we ask you to select the one if for you more attracting.
>
> The questionnaire is anonymous, so we will not know about your identity and choices, but we kindly ask you to be sincere because feedbacks may suggest improvement for the future.[5]

The questions addressed were organized offering firstly a picture of the destination that could tell about the contexts, the ambience, the genius loci, the atmosphere, and secondly two pictures of two different rooms from different, but similar in cost (namely range of 150–180 Euro per night) hotels. A common date was taken, not to depend on seasonality (in specific one night between 14 and 15 October 2021), and the source of costs and pictures have been booking.com, considered one of the most common tourism accommodation engines. The common criteria, beyond the fact of having selected hotels or BnB in renovated traditional buildings, to select pictures have been the wide angle perspective to display as much as possible of the interior and the overall good comfort. The difference of the rooms derives from a precise conservation and design perspective and consists in having traditional and designed furniture or more global one, respect of the historic architectural technologies or not, the use of local crafted materials or not, the connection with the history of the site or not. To the reader/user, the following multiple selection criteria have been proposed:

- *Charme*, but this was not defined at all, leaving to the users their own interpretation;
- *Historic authenticity and integrity*, where only the "historic" character was underlined;
- *Sense of place*, without defining its meaning, leaving to the users their own interpretation;
- *Comfort*, to assess the perception of comfort by the users, expecting both the acceptance by users that the more space is authentic the more it needs our adaptation and to verify if users, however, prefer comfort to the experience;
- *Combination of colours*, as an additional purely aesthetic dimension considering the limited dimension of pictures as displayed in a mobile phone;
- *Type of materials*, to assess the interest by users as well as their valuing of such aspects.

The proposed destinations have been Venice, Siena, Utrecht, La Valletta, Rovinj, Krakow, Ljubljana, Mikonos, the Scottish Highlands, Helsinki, Bruges, the Veneto Countryside, and Provence.

The pictures of Venice show what the respondents were asked to assess visually. The picture of the city (Fig. 1) was selected to transmit the common atmosphere of the city, meant to inspire those who did not have the opportunity to make a real visit in life. On one side, we have considered a hotel that has preserved the terrazzo floor, the original windows (or similar), the original ceiling and part of the traditional furniture, being also less invasive in the installations, for example lighting (Fig. 2), and on the other side, we have considered a comfortable and contemporary room that had a more drastic renovation, recalling the venetian tissues, but in a more globalized perspective (Fig. 3).

[5] https://docs.google.com/forms/d/e/1FAIpQLSfWP8D27o45B9v4vcegD1CZsA0gSt2iSNIAO313sszfb0Ip8Q/viewform.

Fig. 1 Venice, Gran Canal by night in front of La Salute. *Credits* Marco Acri

Fig. 2 Hotel in the historic centre of Venice. *Source* booking.com

The results of the questionnaire (Fig. 4) showed that respondents had a clear attitude in selecting options that are aligned with the expectations of the cultural destination, insisting much on the criteria of *charme*, authenticity, sense of place, and type of materials, giving less importance to comfort. Below is the results for Venice, opting in big majority for the solutions shown in picture 2, in the questionnaire option A.

In general, the questionnaire highlighted a predominance of the voices sense of place and *charme* in the selection of

Fig. 3 Different hotel in the historic centre of Venice. *Source* booking.com

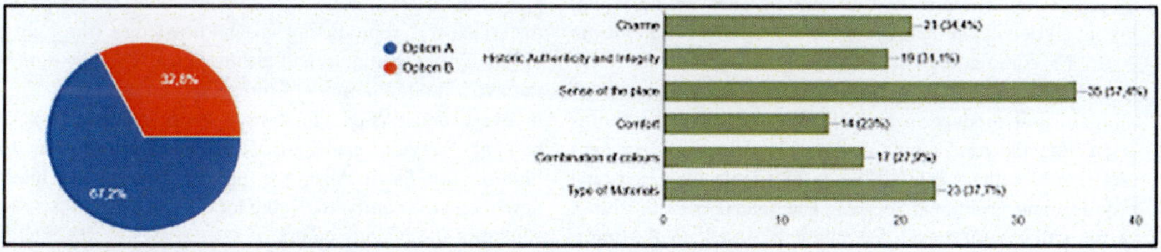

Fig. 4 Results from the case of Venice

the options, which namely put the accent on the desire to have a contact with the *genius loci* and a true cultural experience even in the practical uses.

6 Two Case Studies: Rosa Rosae Guesthouse and the Spirito Santo Palazzo Storico

The further considerations made in our analysis is from a professional viewpoint, selecting two of the examples that were included in the above-described questionnaire, namely the Rosa Rosae Inn in the Veneto countryside, Italy, and the Hotel Spirito Santo in Rovinj. These two case studies were selected, out of the others, firstly for the very strong unanimity in the choice by respondents, and secondly for two different aspects that are debated in this paper, namely conservation and design. The former, Rosa Rosae, for its evident restoration approach based on minimal intervention, relying as much as possible on reuse of materials and objects in line with the circular economy principles. The latter, Spirito Santo, was redesigned starting for two traditional houses in the centre of Rovinj, and the approach taken was of critical restoration but with an enormous emphasis on design inspired by local traditions and history.

The Rosa Rosae Inn is located in the Veneto countryside, nearby Treviso, in the small municipality of Breda di Piave and closed to the famous Piave River (Fig. 5).

The building was a former river mill, used by locals to produce flour. Along the twentieth century, it was never restored by its owner, keeping most of the architectural technologies, structures, materials, and finishing. The new owners in the early twenty-first century carried out a restoration project inspired by the principle of minimal intervention, replacing only when necessary the most

Fig. 5 Picture of the context of the Rosa Rosae Inn: an agricultural plane rich of rivers and with a view to the Dolomites. *Credits* Marco Acri

necessary components, as windows, although reproducing the traditional construction model.

The perceived leading idea of the Inn had been to create an informal place offering restaurant and accommodation services, as well as residence of the two owners, architects by profession. The principle of minimal intervention is visible in all details, in the awareness that comfort and *charme* could be coming together, trying to let guests taste the intimate nature of the building and its history, made of humble rural production and life. Not only are the rooms re-creating the rural conditions of the early twentieth century, but also the meals are made of simple local seasonal products and traditional recipes. The selection of finishing, as the cutlery, furniture, decoration is based on the traditional rural practice of "nothing should be waste" and "everything can be reused" (Fig. 6).

The Spirito Santo Palazzo Storico Hotel is located in the Istrian seaside town of Rovinj, in Croatia (Fig. 9). This hotel went through a different story. By merging two distinct collapsing houses (Figs. 8, 9, and 10) in the centre of the town and carrying out a respectful restoration project based on design, a new small-sized hotel and restaurant opened a new perspective in the urban conservation scenario, previously very much oriented to the modernization and globalization of the offer even in case of historic fabrics (Fig. 7).

The intervention agreed between the architects and the investor had to focus on the preservation of the sense of the place, through a critical restoration able to adopt local materials and reproducing local knowledge, through and upscaling of techniques and technologies. The intention was to have a limited number of rooms, extremely well-designed in every detail (Fig. 11), further accompanied by a keen personal service made of warm welcoming and 0 km high-quality food. Also, the project opened the interior, newly created courtyard, to the local citizens, as a cosy little square in the historic centre.

The Spirito Santo Palazzo Storico was also selected for the important feedbacks by users. We have considered Tripadvisor, with 65 "excellent" out of 66 evaluations, being the remaining "very good", with specific mentions on the design and the architectural detailing respecting history but guaranteeing the comfort, as "*…Very high architecture and interior / exterior….*" And "*…The Hotel is gorgeous! A 150-year-old dwelling that was lovingly and thoughtfully*

Fig. 6 Pictures of the Rosa Rosae Inn showing the exterior, restaurant, and a room detail. *Source* Tripadvisor.com and locandarosarosae.it

Fig. 7 Rovinj from the context of the hotel. *Credits* Marco Acri

Fig. 8 Exteriors before renovation. *Credits* Spirito Santo Palazzo Storico

renovated into a chic and stylish space where no stone was left unturned or space not thought of and used beautifully. The architects are yacht specialists…". These comments,

Fig. 9 Ground floor of the complex before renovation. *Courtesy* Spirito Santo Palazzo Storico

Fig. 10 Internal courtyard of the two houses before renovation. *Credits* Spirito Santo Palazzo Storico

out of others, show that attention on architecture and local genius loci was enabled by design and appropriate conservation.

7 Discussion

The paper has tried to highlight the present lack in the link between the tourism industry and the heritage preservation movement. So far, heritage, especially traditional assets, has been considered simple tools for basic tourism services, without understanding their potential in the cultural and educational dimension. This approach has been allowing the cancellation of many tangible manifestations of the local knowledge, genius loci, history, to give space to presumed more efficient, often globalized, solutions for the tourism experience. Examples have shown that visitors do assign importance to the place of their accommodation, considering

Fig. 11 Few pictures of the hotel after renovation, as the courtyard, the newly created terraces, one room, courtesy of Spirito Santo Palazzo Storico

them an important cultural moment of their visit, and they equally appreciate when proper design stimulates and updates local knowledge though looking to the past and the sense of place. In future, it will be fundamental to improve the connections between heritage, especially traditional and non-listed, conservation, and tourism planning through careful design and aesthetic criteria definition, in the awareness that tourism products, even if secondary and ancillary with an emphasis on the accommodations, are an essential means to communicate the local specificity, its genius loci and strongly contribute to the local sense of place.

This paper is meant to emphasize the need to offer a better tourism experience to users, starting from their accommodation, which is too often considered not contributing to the cultural content of the visit, to the creation of a true cultural experience. The study has several limits, considering that the topic has never been taken seriously in the past, given that the tourism mechanisms have been always treated from an economic perspective and that the cultural conservation stress has been always given to the primary tourism products. However, the potential to find confirmation of the theory expressed in the paper is consistent and many similar or complementary investigations can be carried out and my group at the University of Nova Gorica is already working on it.

8 Conclusion

Thanks to the new understanding of the intrinsic qualities of heritage, from the vernacular and traditional one to the monumental assets, new narratives can be imagined. Additional help to these new narratives is offered by the circular economy applied to the built environment because it highlights the native, rooted circularity of the heritage-making in the past. The stress on circular economy and sustainability in tourism may have an important push also for the design in heritage preservation, considering that the project is a fundamental step in the conservation or adaptive reuse of movable and immovable assets. The integration of design and preservation practice performed in secondary and ancillary tourism products would slowly guarantee a deepen and true cultural experience of visitors, concretizing in an additional form the achievement of the intrinsic qualities of a destination, which is also partly defined as the sense of place.

Funding Project CLIC: Circular models leveraging investments in cultural heritage adaptive reuse. CLIC has received funding from the European Union's Horizon 2020 research and innovation programme under grant agreement no. 776758.

References

Acri, M., Dobričić, S., & Jokilehto, J. (2019). The circular economy in adaptive reuse: Respecting authenticity and integrity. *Patrimonio Culturale in Mutamento Le Sfide Dell'uso*, 319–330. http://www.arcadiaricerche.it/editoria/2019.htm

ARUP. (2016). *The circular economy in the built environment*. ARUP. https://www.arup.com/perspectives/publications/research/section/circular-economy-in-the-built-environment

Bandarin, F., & Oers, R. van. (2012). *The historic urban landscape: Managing heritage in an urban century*. Wiley.

Bandarin, F., & Pereira Roders, A. (Eds.). (2019). *Reshaping urban conservation: The historic urban landscape approach in action* (1st ed.). Springer. https://doi.org/10.1007/978-981-10-8887-2

Bertrand, G. (2008). *Le grand tour revisité: Pour une archéologie du tourisme: le voyage des français en Italie (milieu XVIIIe siècle-début XIXe siècle)*. École française de Rome.

Bourdieu, P. (1985). The social space and the genesis of groups. *Theory and Society, 14*(6), 723–744.

Brandi, C. (2000). *Teoria del restauro* (Rist). G. Einaudi.

Burdet, M. (2018). Primary and secondary tourist resources. GeographyCaseStudy.Com. https://www.geographycasestudy.com/primary-and-secondary-tourist-resources/

Cervellati, P. L. (1991). *La città bella: Il recupero dell'ambiente urbano*. Mulino.

CoE. (2000). *European landscape convention*. Council of Europe. https://www.coe.int/en/web/landscape/text-of-the-european-landscape-convention

Cristinelli, G. (2013). *Saverio Muratori e Egle Renata Trincanato: La nascita del restauro urbano in Italia*. GB EditoriA.

DeSeta, C. (1982). L'Italia nello specchio del Gran Tour. In R. Romano, & C. Vivanti (Eds.), *Storia d'Italia: Annali* (4. rist, pp. 127–263). Einaudi.

Dobričić, S., & Acri, M. (Eds.). (2018). *Creative cities: Which (historic) urban landscape*. Mimesis.

Dobričić, S., Jokilehto, J., & Acri, M. (2019). The circular character of building tradition: Which challenges for the HUL approach. In G. Getzinger (Ed.), *Proceedings of the STS Conference Graz 2019*. TU Graz. https://doi.org/10.3217/978-3-85125-668-0-05

Foster, G. (2020). Circular economy strategies for adaptive reuse of cultural heritage buildings to reduce environmental impacts. *Resources, Conservation and Recycling, 152*, 104507. https://doi.org/10.1016/j.resconrec.2019.104507

Foster, G., & Saleh, R. (2021). The adaptive reuse of cultural heritage in European circular city plans: A systematic review. *Sustainability, 13*(5), 2889. https://doi.org/10.3390/su13052889

Fusco Girard, L. (2019). Implementing the circular economy: The role of cultural heritage as the entry point. Which evaluation approaches? *Bulletin of the Calza Bini Center*, 245–277. https://doi.org/10.6092/2284-4732/7269

Gravagnuolo, A., De Angelis, R., & Iodice, S. (2019). Circular economy strategies in the historic built environment: Cultural heritage adaptive reuse. https://doi.org/10.3217/978-3-85125-668-0-08

Greffe, X. (2012). Creativity: The strategic role of cultural landscapes. In L. Fusco Girard, P. Nijkamp, & T. Baycan (Eds.), *Sustainable city and creativity: Promoting creative urban initiatives* (1st ed., pp. 199–223). Taylor and Francis, Routledge.

Hafstein, V. T. (2004). *The making of intangible cultural heritage: Tradition and authenticity, community and humanity*.

ICOMOS. (1994). *The Nara document on authenticity*. ICOMOS International. https://www.icomos.org/charters/nara-e.pdf

Jokilehto, J. (2006). Considerations on authenticity and integrity in world heritage context. *City and Time, 2*(1), 16.

Laureano, P. (2001). *Atlante d'acqua: Conoscenze tradizionali per la lotta alla desertificazione* (1st ed.). Bollati Boringhieri.

Lefebvre, H. (1974). La production de l'espace. *L'homme Et La Société - Sociologie De La Connaissance Marxisme Et Anthropologie, 31–32*, 15–32. https://doi.org/10.3406/homso.1974.1855

Leroi-Gourhan, A. (1943). *L'homme et la matière. Évolution et techniques, I*. Albin Michel.

MacArthur Foundation. (2021). *The circular economy in detail*. https://archive.ellenmacarthurfoundation.org/explore/the-circular-economy-in-detail

Marconi, P. (1999). *Materia e significato: La questione del restauro architettonico* (1st ed.). GLF Editori Laterza.

Markeviciene, J. (2008). Genius Loci and Homo Faber: A heritage making dilemma. In *Proceedings of 16th ICOMOS General Assembly and International Symposium: Finding the Spirit of Place Between the Tangible and the Intangible* (p. 12).

Mossetto, G., & ICARE. (Eds.). (2001). *Economia del patrimonio monumentale: Documenti e ricerche*. Angeli.

Muratori, S. (1967). *Civiltà e territorio* (Vols. 1–2). Centro Studi di Storia Urbanistica.

Norberg-Schulz, C. (1992). *Genius loci: Paesaggio, ambeinte, architettura*. Electa.

Polanyi, M. (1966). *The tacit dimension*. Routledge & Kegan Paul.

Russo, A. P., & Van Der Borg, J. (2002). Planning considerations for cultural tourism: A case study of four European cities. *Tourism Management, 23*(6), 631–637.

Stovel, H. (2007). Effective use of authenticity and integrity as world heritage qualifying conditions. *City and Time, 2*(3), 16.

Tuan, Y.-F. (2011). *Space and place: The perspective of experience* (7th print). Univ. of Minnesota Press.

UNESCO. (1976). *Recommendation concerning the safeguarding and contemporary role of historic areas*. UNESCO. http://portal.unesco.org/en/ev.php-URL_ID=13133&URL_DO=DO_TOPIC&URL_SECTION=201.html

UNWTO. (2015). *Hotel classification systems: Recurrence of criteria in 4 and 5 stars hotels*. https://doi.org/10.18111/9789284416646

Urry, J. (1990). *The tourist gaze: Leisure and travel in contemporary societies*. Sage Publications.

Education and Professional Training in Tourism and Hospitality: The Case of Portuguese Travel Agents

Lígia Maria Rocha Pinto Ribeiro, Eduardo Cândido Cordeiro Gonçalves, and Ricardo Jorge da Costa Guerra

Abstract

This study focuses on tourism education and training in Portugal and how it has been combined with market trends and cycles. Using the case study method, different tools were used including content analysis and open data collection methods were applied by means of questionnaires. The analysis focuses on the educational offer of the subject of Tourism at higher education level in Portugal. The diversity of the identified courses has led to the conclusion that institutions at a university level should reorganize their educational offer to meet market demands. Within this scope, our study focuses on the educational and professional training reality for the tourism sector with special emphasis on curricular units related to tourism distribution. We have put together a list of all the higher education courses in tourism in Portugal which offer curricular units related to travel agencies and tour operators. At the same time, a questionnaire was administered to a broad sample to characterize the exclusive professional training of travel agents. Our aim was that of assessing a significant number of active travel agents in order to identify their academic profile while at the same time analyzing the distribution of tourism-related components within the curricular plans of the courses. Finally, through an analysis of feelings, the intention of the study was to look into the training needs required to reach optimal performance at work.

L. M. R. P. Ribeiro (✉)
University of Maia, Maia, Portugal
e-mail: ligiarib@hotmail.com

E. C. C. Gonçalves
CEGOT—Centro de Estudos de Geografia e Ordenamento do Território and CEDTUR, University of Maia, Maia, Portugal
e-mail: egoncalves@ismai.pt

R. J. da Costa Guerra
CEGOT—Centro de Estudos de Geografia e Ordenamento do Território, CITUR, and UDI/IPG, Polythecnic Institute of Guarda, Guarda, Portugal
e-mail: ricardoguerra@ipg.pt

Keywords

Tourism distribution • Education in tourism • Professional training • Travel agencies

1 Introduction

This study approaches tourism from the point of view of the qualification of its human resources, especially in what concerns education and training. When taking professional careers into account, attending courses at professional and higher education institutes has become of pivotal importance (Ribeiro, 2019). According to the UNWTO (1996), education and training in tourism is viewed as a phenomenon which entails the transmission of knowledge, concepts, values and techniques with the aim of improving students' skills for qualified performance in their functions and versatility throughout life.

Education and professional training in Tourism had been long forgotten in the past, leading professionals to enter the job market without adequate qualifications, while others revealed a self-taught profile as a result of having to adapt to the needs of their professional activity.

In Portugal, the beginning of training in tourism dates back to 1957 with the emergence of the Hotel and Tourism School of Lisbon which, under the patronage of hotelier Alexandre de Almeida, was essentially a hotel management school. It was in the late 1980s that Portugal expanded its training and educational offer in tourism, especially in large cities such as Lisbon, Porto and Aveiro. Since then, several higher education institutions (public and private) have emerged, offering courses in tourism and subsequently, awarding graduate degrees in bachelors, masters and doctorate courses (Salgado, 2004).

With the development of a "new tourism", the demand for courses related to the area has become inevitable. Higher education institutions have gone to extreme measure to

A. Mandić et al. (eds.), *Cultural Sustainable Tourism*, Advances in Science, Technology & Innovation,
https://doi.org/10.1007/978-3-031-10800-6_14

respond to the demand by trying to create curriculum plans that meet the market's minimal needs which, in turn, are also increasingly demanding (Eurico et al., 2012). In turn, the topic under analysis still reveals a great scarcity of studies and publications, one of the reasons which stimulated the pursuit of this study and led us to identify higher education courses and professional training in the area of tourism, related to tourism distribution, travel agencies and tour operators. From here, the strengths and weaknesses of both types of courses were listed. The study design includes (i) literature review; (ii) methodology; (iii) presentation and discussion of results; (iv) conclusions and proposals for further research.

2 Literature Review

2.1 Travel Agencies and Tourism

Travel agencies are one of the pillars of tourism distribution and of key corporate players in this industry which employ thousands of professionals and generating millions of euros annually in Portugal and the whole world (APAVT, 2018).

The history of travel agencies takes us back to the nineteenth century and to Thomas Cook, who organized the famous first train tour in 1841 for a group of 570 passengers who were going to attend a medical congress (Abranja & Magalhães, 2018; Hunter, 2006). In Portugal, it is important to note the pioneering spirit of the Abreu Agency, established by Bernardo Abreu in 1840 and is still today one of the largest travel and tourism agency networks in Portugal (Abreu, 2021). Over the years, several companies followed in his footsteps, but it was only at the beginning of the twenty-first century that large groups appeared, such as Star Turismo, Top Atlântico and Bestravel (Ribeiro, 2019). It was also at this time that management groups emerged, autonomous organizations comprised independent travel agencies, which provide advice and assistance in various operations areas such as commercial negotiation with the various tourism suppliers (AIRMET, 2018).

For decades, travel agencies played a central role as intermediaries in the tourism distribution system as key elements for the entire distribution channel which bridged the existing gap between suppliers and consumers. At this time, we were in a so-called intermediation stage where it was geographically impossible for suppliers to interact directly with the final consumer (Viljoen & Roberts-Lombard, 2016). The travel agencies, tour operators and GDS (Global Distribution System) were considered intermediaries and the pillar that made this whole process possible (Buhalis & Kalis, 2008; Kracht et al., 2009; Vallespín & Molinillo, 2014). During this period, consumers usually visited the AVT`s where they collected information and bought their travel tickets. It was impossible to find the information obtained in the travel agencies anywhere else (Cheung & Lam, 2009).

However, at the end of the last century and start of this one, the tourism industry underwent changes never seen before with the emergence of the internet and consequently, with the development of Information and Communication Technologies (ICT) (Capriello & Riboldazzi, 2019). There have been so many changes that it is possible to state that there is a *before* and an *after* period insofar as the appearance of ICT in distribution channels is concerned.

The emergence of the Internet made it possible for consumers to resort to digital channels to regularly carry out their search and make their reservations without having to go to the travel agency (Cetin et al., 2016; Cheung & Lam, 2009). But the bargaining power was on their side and it was now possible for them to compare and negotiate prices (Sahut, 2009). Within this new context, the basic intermediary functions no longer made sense with distribution channels becoming scarcer with the disappearance of some of the central intermediaries. Travel agencies have consequently become considered as expensive and dispensable intermediaries for both suppliers and consumers (Viljoen & Roberts-Lombard, 2016) which no longer serve their interests (Granados et al., 2014).

Consumers and suppliers started negotiating final prices, as well as transaction conditions directly (Belleflamme & Neysen, 2009). As such, the concept of disintermediation arises. This is observed when an intermediary is removed from the distribution channel (Cheung & Lam, 2009) or is partially replaced by another competitor (Kracht & Wang, 2009), thus reducing its role (Buhalis & Kalis, 2008).

This new scenario has transformed the tourism industry, allowing for the creation of new products and services, which has not only given rise to new business opportunities, but also to many challenges (Buhalis & Kalis, 2008). One of these opportunities was the appearance of a new type of intermediary which emerged very quickly, namely, the Online Travel Agency (OTA). This new intermediary, which is much more powerful than its previous version (Cetin et al., 2016), presents a new business model and the role of mere intermediaries to have other functionalities as business partners. They are considered a virtual intermediary that connects the consumer to the various suppliers, offering various services (Lee et al., 2013). These new players came to fill a gap that existed in e-commerce and consequently liberalized the global travel market (Weigert, 2017).

According to Munikrishnan et al. (2019) travel agencies were already aware that their business model was obsolete since the end of the last century and knew that they would have to innovate to be present in the virtual world. OTAs use marketing strategy associated with the publication of ads that give many users the possibility of browsing and buying

(Tekin & Cebi, 2020), but at the same time, creating the possibility of elaborating tailor-made packages that are customized by the consumer himself (Granados et al., 2014). This new panorama takes consumers to a new level, making them more informed, more demanding while giving rise to new consumption habits. OTA's have profoundly changed traditional tourism distribution, distribution channels and the intermediation structure itself, making it more digital (Boukherouk & Marzoug, 2019; Capriello & Riboldazzi, 2019; Weigert, 2017).

Disintermediation did not mean the end of intermediaries but rather a fresh start "leading them" to reorganize and reinvent themselves (Boukherouk & Marzoug, 2019; Capriello & Riboldazzi, 2019). This new phase, called re-intermediation, has allowed companies to adapt to this new context by reformulating their strategies and business models (Sahut, 2009). Many of the travel agencies that worked exclusively in the offline mode also started working in the online mode, consequently offering services and products through digital tools. This mix allowed companies to work in a hybrid context (offline and online), thus improving the relationship with the customer and creating greater proximity (Casais et al., 2020; Weigert, 2017).

All these new events have led to the mandatory retraining of travel agents that aims to increase their skills so that they may cope with new challenges. As such, they will be able to attend professional or higher education in tourism, and/or professional training provided by various public and private entities.

2.2 Education in Tourism

The entire educational system has undergone several changes over the years due to political and economic changes. From 1950 to 1980, education and training in tourism in Portugal was offered at only a few educational establishments that took interest in these subjects. It was only in 1986 (Table 1) that three private educational institutions began offering courses in the area of Tourism (Salgado, 2004).

Due to the abrupt evolution of tourism and its growing economic importance, the demand in this area by students has increased, leading educational institutions to expand their educational offer with the creation of new courses in tourism (Costa et al., 2012; Salgado, 2004). Today, it is possible to find tourism courses in professional secondary education and in public or private higher education institutions (Cravidão et al., 2010).

The Portuguese higher education structure is organized in a system known as binary, that is, a system which integrates university and polytechnic education. While the university system is aimed at promoting research and the creation of knowledge, the polytechnic is geared toward understanding and finding practical solutions to meet the demands of a particular field of studies. This solution has existed since the 60s in both public and private education establishments (DGES, 2019a).

Table 1 First higher education courses created in tourism both public and private in Portugal

Start year	Sector	No. of courses
1986	Private	3
1988	State	1
1989	Private	3
1990	Private	3
1991	State and private	4
1992	State and private	5
1994	State	2
1995	State and private	4
1996	State	2

Source Adapted from Salgado (2004, p. 106)

Salgado (2007) considers that today's education system should be changed to one that also includes professional training. Nowadays, the higher educational system is divided into public, private and cooperative, as well as concordat (Fig. 1). Public higher education includes public universities, public polytechnic institutes and academies, institutes and police and military schools. Private and cooperative higher education, in turn, comprises private and cooperative universities and polytechnic institutes. The only higher education institution under Concordat law is the Catholic University of Portugal (DGES, 2019a, 2019b).

The higher education institutions mentioned above are divided into three cycles of studies. The first cycle refers to bachelors, the second to master's degrees and finally, the third to Ph.D. degrees (Table 2). The latter can only be offered at universities.

In the 2018/2019 academic year there were 114 higher education institutions in Portugal, 40 of which were public, 72 were private and 2 concordats. A total of 4745 courses were available, of which 3674 belong to public education and 1071 to private education (DGES, 2019a). Table 3 shows the number of higher education institutions that exist by district, as well as by type of institution.

The tourism sector follows the changes in the real world, forcing its professionals to adapt and prepare for these variations. As future professionals, students need to acquire skills and knowledge in order to cope with the growing need of the sector's job market (Costa et al., 2012; Wallis & Steptoe, 2006).

It is essential to guarantee the skill management of future professionals by implementing educational systems that come as close as possible to the existing reality (Cravidão et al., 2010; Wallis & Steptoe, 2006). Education in the area

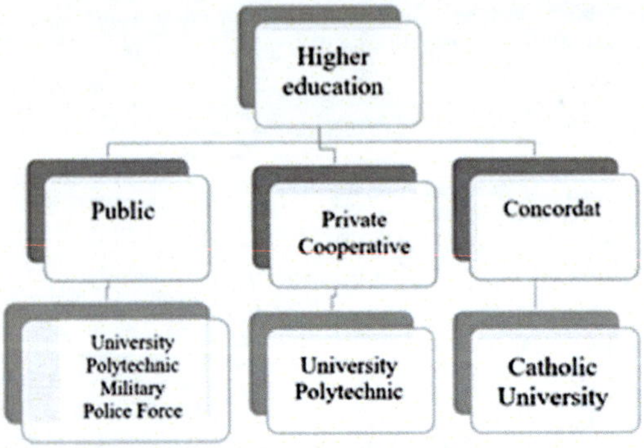

Fig. 1 Division of the Portuguese higher educational system. *Source* Adapted from DGES (2019a, 2019b)

Table 2 Different cycles of the higher education level

Cycles	Degree
1st cycle	Bachelors
2nd cycle	Masters
3rd cycle	Ph.D.

Source Adapted from DGES (2019b)

of Tourism has widely contributed to the global success of this sector especially since technology and innovation have become more present in the Tourism industry (Fuchs, 2021).

Future professional travel agents and tour operators need to acquire specific skills in this subsector by means of various tourism courses that offer educational plans with specific curricular units for this area. Data collection for this study was carried out at all higher education institutions with courses in tourism at bachelor level that were either active in the 2018/2019 academic year or had a curricular unit (CU) linked to travel agents (TA) and tour operators (TO) (Table 4).

The results presented in Table 4, show that the vast majority of courses have a curricular unit related to travel agencies and/or tour operators. However, after analyzing the study plans, we concluded that many of them do not teach the Global Distribution System (GDS) program, which better prepares students for future professional challenges by providing them with knowledge about airline reservations and availability.

Turismo de Portugal is another entity which has developed a training offer in its network of Hotel and Tourism Schools by presenting courses adapted to new trends and market demands. An example of this are the on the job courses which

Table 3 Number of institutions of higher education by district and by type in Portugal

Districts	No. of institutions	Public	Private	Concordat
Aveiro	6	2	3	0
Braga	5	2	4	0
Bragança	1	1	0	0
Beja	1	1	0	0
Castelo Branco	2	2	0	0
Coimbra	5	3	2	0
Évora	1	1	0	0
Faro	3	1	2	0
Guarda	1	1	0	0
Leiria	3	1	2	0
Lisboa	38	11	28	1
Portalegre	1	1	0	0
Porto	27	3	26	1
Santarém	3	2	1	0
Setúbal	6	3	2	0
Vila Real	2	1	1	0
Viana do Castelo	2	1	1	0
Viseu	3	1	2	0
Azores	1	1	0	0
Madeira	3	1	2	0
Total	114	40	72	2

Source Adapted from DGES (2019a)

Table 4 List of tourism bachelor degrees in Portugal in the 2018/2019 academic year

Course name	Number of courses	Number of courses with CU of TA/TO
Tourism	25	20
Tourism and leisure	1	1
Tourism, territory and patrimony	1	0
Tourism and hotel management	1	1
Tourism companies management	1	1
Tourism in rural and natural spaces	2	0
Tourism + marketing and advertising	1	1
Marketing and advertising + tourism	1	1
Nature sport and active tourism	1	0
Environmental education and nature tourism	1	0
Tourism management	3	2
Cultural and tourism management	1	1
Tourism and hotel management	1	1
Management of tourism activities	2	2
Leisure and tourism animation management	1	0
Cultural and equity tourism management	1	1
Tourism information	1	0
Tourism animation	1	0
Total	46	32

Source Adapted from DGES (2019a)

aim to qualify trainees with concrete knowledge of the business reality, allowing them to have cooperative education programs which combine classroom learning with curricular internships (Turismo de Portugal, 2021).

2.3 Professional Qualification in Tourism

Training highlights the more practical and hands-on approach to learning. It is aimed at trainees who wish to develop more technical and specific skills. Its duration is much shorter, compared to educational courses considering that the objective focuses on obtaining results within a shorter period of time (Fragoso, 2015).

It is believed that professional training in tourism first took place in 1893 at the renowned Hotel School of Lausanne in Switzerland. This first professional training was related to hotel management. Over the decades, the number of training courses has increased significantly, leading many researchers to agree that the development of this type of training is closely associated with Higher Schools of Tourism and Hospitality (Cunha, 1997).

In recent years, there has been a structural change in relation to the concept of professional training, which has become broader and makes it possible to view training as a teaching instrument that allows trainees to obtain different qualifications and skills, including cultural and social ones (Vasco, 2014).

Portuguese travel agents need professional training to stay abreast of the latest updates in the sector. Many of these courses are delivered by different suppliers, management groups, including the companies themselves (Ribeiro, 2019). Table 5 shows the main training courses offered to professionals in this subsector. There is an enormous difficulty in obtaining information on this subject, although it is possible to find some information in specialist magazines or on the internet pages of the management groups. All the information presented in Table 5 is only shared on the network of Portuguese travel agencies, which attests to the lack of information.

In addition to the aforementioned entities, the Schools of Turismo de Portugal must also be included considering that their courses also offer ongoing and professional training with the aim of providing added-value to the companies and enhancing the skills of their human resources. This training includes the areas of catering and drinks, hotels, travel agencies, tour operators and other related activities with tourism companies (Turismo de Portugal, 2021).

In order for tourism companies to maintain their service of excellence, it is essential to invest in the continuous

Table 5 Professional training by entities

Entities	Companies	Training
Airlines	TAP, Air France; KLM; Lufthansa	Presentation of new routes, fares
Global distribution system (GDS)	Galileo; Amadeus	Basics of booking and fares; Advanced of bookings, re-issuance, conversion
Travel agencies groups	Abreu, Geostar, Top Atlântico	Destinations, GDS, customer service; claims; invoicing; languages
Management groups	Airmet; Dit Portugal; GEA; RAVT	Invoice system, sales techniques, successful communication, conflict management, customer service, agency training, keeping and retaining customers, leadership and team motivation, new technologies, destinations, emotional sales; airlines; data protection policy; GDS
Tour operators	Sóltropico; Viajartours; Jolidey; Jadetravel; Lusanova; Nortravel; Sólferias; Soltour; Travelplan; Viajartours	Destinations
Cruise companies	MSC; Costa Cruzeiros; Melair	Routes, bookings and websites

Source Adapted from Publituris (2015), Ambitur (2016), Opção Turismo (2017), Presstur (2017), Publituris (2017), RAVT (2018), Publituris (2018a), Presstur (2018), Airmet (2018), Dit Portugal (2019), Sulforma (2019), Go4travel (2017), Amadeus (2019), Ribeiro (2019), and Ribeiro et al. (2020)

training of their employees. However, there are still many companies that see training as a process that only entails costs, without taking its benefits into account. Also linked to this situation, is the employees' lack of motivation and difficulties in working with new information technologies, which makes training considered as an obligation rather than a benefit (Fragoso, 2015).

3 Methodology

The object of study of this work is the Portuguese travel agents. For the study sample, we selected travel agents who work with the GDS Galileo. A questionnaire survey was carried out based on a descriptive quantitative methodology. This data collection tool was delivered in paper format (during the GDS—Galileo trainings sessions, exclusive to travel agents) and through the Portuguese Travel Agents' Facebook Page between the months of December 2018 and February 2019.

The validation of the questionnaire was subjected to a pre-test with 15 agents, with the aim of testing the perception of the questions presented. After collecting the answers and verifying that the respondents had a good understanding of the questionnaire, the test was validated and applied to the selected sample. All participants in this study contributed voluntarily and with no external interference. The collected data was analyzed using the Statistical Package for Social Sciences (SPSS) software (version 26). A total of 600 questionnaires were received, but only 512 were eligible for the sample.

The analysis of the sample under study includes some factors that reveal educational offer in scientific area of tourism. Along with this information, it was possible to assess the current status of professional training that is offered exclusively to Portuguese travel agents. These factors made it possible to establish 3 profiles: sociodemographic and professional; academic and professional training. Descriptive and univariate analysis were performed using relative and absolute frequencies to characterize the sample.

4 Results

4.1 Sociodemographic and Professional Profile of Travel Agents

In this sample with a total of 512 respondents, it was found that the vast majority are female (82.4%), and only 17.6% are male (Table 6). These results are in line with the study conducted by UNWTO in 2019 on women in tourism. This study showed that the majority of the tourism workforce is globally comprised women. They represent 54% of people employed in tourism compared to the general economy which is only 39% (UNWTO, 2019).

Regarding age, the age group between 31 and 40 years is the one that is more represented, with 34.8% of the answers, followed by 21–30 years with 29.1% and 41–50 years with 25.6%. These three age groups represent almost all of the respondents (89.5%). With these results, we can see that we have a group of workers who are still quite young.

Table 6 Sociodemographic and professional profile of travel agents

Gender	(*N*)	(%)	Time working as an agent	(*N*)	(%)
Male	90	17.6	Less than 1 year	50	9.8
Female	422	82.4	From 1 to 5 years	136	26.6
			From 6 to 10 years	67	13.1
			From 11 to 15 years	93	18.2
			From 16 to 20 years	82	16.0
			More than 20 years	84	16.4
Age			No. of branches TA		
18–20 years	11	2.1	1 branch	187	36.5
21–30 years	149	29.1	2–4 branches	186	36.3
31–40 years	178	34.8	5–9 branches	36	7.0
41–50 years	131	25.6	More than 10 branches	107	20.2
51–60 years	41	8.0			
More than 60 years	2	0.4			

Source Based on questionnaires administered to travel agents in Portugal

The sample does not show a discrepancy with regard to years as workers in the travel agencies subsector. There are the travel agents with 1–5 years of experience that are most represented in this study with 26.6%, followed by 11–15 years with 18.2%, over 20 years with 16.4% and 16–20 years with 16%. The results presented allow us to assert that travel agencies have experienced workers, which allows them to give more security to this sector.

The vast majority of agents work in travel agencies with up to four branches, totaling 72.8% of responses.

Of the inquired travel agents, 24.6% work in travel agencies that are part of a travel agency management group known as GEA (Table 7). This result is to be expected, as it is one of the largest independent travel agency networks in Portugal (Ribeiro, 2019).

The remaining travel agents are distributed across the different networks, with the exception of those that do not belong to any group, which represents 14.1% of the total.

4.2 Academic Profile of Travel Agents

When travel agents were asked about the type of higher education institution they had attended, only 326 answered this question. The vast majority of respondents have higher education qualifications, comprising a total of 66.3%. It should be noted that 62.30% of the professionals in this subsector have higher education qualifications and are from the tourism area (Table 8). The majority of qualifications are centered on Bachelor degrees in tourism and secondary education in tourism with 36.1% and 13.7%, respectively. These results are in line with the studies carried out by APAVT (2018) and Turismo de Portugal (2017) which state that travel agents are among the tourism professionals with the highest level of higher education qualifications.

The vast majority of travel agents attended public higher education, corresponding to 61%, leaving only 39% who attended private institutions. According to Ribeiro (2019), the main reason for this result could be related to the in lower tuition fees in public education in comparison to private ones. Regarding the question whether the course they attended had prepared them for professional life, 61% of respondents answered no and only 39% answered yes. The vast majority of courses 69% had curricular units linked to TA and/or TO, in line with the survey carried out and presented previously. These results lead to the conclusion that it is necessary to review the educational systems related to tourism, so that they could include curricular units that are closer to the current reality.

4.3 Professional Training Profile of Travel Agents

When the travel agents were asked if they felt lack of professional training when they first started working, 81% answered yes and only 19% answered no (Table 9). This information leads us to conclude that some training entities do not provide adequate training for those who are starting this activity. It is of pivotal importance for this data to reach the competent entities, so that they can provide training focused on these latent needs. Travel agents were also asked which type of training they felt they required the most, with 49% making reference to GDS and 37% referring to destination information, which made up 86% of the total responses. The results presented are not in line with previous

Table 7 Network of travel agencies where the companies of the respondents are associated

Agencies by network	(*N*)	(%)
Abreu	49	9.6
Airmet	54	10.5
Bestravel	14	2.7
Gea	126	24.6
Geostar	36	7
Go4travel	68	13.3
RAVT	23	4.5
Top Atlántico	8	1.6
Other	72	14.1
None	62	12.1

Source Based on questionnaires administered to travel agents in Portugal

Table 8 Academic profile of travel agents

Education level	(*N*)	(%)	What type of higher education institution did you attend?	(*N*)	(%)
Basic level	5	1.0	Private	198	39
High school (other)	107	20.9	Public	128	61
High school in tourism	70	13.7	Did the course prepare you for your professional life?	(*N*)	(%)
Bachelor degree (other)	6	1.2	Yes	113	61
Bachelor degree in tourism	43	8.4	No	213	39
Graduation (other)	60	11.7			
Graduation in tourism	185	36.1	Degree with curricular units related to TA and TO	(*N*)	(%)
Postgraduate studies (other)	5	1.0	Yes	224	69
Postgraduate studies in tourism	8	1.8	No	102	32
Master (other)	10	2.0			
Master in tourism	12	2.3			
Ph.D. (other)	0	0			
Ph.D. in tourism	1	0.2			

Source Based on questionnaires administered to travel agents in Portugal

data which shows that GDS has basic training on booking and fares for its trainees. This may indicate that when travel agencies hire new employees, they do not offer these training courses immediately. This could be due to a series of factors: monetary, lack of staff or even lack of knowledge of the GDS training calendar.

The vast majority of travel agents responded that they attend professional training 1–2 times a year (43%), or 2–4 times a year (27%). However, there is still a significant number of agents who do attend any type of professional training annually (7%).

When asked whether they currently receive sufficient training to perform their duties, the vast majority 69% answered yes. The main entities that offer this type of professional training are the tour operators with 45% of the responses and the GDS with 42%. These results may contradict what Travelport (2021) and Ambitur (2018) state when referring that training is offered in several cities in Portugal, making it possible for all agents to have access to them. At the time, face-to-face training was the most attended, corresponding to 47% of the responses, followed by training in e-learning format with 36% and b-learning with 17%.

The vast majority of professional training courses were delivered on the suppliers' premises, with 47% (Table 10). Travel agents were unanimous in answering that they know that participation in professional training is an asset added-value to their qualifications.

From the results obtained through the conceptual and empirical components, it was possible to draw up a table listing the weaknesses and strengths of the Portuguese educational and training system (Table 11).

Table 9 Professional training profile of travel agents (1)

Did you sense that at the beginning of your career there was a lack of training?	(*N*)	(%)	Nowadays do you think you have sufficient training to perform your profession?	(*N*)	(%)
Yes	415	81	Yes	355	69
No	97	19	No	157	31
What kind of training do you think is missing?	(*N*)	(%)	What entities offered the training courses you have participated in?	(*N*)	(%)
Destinations	154	37	Tour operators	215	45
Airlines	37	9	Airlines	20	5
GDS (Galileo and/or Amadeus)	203	49	GDS (Galileo or Amadeus)	200	42
Insurances	5	1	Insurances	5	1
Rent-a-car	6	2	Rent-a-car	5	1
Touristic information	10	2	Tourism destination representatives	29	6
Training frequency during the civil year	(*N*)	(%)	What kind of training did you attend?	(*N*)	(%)
More than 6 times	41	8	Classroom	350	47
4–6 times	67	13	E-learning	74	36
2–4 times	139	27	B-learning	50	17
1–2 times	277	43			
None	38	7			

Source Based on questionnaires administered to travel agents in Portugal

Table 10 Professional training profile of travel agents (2)

Where did the presential training take place	(*N*)	(%)
Workplace	60	17
Supplier's classroom	165	47
Other	125	36
Is training an asset to your educational background?	(*N*)	(%)
Yes	512	100
No	0	0

Source Based on questionnaires administered to travel agents in Portugal

Table 11 Strong and weak points of the educational and training system in Portugal

Educational system	*Strong points*: • Great diversity of courses • *Vast* geographical network • Variety of courses with curricular unit related to TA and TO • 3 cycles of studies • Offered by public and private entities
Training system	*Weak points*: • Some higher degree courses do not provide adequate preparation for students to face the work market • Greater need of articulation between the educational institutions and the sector companies
	Weak points: • Lack of training to newcomers to this sector • Low qualification training entities as GDS and tour operators to provide training to newcomers starting their professional careers • Low diversity on the types of training (almost exclusively in the classroom) • Lack of information available to the scientific community

Source Based on questionnaires administered to travel agents in Portugal and DGES (2019a, 2019b)

5 Conclusions

The importance of education and professional training is increasingly highlighted by the major world and national tourism entities. An example of this was the document prepared by Turismo de Portugal and the Portuguese Ministry of Economy, entitled Tourism Strategies 2027. This document stresses the importance of professionals in this industry in obtaining more educational qualifications and actively participating in congresses, workshops, seminars, and debates on how to keep up to date with the constant changes in the market (Turismo de Portugal, 2017).

According to the UNWTO (2020), human resources in tourism are the main pillar for its proper functioning. However, they have undergone some changes in recent decades due to commercial, social, legislative, ethical and technological issues (Baum, 2015). These changes are related to the creation of new business models, new consumption patterns, changes in supply and demand dynamics and in the distribution of tourism itself (Baum, 2015; UNWTO, 2015, 2020). Educational institutions and professional training provided by private and public entities are the best equipped organizations to train professionals in this sector with the best skills.

By carrying out this study, it was possible identify the strengths and weaknesses of the Portuguese educational and training system and draw some conclusions. The existence of a great variety of tourism courses is evident. However, not all of them have study programs that better prepare their students by preparing them with the required qualifications for the job market, leading to higher employability rates (Forte, 2017). It is essential for curriculum plans to be focused on responding to market needs and allowing these future professionals to add some value to their future potential employers, thus responding to the new demands and challenges of this industry (Eurico et al., 2012; Forte, 2017; Marques, 2019). According to Forte (2017), the student only truly completes his course when he enters the job market. Within this scope, we have professional training that needs to provide more training for professionals who are starting a career in this sector.

To address this issue, it is necessary to implement greater articulation between higher education institutions and private and public entities that provide professional training with the sector's business, in order to create content that meets existing needs and thus prepare future professionals for such a demanding market (Eurico et al., 2012). On the other hand, it is clear that the educational and training service must be evaluated in order to obtain feedback from the different educational and training plans and their execution. Monitoring students' and/or trainees' satisfaction will allow for the assessment of the quality of education and training (Fuchs, 2021).

Due to the scarcity of studies on this topic, it would be relevant for future research to focus on professional education in tourism, considering that until now no survey or study of this nature has been carried out. Additionally, research into professional training in the travel agency subsector would also be relevant.

References

Abranja, N., & Magalhães, C. (2018). *Gestão de agências de viagens e turismo* (2nd ed.). LIDEL.

Abreu. (2021, September 14). *A nossa história*. https://www.abreu.pt/a_nossa_hist%C3%B3ria-30979.aspx

Airmet. (2018, May 11). Airmet. https://www.airmet.pt/index.php

Amadeus. (2019, November 15). *Amadeus training*. https://amadeus.com/en/training

Ambitur. (2016, October 13). *Travelport e TAP em parceria junto dos agentes de viagens*. https://www.ambitur.pt/travelport-e-tap-em-parceria-junto-dos-agentes-de-viagens/

Ambitur. (2018, January 27). *Roadshow "Os especialistas" tem início a 5 de fevereiro*. https://www.ambitur.pt/roadshow-os-especialistas-tem-inicio-a-5-de-fevereiro/

APAVT. (2018). *O valor económico da distribuição turística em Portugal*. Associação Portuguesa de agentes de viagens e turismo.

Belleflamme, P., & Neysen, N. (2009). Coopetition in infomediation: General analysis and application to e-tourism. In: Á. Matias, P. Nijkamp, & M. Sarmento (Eds.), *Advances in tourism economics: new developments* (pp. 217–234). https://doi.org/10.1007/978-3-7908-2124-6_14

Baum, T. (2015). Human resources in tourism: Still waiting for change? A 2015 reprise. *Tourism Management, 50*, 204–212. https://doi.org/10.1016/j.tourman.2007.04.005

Boukherouk, M., & Marzoug, M. (2019). Enjeux et impacts des réseaux sociaux pour les agences de voyages marocaines. *Communication Et Management, 16*, 83–98. https://doi.org/10.3917/comma.162.0083

Buhalis, D., & Kalis, K. (2008). eEnabled internet distribution for small and medium sized hotels: The case of Athens. *Tourism Recreation Research, 33*, 67–81. https://doi.org/10.1080/02508281.2008.11081291

Capriello, A., & Riboldazzi, S. (2019). Exploring service innovation in a network of travel agencies: The Robintur case. *Journal of Hospitality and Tourism Technology*. https://doi.org/10.1108/JHTT-02-2019-0014

Casais, B., Fernandes, J., & Sarmento, M. (2020). Tourism innovation through relationship marketing and value co-creation: A study on peer-to-peer online platforms for sharing accommodation. *Journal of Hospitality and Tourism Management, 42*, 51–57. https://doi.org/10.1016/j.jhtm.2019.11.010

Cetin, G., Cifci, M. A., Dincer, F. I., & Fuchs, M. (2016). Coping with reintermediation: the case of SMHEs. *Information Technology Tourism, 16*(4), 375–392. https://doi.org/10.1007/s40558-016-0063-2

Cheung, R., & Lam, P. (2009). How travel agency survive in e-business world? *Innovation and Knowledge Management in Twin Track Economies Challenges and Solutions—Proceedings of the 11th International Business Information Management Association Conference* (Vol. 10, pp. 900–907). https://ibimapublishing.com/articles/CIBIMA/2009/884409/884409.pdf

Costa, C., Umbrelino, J., & Amorim, E. (2012). A relação entre a formação superior em Turismo e o planejamento turístico – o caso português. *CULTUR—Revista de Cultura e Turismo, 2*, 43–59.

https://www.researchgate.net/publication/260348707_A_RELACAO_ENTRE_A_FORMACAO_SUPERIOR_EM_TURISMO_E_O_PLANEJAMENTO_TURISTICO_-_O_CASO_PORTUGUES
Cravidão, F., Santos, N., & Cunha, L. (2010). Educação e Turismo: formação e capacitação. A evolução da formação em Turismo: o caso português. 4º Congresso latino Americano de investigação Turística. https://estudogeral.uc.pt/bitstream/10316/13836/1/155.pdf
Cunha, L. (1997). *Economia e política do Turismo*. McGraw-Hill
DGES. (2019a, March 11). *Cursos e instituições*. https://www.dges.gov.pt/pt/pesquisa_cursos_instituicoes?plid=372
DGES. (2019b, March 12). *Quadro de Qualificações*. https://www.dges.gov.pt/pt/quadro_qualificacoes
Dit Portugal. (2019, April 02). *Ditgestion*. http://ditportugal.com/
Eurico, S., Silva, J., & Valle, O. (2012). Recursos humanos qualificados no turismo: factores de diferenciação nos polos de desenvolvimento turístico do Oeste e de Leiria-Fátima. *Tourism and Management Studies, 8*, 30–40.
Forte, A. (2017, May 06). *Turismo: Educar, Formar e Empregar*. Turisver. https://www.turisver.com/iturismo-educar-formar-empregar-atilio-forte/
Fragoso. A. (2015). *Contributos para o debate sobre as políticas de formação em turismo na União Europeia e em Portugal* (Master's dissertation, Faculdade de Letras da Universidade do Porto). Porto University Research Repository. https://sigarra.up.pt/fbaup/en/pub_geral.pub_view?pi_pub_base_id=37176
Fuchs, K. (2021). How do undergraduate tourism students perceive the service quality in their educational institution? Na assessment using the SERVQUAL framework. *Journal of Environmental Management and Tourism, 12*(4), 1076–1084. https://doi.org/10.14505//jemt.v12.4(52).20
Granados, N., Kauffman, R., Lai, H., & Lin, H. (2014). Decommoditization, resonance marketing, and information technology: An empirical study of air travel services amid channel conflict. *Journal of Management Information Systems, 28*(2), 39–74. https://doi.org/10.2307/41304620
Go4Travel. (2017, January 12). *Go4travel promove formações entre colaboradores*. https://www.go4travel.pt/www/noticias_view.php?id=14
Hunter, F. (2006). Tourism and empire: the Thomas Cook & Son Enterprise on the Nile, 1868–1914. *Middle Eastern Studies, 40*(5), 28–54. https://doi.org/10.1080/0026320042000265666
Kracht, J., & Wang, Y. (2009). Examining the tourism distribution channel: Evolution and transformation. *International Journal of Contemporary Hospitality Management, 22*(5), 736–757. https://doi.org/10.1108/09596111011053837
Lee, H., Guillet, B. D., & Law, R. (2013). An examination of the relationship between online travel agents and hotels a case study of choice hotels international and Expedia.com. *Cornell Hospitality Quarterly, 54*, 95–107. https://doi.org/10.1177/1938965512454218
Marques, A. (2019). *A importância da qualificação dos recursos humanos no Turismo: O caso dos hosteis da cidade do Porto* (Master's dissertation, Escola Superior de Hotelaria e Turismo do Politécnico do Porto). Instituto Politécnico do Porto Research Repository. https://recipp.ipp.pt/handle/10400.22/14870
Munikrishnan, U. T., Imm, N. S., Ann, H. J., & Yusof, R. (2019). Disintermediation threat: Do small medium traditional travel agencies in Malaysia embrace ICT adequately? *Pertanika Journal of Social Sciences and Humanities, 27*, 707–728. http://psasir.upm.edu.my/id/eprint/67813/1/47%20JSSH-2617-2017.pdf
Opção Turismo. (2017, December 09). *Gea promove novas formações*. http://opcaoturismo.pt/wp/grupo-gea-promove-novas-formacoes/
Presstur. (2017, May 12). *Ultratur promove roadshow para agentes de viagens até ao fim do mês*. http://www.presstur.com/empresas—negocios/agencias—operadores/ultratur-promove-roadshow-para-agentes-de-viagens-ate-ao-fim-do-mes/
Presstur. (2018, December 11). *Costa Cruzeiro, lança nova plataforma de vendas e formação para agentes de viagens*. http://www.presstur.com/empresas—negocios/cruzeiros/costa-cruzeiros-lanca-novas-plataformas-de-vendas-e-formacao-para-agentes-de-viagens/
Publituris. (2015, November 04). *Melair anuncia formações em todo o país*. https://www.publituris.pt/2015/09/30/melair-anuncia-formacoes-em-todo-o-pais/
Publituris. (2017, March 30). *Soltrópico aposta na proximidade aos agentes de viagens com lançamento de Academia*. https://www.publituris.pt/2017/02/23/251652/
Publituris. (2018a, April 22). *Inscreva-se no webinar Publituris – Costa Rica com a Jolidey*. https://www.publituris.pt/2018a/01/31/inscreva-no-webinar-publituris-costa-rica-jolidey/
Publituris. (2018b, July 29). *Viajar Tours Promove webinar sobre Agadir e Creta*. https://www.publituris.pt/2018b/07/02/viajar-tours-promove-webinar-agadir-creta/
Publituris. (2018c, July 29). MSC Cruzeiros promove roadshow de formação em 14 cidades. https://www.publituris.pt/2018c/07/24/msc-cruzeiros-promove-roadshow-de-formacao-em-14-cidades/
RAVT. (2018, May 16). *Formação*. https://www.ravt.pt/optitravel/www/ravt/details.php?content_id=56920281
Ribeiro, L. (2019). *Educação e formação profissional em Turismo, na área da intermediação – Uma análise aos agentes de viagens portugueses* (Master's dissertation, ISMAI).
Ribeiro, L., Gonçalves, E., & Guerra, R. (2020). The importance of qualification of the human resources in tourism: A study applied to travel agents in Portugal. *PASOS. Revista de Turismo y Patrimonio Cultural, 18*(4), 669–684. https://doi.org/10.25145/j.pasos.2020.18.047
Salgado, M. (2004). As organizações de Ensino Superior em Turismo. *Revista Turismo e Desenvolvimento, 2*(1), 105–109.
Salgado, M. (2007). *Educação e Organização Curricular em Turismo no Ensino Superior Português* (PHD Thesis, University of Aveiro). Aveiro University Research Repository. https://ria.ua.pt/bitstream/10773/1842/1/2008000400.pdf
Sahut, J. M. (2009). The impact of internet on pricing strategies in the tourism industry. *Journal of Internet Banking and Commerce, 14*. file:///C:/Users/Ligia/Downloads/The_Impact_of_Internet_on_Pricing_Strategies_in_th.pdf
Sulforma. (2019, May 22). *Formação Travelport*. https://sulforma.com.pt/pt/formacao/formacao-travelport-sulforma
Tekin, A. T., & Cebi. F. (2020). Click and sales prediction for digital advertisements: Real world application for OTAs. In: C. Kahraman, S. Cebi, S. Cevik Onar, B. Oztaysi, A. Tolga, I. Sari (Eds.), *Intelligent and Fuzzy Techniques in Big Data Analytics and Decision Making. INFUS 2019*. Advances in Intelligent Systems and Computing (Vol. 1029, pp. 205–212). Springer. https://doi.org/10.1007/978-3-030-23756-1_26
Travelport (2021, September 25). *Formação em Galileo*. https://pt.travelportservices.com/extranet/24/Homepage/FORMACAO
Turismo de Portugal. (2017, December 11). *Estratégia Turismo 2027*. http://www.turismodeportugal.pt/SiteCollectionDocuments/estrategia/estrategia-turismo-2027.pdf
Turismo de Portugal. (2021, March 05). *Formação*. http://business.turismodeportugal.pt/pt/Gerir/Formacao_Recrutamento/formacao/Paginas/default.aspx
UNWTO. (1996). *El capital humano en la indústria turística del siglo XXI*. https://www.e-unwto.org/doi/pdf/10.18111/9789284402106
UNWTO. (2015). *Tourism and trade: A global agenda for sustainable development*. https://www.intracen.org/uploadedFiles/intracenorg/Content/Publications/Tourism_and_Trade__low%20res_2014-2015-335.pd
UNWTO. (2019). *Global report in women in tourism* (2nd ed.). https://www.e-unwto.org/doi/epdf/10.18111/978928442038

UNWTO. (2020). *El futuro del trabalho en el turismo y el desarrollo de competências*. https://www.e-unwto.org/doi/epdf/10.18111/9789284421404

Vallespín, M., & Molinillo, S. (2014). El futuro de la intermediación en el sector turístico. *Revista de Análisis Turístico* (17), 13–25. https://dialnet.unirioja.es/servlet/articulo?codigo=478412

Vasco, A. (2014). *Formação profissional e turismo – Caracterização da oferta de uma escola profissional e perceção dos empregados* (Master's dissertation, University of Coimbra). Coimbra University Research Repository. https://estudogeral.uc.pt/handle/10316/27793

Viljoen, K., & Roberts-Lombard, M. (2016). Customer retention strategies for disintermediated travel agents: How to stop customers from migrating to online booking channels. *Journal of Applied Business Research, 32*(3), 681–694. https://doi.org/10.19030/jabr.v32i3.9669

Wallis, C., & Steptoe, S. (2006, December 10). How to bring our schools out of the 20th century. *Time Magazine*. http://www.time.com/time/magazine/article/0,9171,1568480,00.html

Weigert, M. (2017). Jumia travel in Africa: Expanding the boundaries of the online travel agency business model. *Tourism Review, 74*(6), 1167–1178. https://doi.org/10.1108/TR-04-2017-0073

The Future of Historic Re-enactments in Portugal—The Case of Santa Maria da Feira

Roberto Carlos Reis

Abstract

The aim of this paper is to share the future of historic re-enactments in Portugal, the case of Santa Maria da Feira on historical reconstruction and recreation, and analyze the consequences of cancelation of this types of events. Recently, there has been a great expansion in the re-enactment of history in Portugal, so it is well worth debating the phenomenon in order to be better acquainted with it and to disseminate the associated instruments. The renewal of the State of Emergency in Portugal, on April 2020 and 2021, forced the cancelation of the largest historical re-enactment in Europe, which takes place in Santa Maria da Feira. In the list of the exceptional and temporary measures adopted by the government, in the context of the combat of COVID-19, is included the non-holding of mass events. In thousands of localities, like Santa Maria da Feira, there has been a suspension of many business operations which directly or indirectly depend on the economy fueled by visitors, and it is by no means certain that they will all reopen. The consequences of global immobility induced by The COVID-19 crisis resulted in worldwide immobility, which rather all confused those involved in the promotion and research of tourism. Probably, this pandemic will create a psychological change and during the first year many tourists will prefer to travel to nearby destinations in their own country, or to countries where the pandemic has not been too intense (if there is such a country in a few months…). However, the tourism industry has recovered before from such periods of crisis, epidemics included, which shows how resilient it is and capable of mitigating abrupt shifts in demand or supply. One should create the conditions for small local businesses to adapt, through a flexible approach to the demands of the market.

R. C. Reis (✉)
University of Maia - ISMAI, Maia, Portugal
e-mail: robertoreis@umaia.pt

Keywords

Re-enactment • Covid-19 • Tourism • Cancelation of events • Medieval Journey • Santa Maria da Feira

1 Introduction the Normal Flow of a Crisis in the Tourism Industry

The behavior of the economy of tourist destinations and the growth in the number of individuals infected by COVID-19 are inversely proportional situations; it is clear that the greater the number of infected, the greater the impacts on the economy. As established by McKinsey & Company in its report entitled "COVID-19 Global Health and Crisis Response," a crisis of this type will develop in three different phases, in which, for each of them, special actions will be required, responding to the most prevailing needs of each moment.

According to the last press release of World Tourism Organization (UNWTO) published in Jan. 18, 2022, *Global tourism experienced a 4% upturn in 2021, compared to 2020 (415 million versus 400 million). However, international tourist arrivals (overnight visitors) were still 72% below the pre-pandemic year of 2019, according to preliminary estimates by UNWTO. This follows on from 2020, the worst year on record for tourism, when international arrivals decreased by 73%.*

The same document adds that the first 2022 issue of the UNWTO World Tourism Barometer points out toward a release in pent up demand, as a result of positive vaccination rates, together with less restrictions regarding traveling due to more international cooperation. The world's tourism has shown a moderate rebound during the second semester of 2021, with a 62% decrease in international arrivals in the third and fourth quarters. There was a 65% decrease in international arrivals in December, as compared to 2019

A. Mandić et al. (eds.), *Cultural Sustainable Tourism*, Advances in Science, Technology & Innovation,
https://doi.org/10.1007/978-3-031-10800-6_15

levels. The full impact of the Omicron variant and surge in COVID-19 cases is yet to be seen.

As can be seen in Fig. 15.1, for a first moment "A", which is where we find ourselves in most countries of the world (increase in the number of infections and reduction in consumption), the imperative actions are all those that are, on the one hand, linked to the control of the growth of the virus and, on the other hand, those related to the support of people and businesses that are being affected by the restrictions generated by the actions focused on the control of the virus.

This is surely the most difficult moment for all companies in the tourism sector, and at this point in the process, all those actions that are linked to the protection of our collaborators and suppliers will be the pillar that will support the future strategies that are developed for the reactivation of tourist activity in the world.

As in all the crises that have impacted tourism in the past, to name only those that occurred in the last two decades, the tourism industry had to face with aplomb the attack on the twin towers in 2001, SARS in 2003, and the economic crisis of 2009. Situations in which, despite the reduction in the number of travelers, the drop in average rates and the imminent need to increase operating costs to face the new fears of travelers, at the end of everything in this reinvention process, the global tourism figures were restored to their normal state.

Sluggish and Unbalanced Recovery

Recovery is still sluggish and unbalanced around the world because of the various local degrees of mobility restrictions, penetration of vaccines, and traveler confidence. European and American obtained the best results from 2020 to 2021 (+ 19% and + 17%, respectively); still, they represented a 63% decrease as compared to the period before the pandemic (Fig. 15.2).

The Caribbean performed the best, with 63% above 2020, but 37% below 2019, and specific destinations came close to, or even exceeded, the levels registered before the pandemic.

Other sub-regions that significantly rebounded include Southern Mediterranean Europe, with a 57% increase and Central America, with 54%, while remaining 54% and 56% below 2019, respectively. North America, with a 17% increase, and Central Eastern Europe, with 18%, also improved on their 2020 performance.

The UNWTO Panel of Experts reports that the majority (61%) of tourism professionals are positive about 2022. Fifty-eight percentage are expecting 2022 to provide them with a rebound, mostly in the third quarter, while 42% expect it to happen only around 2023. Sixty-four percent of experts are not counting on the return of the 2019 levels of international arrivals before 2024. The number in the September survey was 45%.

UNWTO conducted a global survey among its UNWTO Panel of Tourism Experts on the impact of COVID-19 on tourism and the expected time of recovery. At the question *"When do you expect international tourism to return to pre-pandemic 2019 levels in your country?"* The UNWTO Confidence Index slightly declined between January and April 2022. Experts identify the improved vaccination process, less travel restrictions, and more efficient travel as the main factors behind the rebound of international tourism. International tourist arrivals are expected to grow between 30 and 78% as compared to 2021, yet 50–63% less than

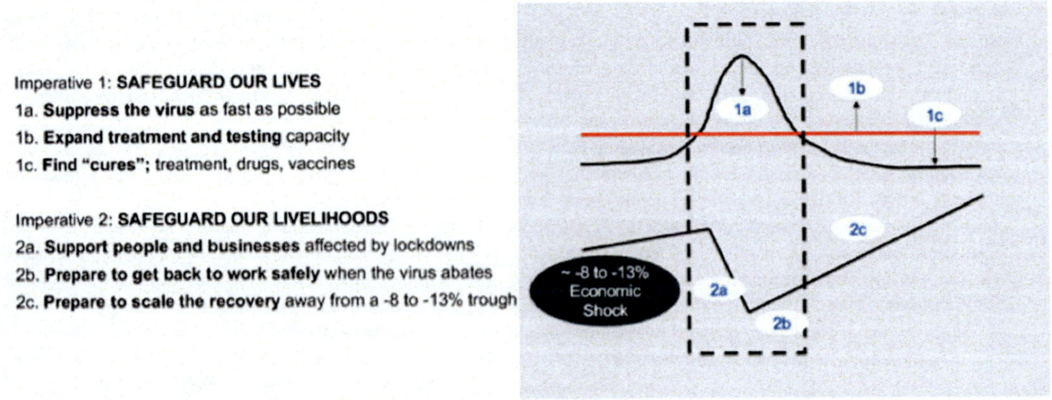

Fig. 15.1 Virus and the economic shock. *Source* McKinsey & Company

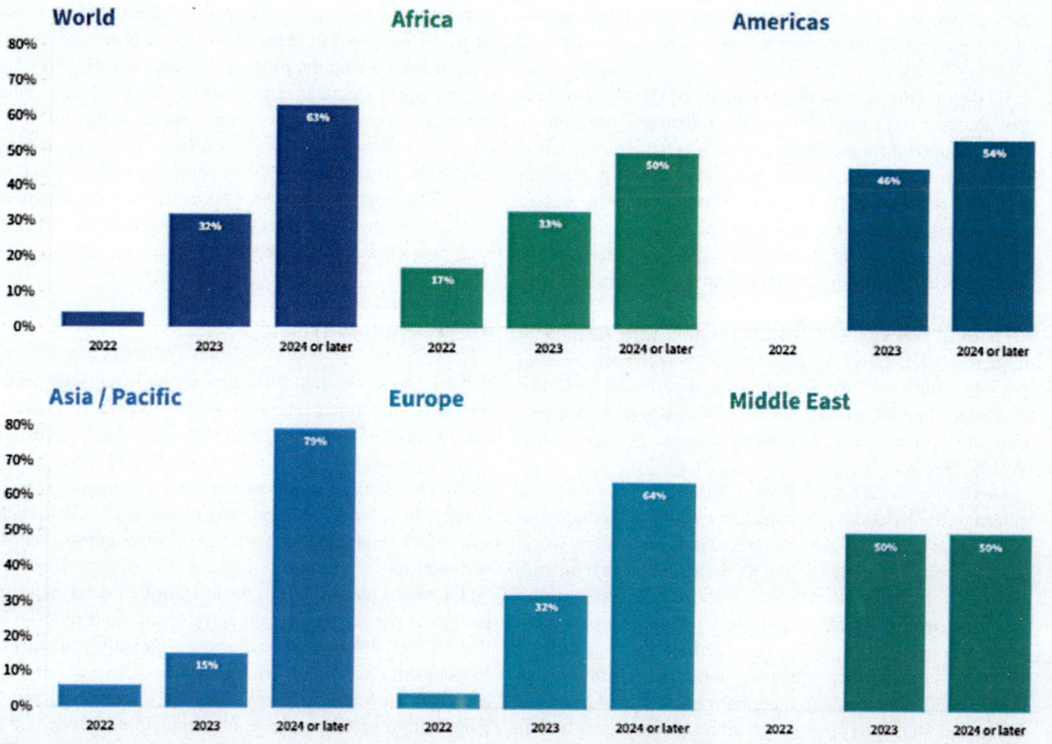

Fig. 15.2 Expect international tourism to return to pre-pandemic 2019 levels in your country. *Source* World Tourism Organization (UNWTO)

before the pandemic. The Omicron-fueled increase in cases should affect the recovery and confidence early this year, and some nations are reintroducing some level of bans and restrictions regarding travel, while the vaccination process remains uneven and many borders remain closed, most of which in the Pacific and Asia. The recovery of international tourism is threatened by the economy, namely high oil prices, inflationary pressure, rising interest rates, high debt, and disrupted supply chains. Still, in 2022, consumer confidence could be restored and the recovery of tourism accelerated by more vaccination and less travel restrictions, namely in the Americas and Europe.

International tourism is rebounding, while new tendencies point toward domestic tourism as a leading factor in the recovery of the sector in many destinations, mostly the ones whose domestic markets are significant. Among 2022s, main travel trends are domestic tourism, traveling near home, open-air activities, nature and rural tourism experts predict.

2 Historical Re-enactments in Portugal—What Future?

As we defended historical re-enactments began to play a significant role in the cultural programming of parishes, municipalities, regions, and some countries to create an impact in the media, develop tourism activities, and enhance the image of destinations. Recently, historical re-enactments have been increasingly debated and criticized regarding their worth for education, as well as how they contribute to the way cultural heritage is understood in our post-modern consumer societies." (Carnegie & Mccabe, 2008).

It is emphasized that "historical re-enactments achieve a variety of purposes and provide examples from different perspectives: public policy and event organizers, reenactors and academics. The heritage industry contributes to making these events meaningful and as such represents unique frameworks through which to understand issues of authenticity and identity in the production and consumption of post-modern cultural heritage attractions." (Carnegie & Mccabe, 2008).

Today, taking sustainable advantage of this form of culture and its various applications—in tourism or citizenship—can be important for the development of certain geographical areas that are rich in history but have a low population density, such as Trancoso, Castro Marim, Almeida, among others. Making historical re-enactments is worth its weight in gold, as they bring the Middle Ages or other eras of history to life with fun, entertainment and music, sparking our imaginations.

There is clearly a pervasive fascination with attempts to remember and reflect on past events, whether by wearing period costumes or taking on the role of particular historical characters. However, there is also skepticism and, at times, outright hostility toward these practices. (Crang, 1996; Walsh, 1992).

According to Warwick Frost, "Recreation events are a subset of the living history approach to cultural heritage. As has been suggested by several researchers, our interest in the past is fuelled directly by our concerns about the intense pressures of modern society."

Immersion in a certain historical period can be understood as a return to a simpler, better and more authentic time, being an effective and attractive antidote, constituting an escape from alienation and marginalization to modern times. To escape modernity, people are increasingly drawn into an idealized past (Laing & Frost, 2012; Lennon & Foley, 2000; Lowenthal, 1985). Re-enactment events allow participants and audiences to suspend their disbelief, disconnect from the problems of the present, and step into a version of the past for a short time. This version of the past is, of course, a fantasy; but it is also a seductive fantasy."

When a historical event is re-enacted, we are always faced with the problem of truth and fact. Some recreations are better than others. Many are careless and imperfect. So, how can we solve this problem? For starters, we need to demystify the story. Historians write history, but we are not evoking an exact replica of the past, simply because that will never be possible. Historians "invent" the past. Of course, we should not write irresponsibly, without using the correct fonts, we cannot write whatever comes to our mind, just because it sounds fashionable. Historians must "present" past events. However, the history we display is not so different from the show we provide with these re-enactments. It is important to communicate and share ideas, build bridges, but these bridges must allow for the crossing of information in both directions.

Organizers of these events should discuss the finer details with historians. For example, what did medieval wine taste like? What kind of weapons did a knight use in a Portuguese joust? There are people in Portugal who study these details, and therefore, we should encourage these researchers to keep in permanent contact with the organizations and become more involved with the progress of these events. They must express their opinions and suggest improvements where necessary. We all know, of course, that the main objective of this kind of events is entertainment, the involvement of thousands of people, but the economic return is what prevails. The hope is that the public will be interested and curious to know more, and if people have fun, they will have time to learn more and spend more...

3 Methodology

For this study, we shall first carry out an extensive literature review on the subject of re-enactments, its origin and history, the difficulty of forming a definition of the field, the dangers posed by globalization as well as the opportunities brought up by globalizing forces and tourism for the preservation of history. In order to carry out this research, we used articles, scientific journals, and books on the pertinent areas related to re-enactments.

Guerreiro and Mendes (2011) identified what motivated people to participate in a medieval event, and novelty was the main drive: "Socialization, Novelty, Intellectual Enrichment, Rest and Relaxation, and City Identity. The search for novelty, to know the roots of the city, and the contribution of this event to intellectual enrichment are the most important factors" (Guerreiro & Mendes, 2011). Formica & Uysal also identified novelty as the main motivation. Roche concluded that various reason underlies the creation of a festival or fair, such as celebration, taking pride in one's community, and preserving one's cultural identity (Roche, 2000 cited by Carnegie & McCabe, 2008). The motivation for organizing a medieval festival is organized mainly in order to benefit the host venue in tourism terms. According to Ritchie, special events, such as medieval fairs are "large one-time or recurring events of limited duration, developed primarily to increase the awareness, attraction and profitability of the host site" (cited by Guerreiro & Mendes, 2011). Of course, this only applies if they are well organized.

Medieval events reproduce the past in our days through the use of theatrical devices. "Theatre is used here [in re-enactments] as a technology for organization, creating its own time and place, allowing realities to be parallel" Taalas (2006): To a large extent, we deal with reconstructions of history when we engage with medieval and heritage events: "People not only live in society and occupy social space, we are producing history in order to live (...) it [a medieval event or re-enactment] is not just the construction of a new history, but a reorganization and renegotiation of the past" (Godelier, 1984/1987 cited by Taalas, 2006)."

There are studies that seem to indicate that the more fun in its nature an event is, the more likely it is that it will be re-enacted. Regarding the replication of events of a more fun nature: "the researchers highlight that those who feel the greatest pleasure and excitement are also the most satisfied and loyal to the event". If an event produces happy participants, that will lead to it being re-enacted and replicated more than another event that is more honest in the re-enactment of the respective historical event. As Pine & Gilmore stated in their article on marketing: "Customers buy an experience, pay to spend time enjoying a series of memorable events". Individuals are unlikely to want a tedious re-enactment down to the last detail of history. They want an entertaining experience as consumers, being provided with a memorable experience, they wish to purchase a series of memorable events.

It becomes increasingly difficult to distinguish between truth and mere interpretation in the world: "Scholars argue that it is too compelling to determine what is authentic [medieval culture] within increasingly fragmented communities where such points of reference are lost or altered" (Cohen-Hattab & Kerber, 2004). Our modern globalized world is slowly undermining our understanding as to how these festivals and events can represent medieval culture in an authentic way.

Other investigations cited by Fernández et al. (2016) point to contradictory studies that state that "Historical re-enactment encompasses several themes and genres, where authenticity and care in the re-enacted elements demand an utmost attention to detail (Hart, 2007). This particularity can make the difference between different events, where the value of authentic experience is increasingly valued. (…)." More than mere theatrical performances, such events are unique for their historical setting and the committed participation of the locals, who thus seek to maintain and defend their heritage (Agnew, 2005). The opposing research asserts that tourism and globalizing forces are not only not destructive, but can play a positive role, as they help communities that are proud in their heritage and its preservation. An example is living history, which turns the interpretation and construction of the past into an important educational tool and a part of modern life's leisure for all involved, as well as for historians and educators (Carnegie & McCabe, 2008). Cohen-Hattab & Kerber, in the same vein as most researchers, state: "people are able to create new communities in the present, creating or building shared ideas about the past and taking advantage of material culture in order to meet their present needs/interests" (Carnegie & Mccabe, 2008). It is a means by which society seeks to avoid the loss of authenticity. It should be noted that Granja states that "Some visitors are extremely interested in the authenticity of the sites and may be disappointed if they are not 'real'".

It is often argued that the people that most damage medieval heritage and culture are the ones that seek to preserve them, as tourism and commodification degrade authenticity. Carneiro, Eusébio & Santos point out: "cultural events have been widely studied in the anthropological literature, where tourism is often seen as an agent of change, contributing to the decline of cultural authenticity. Cultural events can be seen as a contribution to development, but also as a risk that imposes a negative effect on the community in terms of cultural change or commodification" (Carneiro, et al. 2016). Living history is criticized by some researchers as an educational tool as being an illusion and a sham (although it allows human contact which is a benefit): "… contact with the living past is impossible and those interpretations of living history in museums or historic sites are as illusory as any other form of populating historic space. Their main benefit is that they allow us to explore our relationship to the past through human contact'." (Goodacre & Baldwin, 2002: 59 cited by Carnegie & Mccabe, 2008). In the midst of the debate, we often find festivals with this so-called "staged authenticity": "Ethnic groups hold festivals to promote nostalgia for the past, as well as to reinforce their present cultural ties… the staging does not need to exclude authenticity. In other words, what is staged is not superficial, as it contains elements of the original tradition."

However, Carnegie & McCabe (2008), in the context of re-enactment events, state that "Cohen-Hattab & Kerber (2004) argue that academics are more concerned with authenticity than tourists, for which experience is key." Once again, minimizing the issue of the effects of tourism, globalization, and cultural exchanges on the authenticity of heritage events, that is, medieval tourist events.

4 New Tendencies

Evidently, new trends in tourism have emerged, as well as in the frequency of historical recreation events, as people have changed their behavior patterns:

Increased demand for less crowded and more sustainable destinations.
The most sought-after tourist products are those that offer greater flexibility for changes and cancelations.
Tourists are more demanding of tourism providers in terms of hygiene and sustainability.
Some people will avoid traveling if they have a cold or the flu, or because of the attention they will get from people traveling with them.
There is a greater demand for hygiene measures (both in the facilities and in the employees).
Recommendations are requested from travelers to minimize the risk of contagion.
Travelers will be more concerned about health safety, both in facilities and in food.

We found that, in the particular case of Santa Maria da Feira, several factors contributed to the consolidation of projects for historical re-enactments, and the city knew how to use them to enhance the regeneration of its historic center, Quinta do Castelo and the banks of the river. Together, these prerogatives make us dream of an open-air theme park, in spaces that allow the control of entrances and that encompass the different periods of history.

This possibility is undoubtedly among the latest trends in tourism, as Newton (2020) emphasizes, "the restrictions imposed by COVID-19 have caused an increase in visits to parks and forests."

Feira's tradition, history, heritage, and experience in the organization of historical re-enactments, and consequent urban renewal, make it a prime location for a theme park.

However, it is not dissociated from other factors and trends that appear every day in the world, because of COVID-19. Tourism undergoes transformations at the same speed that societies also change.

5 Discussion

In view of this, several questions must be asked: It is safe to say that investing in a theme park is a good option, as it tends to potentiate synergies where it is inserted and contributes to the rejuvenation of traditional tourism, by attracting unexplored market segments like Puy du Fou, Guédelon, or Open Air Museum—Poggibonsi Archaeodrome or other product?

Puy du Fou

Puy du Fou (Fig. 3) is regarded as the world's best theme park, with four themed villages one can visit to become more immersed in history: The Royal Village, the Andalusian Askar, La Venta de Isidro and El Arrabel, a medieval market at the feet of the great wall.

At sunset, it will be time to return to The Toledo Dream, the outdoor evening show that takes spectators through more than 1500 years of history on a historic dramatized journey, with 200 actors and incredible special effects, ranging from the reign of King Recaredo I in the sixth century to the arrival of the railway, as well as the Navas de Tolosa battle or the discovery of America.

A number of shows are held throughout the day, each lasting approximately 30 min. They can each seat between 1800 and 2000 people. The program will be adapted depending on the number of spectators and on the current government guidelines in order to avoid queues and long waits.

In the four historic villages, there are more than 20 food outlets as well as craft workshops and boutiques with handmade, local, original and sustainable products. Furthermore, to cope with the heat of the Toledo summer, 15 fountains and water points have been installed throughout the park, as well as 4750 m^2 of pergolas and awnings, and 80% of the paths have rest areas in the shade, benches and misting points.

Guédelon

Guédelon (Fig. 4) is a castle recently built from scratch near Treigny, France, and the main component of an archeology project that aimed at recreating a thirteenth-century castle using period dress, materials, tools, and techniques.

The stone and wood used in the construction are all local. The project's chief architect for the project designed the fortification in accordance with the twelfth and thirteenth centuries model promoted by Philip II of France.

Construction began in 1997 on a site with plenty of construction materials nearby: a stone quarry, a forest, a pond.

The promoters of the project, a fortification expert and a castellologist, following a study of Saint-Fargeau castle, which revealed a stone castle beneath the more recent red-brick walls, had the idea to recreate it. Together with a small group of enthusiasts, they made it happen.

Poggibonsi Archaeodrome—Open Air Museum

Part of the Park of Poggio Imperiale in Poggibonsi (Tuscany, Italy), the Archaeodrome (Fig. 5) is an open-air museum, still in construction, that reproduces a ninth–tenth century village that was discovered by archeologists from the University of Siena.

Its larger building reproduces the proprietor's dwelling, who was probably a member of Carolingian military. Around it, one can find areas devoted to crafts. The other, significantly smaller building, was the abode of a family of peasants. Also, part of the complex: blacksmith's forge, a hen houses a bread oven and a space for household chores.

Creation of New Tourist Products, Such as Dias do Burgo Does the future lie in the creation of a park alluding to the various periods of the history of Portugal in the municipality of Feira, and attracting new customers/tourists, like the so-called digital nomads?

As we have already seen, the impacts on professionals who depend on tourism have been terrible, as many jobs related to the sector are precarious. (Chanel, 2020; see also, UNWTO, 2022). There are numerous locations, such as Santa Maria da Feira, where businesses of various sizes that in some way depend on tourist activity are in a situation of

Fig. 3 Puy du Fou. *Source* Puy du Fou

Fig. 4 Guédelon, a Medieval Castle in the Making. *Source* https://www.guedelon.fr/en

suspension, with indications that many of them will probably never reopen.

The global immobility caused by the pandemic crisis has greatly confused tourism professionals, policy-makers, and researchers (Miles & Shipway, 2020). Relatively recently, Hall (2015) predicted that a similar health crisis could produce such manifold effects.

But we know that following past crises, including epidemics, the tourism managed to recover, thus proving its resilience as it mitigates fluctuations in supply and demand (Novelli et al., 2018; Papatheodorou et al., 2010). Small-scale, local companies targeting niches should be better prepared to face the challenges of market demands.

Fig. 5 Archeodromo live. *Source* https://www.archeodromopoggibonsi.it/

6 Results the Future of These Events: Historical Re-Enactment and Heritage Education

Several authors have pointed out the educational potential that historical re-enactments have as a methodology for disseminating History, especially in the field of Museology and Heritage, since direct contact with the public favors their interactivity (fundamental in didactic and educational actions) and incorporates resources such as empathy, surprise or emotion, elements necessary to achieve meaningful learning.

With regard to the creation of characters and their costumes, Llonch highlights that costumes are "an object of unquestionable value for the didactics of history and other social sciences, as it has all the qualities of the didactics of the object, with an attractive additional closely linked to its appearance, aesthetic functionality and its visual appeal [...]" (Llonch, 2010).

In the same way, living museums and historical re-enactments or theatrical narrations are a powerful pedagogical resource in the transmission of knowledge of history, as highlighted by Gregorio and Garcia (2014).

The use of this type of resources is now highly recommended in the dissemination and interpretation of our heritage; since, in a playful and attractive way, it transmits knowledge through pedagogical capacity, seeking emotional interactivity and giving rise to positive attitudes, creating a unique experience (Gregorio & Garcia, 2014, p. 96).

In recent years, many museums and interpretation centers have started to offer their visitors didactic content based on tactile experience, with the use of replicas as the main element of effective learning. In response to this new situation, the demand for reproductions of reconstructed archeological materials and scenarios has increased. The professional elaboration of archeological replicas (objects, fabrics, clothes, etc.) is usually done by hand, the result of a process in which the knowledge derived from a specific artistic training and experimental studies of archeology must converge, including the recovery of artistic, traditional techniques. One of the most interesting types of historical recreation is the material reconstitution of historical–archeological structures and contexts called archeodromes.

They consist of a full-scale reproduction of an architectural–archeological ensemble previously documented with precision. They respond to ambitious experimental archeology projects (as we have seen before, the Guédelon Project, among others) that incorporate construction techniques and provide new data on old technologies during their execution. In the same way, they represent an excellent communication channel and constitute an informative medium of great impact, translating into a large amount of scientific data in a visual summary of easy access to the viewer, in addition to leading him to an immediate understanding of how it is done.

Therefore, in the last decades, the concept of the archaeodrome, associated with the concept of open-air museum (OAM), implemented in Skansen (1891), a large number of open-air archeological museums (AOAM) emerged. This type of interpretive center is supported, as Valenti (2018) stressed, on the following pillars:

Reconstruction of structures, supported by experimental archeology, using ancient technology, recreating everyday life associated with narration and interpretation. They are not, therefore, mere large-scale dioramas;

AOAMs are live recreations that usually include live museum actions, in which the storytelling is given in the "first person' by the locals themselves, represented by different characters that are brought to life by the reenactors which brings the story back to life (re-enactment)? This type of teaching is a didactic strategy that uses dramatizations, making it unnecessary for the audience to have to imagine some of the more abstract historical concepts. However, it implies the need for prior and exhaustive documentation work, since the information provided must be as objective as possible, distinguishing between what is plausible and what we know for sure, which unfortunately in most cases is not what happens.

One of the paradigmatic examples of archeological open-air museum, AOAM is the Archeodromo Live in Poggibonsi (Siena, Italy).

The successive archeological excavations commissioned by the University of Siena from 1993 until now prove the medieval occupation of Poggibonsi from the fifth to the fifteenth century. From the Carolingian phase (ninth–tenth centuries) known as Poggio Imperiale (excavated between 1993 and 2009), there is a large number of elements identified with various structures, corresponding to: 6 houses (one of them large, longhouse type), 1 barn, 1 warehouse, 1 blacksmith's workshop, 1 place for slaughtering animals, 2 enclosures (multifunctional yard and orchard), 1 chicken coop and 2 haystacks. These structures would be an important structure of the time, probably ruled by a landowner or Dominus related to the Carolingian court and its militias (Fronza, 2018, p. 67).

The large-scale reconstruction project of the archeological context related to the Poggio Imperiale in its Carolingian phase began to materialize in 2014. However, the preliminary project began to be developed a few years earlier by Marco Valenti & Vittorio Fronza (Fronza, 2018, p. 72). After evaluating different proposals, it was decided (2013) that its location should be close to the archeological site, at the top of the hill. The site offers three artificial esplanades that modify the original morphology of the city, making the spatial relationships remain intact.

According to Fronza, a pragmatically adapted experimental archeology protocol was established for the reproduction of structures, but in no case did it lead to less data collection. In this way, wherever possible, replicas of tools and materials were used (axes, knives, scrapers, spatulas, wooden needles to fix the reed beams on the roof, etc.), although modern machines and tools were occasionally used, (excavators, cranes, scaffolding, harnesses, pulleys, chainsaws), to comply with the safety regulations in force and the project schedule (only 4 months for the longhouse and between 2 and 3 months for smaller housing structures.

As the author points out, "Overall, the production of the reed bunches is a good example of the compromise we took between a close experimental protocol and the pragmatic approach dictated by external needs" (Fronza, 2018, p. 81). Likewise, all the interior furniture was carried out according to traditional procedures.

The main objective of heritage didactics is, once again, to establish an inclusive communication channel, capable of winning the interest of the general public and moving it "*senza ricorrere a quel sensationalismo*" (Salzotti, 2018, p. 0.101).

However, in the current period of pandemic, we live in the age of technology, a reality in which digitization is becoming increasingly wild and extreme. Therefore, and paradoxically, the fact that we resume direct contact with the public through verbal and primitive communication proves to be original and innovative (Salzotti, 2018). Therefore, the creation of story environments contributes to capturing the user's attention, especially the younger ones.

The first-person narrative of an inhabitant of a village (wearing rigorous period costumes) breaks the psychological and reverential barrier between the guide and the consumer/user (teacher–student within the scope of regulated education) between "learned and ignorant" (Salzotti, 2018, p. 103), creating a safer and more conducive climax for the user/student to feel comfortable and free to ask questions without fear of sounding ridiculous (because they feel like they belong). In addition, it allows establishing a relationship between the two from a supposed reciprocal curiosity, typical of characters belonging to two different eras, which can make it a guiding thread for dealing with everyday discourse. The use of a simple and mainly close language is fundamental to stimulate the user's curiosity (tourist, visitor or student) to deepen what is being told, because using understandable language does not make you feel excluded. In the same way, the use of pedagogical strategies related to experiential archeology and object didactics are the order of the day in this type of interpretation center.

Thus, considering the different nuances, it seems to me that historical recreations can integrate the archeological open-air museum model (adapting the Dias do Burgo project, for example integrating the uniqueness and spectacularity of the Puy du Fou de Toledo project), but addressing the different periods of history, thus preventing the theme of the Middle Ages from being exhausted.

7 Conclusions

This new paradigm could take place in the refurbished Quinta do Castelo of Santa Maria da Feira, which represented a total investment budgeted at 1,822,138.46 euros, and in other areas of the city.

The project included interventions in landscape architecture, architecture, foundations and structures, hydraulic infrastructure, electrical infrastructure, telecommunications infrastructure, installations, and mechanical equipment, in a total area of four hectares of Quinta do Castelo.

Among the interventions, we highlight the restoration of the lake and the grotto, the construction of specific areas to host temporary camping and parking facilities, the restoration of an old building, currently in ruins, for a support building for activities in the park, restoration the vegetation structure, planting trees and shrubs, defining new paths or moving the Porta da Cruz, also called "Porta da Cerca," to the main entrance of the area. The project also included important interventions in terms of irrigation systems, drainage infrastructure, sanitation and water supply, scenic lighting, electrical supply, Wi-Fi network and public address system. These large areas of intervention are broken down into 28 requalification actions foreseen in the project,

Quinta do Castelo, in Santa Maria da Feira, opened its doors to the public on July 6, 2019, completely renovated.

In addition to this important requalification of Quinta do Castelo, for the next four years, the municipality favored the construction of the Feira Cycle Route—Urban Path of Cáster, since cycling is not just a fad. It has unquestionable benefits for the improvement of the quality of life of populations and the environment, also allowing cities to be fully experienced. It is from the perspective of encouraging the use of mobility in soft modes and valuing the public space that the Santa Maria da Feira City Council will build the Feira Cycle Route—Urban Path of Cáster, in a total of 2665 m, favoring a natural channel, the river Caster.

The Feira Cycle Path—Cáster Urban Route will unify the eastern part of the city, with the historic core as its main centrality, and with the western zone, of recent urban expansion, always taking advantage of the Cáster River for its implementation.

This clickable/pedestrian route will connect the historic center, the core of schools, the sports facilities (Municipal Swimming Pool and Lavandeira Gymnasium Pavilion), Holy House of Mercy, social equipment, the railway station and the future transportation coordination center, and will benefit from the proximity to parking lots and the main access to the city (N223).

An interesting proposal for attracting new visitors could involve capturing digital nomads, providing some of the spaces mentioned above (Quinta do Castelo, Guimbras and the Urban Park) with structures that would allow "new customers" to stay in the region for longer periods of time. The pandemic has brought work into our homes, but these "homes" can be anywhere (as long as they are connected to the Internet). Digital nomads are proof of this, and there are several countries creating conditions to attract these "tourists–workers," who can stay for a few months or more than a year—Portugal is no exception and is seen as "a country of excellence" for this.

COVID-19 has made teleworking the rule for many. In the last year and a half, employees and employers have realized that with a computer and Internet access, the office can be divided into tens or hundreds of homes. But these "homes" don't always have to be the same: and this is where digital nomads come in, who end up roaming the world while working.

One of the advantages of this professional nomadism is the possibility of exploring a local culture as a semi-resident, since the period of stay as a worker extends beyond what would be the short visit that the holidays allow. In this sense, many countries have created visas for digital nomads—which, in times of a pandemic, makes up for the shortfalls in tourism.

According to a top from InsureMyTrip, and contrary to what would be expected, it is not a country full of sun that occupies the first place in the table of preferences of digital nomads, but one where there is plenty of cold: Norway. This is where these hardworking tourists find themselves staying for a year or more. The other places on the podium are occupied by Mexico and Germany. But Portugal appears right behind, in fourth place. The remaining top 10 includes Iceland, Greece, Costa Rica, Jamaica, Spain, and Bermuda. Our country has developed a temporary resident visa that can be used by freelancers and entrepreneurs. This way, the holder can stay in the country for one year, with the possibility of renewal for successive periods of two years. Another outstanding initiative is the creation of the e-Residency Program, which was included in the Simplex Program 2019 with the designation "Digital Identity." But specifically for digital nomads, there is no doubt about which is the main project so far: The Digital Nomads Madeira 85, which seeks to attract digital nomads to the region. However, other places are also attractive for this way of life, as is the case of Porto, Lisbon, Ericeira, and Lagos.

However, in our perspective, the city of Santa Maria da Feira is clearly able to develop, in a creative and innovative way, the permanence of these nomads, perfectly integrated in our events of historical re-enactment events, which are differentiating and capable of attracting tourists.

References

Carnegie, E. & Mccabe, S. (2008). Re-enactment events and tourism: meaning, authenticity and identity. *Current Issues in Tourism*

Carneiro, M. J., Eusébio, C., & Santos, A.C. (2016, July). The emotions generated by re-enactment events: A market segmentation study of the Medieval Journey in the Land of Saint Mary. *Journal of Convention & Event Tourism, 17*(3), 192–219.

Chanel, S. (2020, April 15). 'It's catastrophic': Fiji's colossal tourism sector devastated by coronavirus. The Guardian. https://www.theguardian.com/world/2020/apr/16/its-catastrophic-fijis-colossal-tourism-sector-devastated-by-coronavirus [Google Scholar]

Cohen-Hattab, K., & Kerber, J. (2004). Literature, cultural identity and the limits of authenticity: A composite approach. *International Journal of Tourism Research, 6* (2), 57–73

Crang, M. (1996). Magic kingdom or quixotic quest for authenticity.

Fronza, V. (2018). «From excavation to reconstruction of timber buildings at the Archeodromo of Poggibonsi (Siena, Italy) », en M. Valenti, S. Ricci, V. Fronza (coords.): *Dalle fonti alia narrazione. Ricostruzione storica per il racconto delia quotidianità,* Quaderni dei dipartimento di Archeologia e Storia delle Arti. Sezione Archeologia, Universita di Siena. Florencia: Alfinsegna dei Giglio, pp. 67–84.

Gregorio, M. A. C. D., & García, Y. J. (2014). «Living History, teatralizaciones y demostraciones como recurso didáctico, lúdico y pedagógico en la interpretación de yacimientos arqueológicos y espacios patrimoniales», en P. J. Lavado Paradinas y V M. Lacamba Gambau (coords.): VII Jornadas Nacionales de Ludotecas: ponencias y comunicaciones: juegos romanos, juegos de agua, Teruel: Comarca de la Sierra de Albarracín, pp. 95–119.

Guerreiro, M., Valle, P. & Mendes, J. (2011). What motivates visitors of a medieval fair? *Tourism Today*

Laing, J. & Frost, W. (2012). *Books and travel: Inspirations, quests and transformation.* Channel View.

Lennon, J. J., & Foley, M. (2000). *Dark tourism: The attraction of death and disaster.* Casse.

Lowenthal, D. (1985). *The past is a foreign country.* Cambridge University Press.

Llonch, N. (2010). «La indumentária como fuente para la didáctica de la Historia: problemática y estado de la cuestión», Didáctica de las ciências experimentales y sociales, *24,* 63–72

Miles, L., & Shipway, R. (2020). Exploring the COVID-19 pandemic as a catalyst for stimulating future research agendas for managing crises and disasters at international sport events. *Event Management.* https://doi.org/10.3727/152599519X15506259856688

Novelli, M., Burgess, L. G., Jones, A., Ritchie, B. W. (2018). "No Ebola… still doomed"–The Ebola-induced tourism crisis. *Annals of Tourism Research, 70,* 76–87.https://doi.org/10.1016/j.annals.2018.03.006

Papatheodorou, A., Rosselló, J., & Xiao, H. (2010). Global economic crisis and tourism: Consequences and perspectives. *Journal of Travel Research, 49*(1), 39–45. https://doi.org/10.1177/0047287509355327

Salzotti, F. (2018): «La comunicazione alT Archeodromo di Poggibonsi (SI): Accessibi- lità, esperienzialità, partecipazione attiva e dialoghi immersivi», en M. Valenti, S. Ricci, V. Fronza (coords.): *Dallefonti alia narrazione. Ricostruzione storicaper il racconto delia quotidianità,* Quaderni del dipartimento di Archeologia e Storia delle Arti. Sezione Archeologia, Universita di Siena. Florencia: AlFInsegna del Giglio, pp. 101–116.

Taalas, S. L. (2006). Consumption of a medieval event: Employing theatre in the construction of local and spatial organization. *Management & Organizational History.*

UNWTO. Published: 18/01/2022

Valenti, M. (2018): «Dalle fonti alia narrazione. Ricostruzione storica per il racconto delia quoti- dianità. Problemi e prospettive», en M. Valenti, S. Ricci, V. Fronza (coords.): *Dallefonti alia narrazione. Ricostruzione storica per il racconto delia quotidianità,* Quaderni dei Di- partimento di Archeologia e Storia delle Arti. Sezione Archeologia, Universita di Siena. Florencia: AlFInsegna dei Giglio, pp. 7–45.

Walsh, K. (1992). *The representation of the past in the post-modern world.* Routledge.

Assessing On-Water Recreation Experiences in Midwestern U.S. Glacial Lakes: Identifying Areas of Conflict and Agreement

Rebecca L. Kauten and Olivia J. Calvin

Abstract

Lake-based recreation resources are bound by physical, social and economic qualities which often lead to conflicting experiences and perspectives on personal experiences. What one person may perceive as a highly positive recreation experience may in fact lead to a strong, negative sentiment for others, particularly when activity is confined to a finite space. Conversely, unintended consequences of policy to limit one form of recreation may introduce other, new challenges. This study aims to assess perceptions and levels of past and present on-water activities by current users of the Iowa Great Lakes as a recreation resource to generate a general profile of the local tourism community. Qualitative survey data and observational activity assessments were collected from May to September during the 2020 recreation season. Both approaches test a central hypothesis that all community members use and value local water recreation resources equally. Observational data also provide foundational knowledge of regular activity throughout the season in a designated "no-wake zone" known for high levels of anchored on-water activity due to restricted motorized activity. Results indicate a divide between property owners and others recreating in the region, as well as diverging satisfaction levels for recreation using motorized and non-motorized watercraft. The study builds capacity for researching broader water recreation topics related to environmental impact, recreation conflict and the capacity for locally led comprehensive usage plans for lake-based resources.

R. L. Kauten (✉)
Iowa Lakeside Laboratory, Milford, IA, USA
e-mail: rebecca-kauten@uiowa.edu

O. J. Calvin
Coe College, Cedar Rapids, IA, USA
e-mail: ojcalvin@coe.edu

Keywords

On-water recreation • Recreation conflict • Sustainable tourism • Glacial lakes

1 Introduction

This study aims to assess perceptions and levels of past and present on-water activities by northwest Iowa residents regarding regular usage of the Iowa Great Lakes as a recreation resource. Data collected test a central assumption that all residents use and value local water recreation resources equally. It launches research on broader water recreation topics related to environmental impact, recreation conflict and the capacity for locally led comprehensive usage plans for lake-based resources. As a pilot study, this research gathers a combination of qualitative and quantitative data to assess conditions regarding recreation-based use of local water bodies and prioritization of activities based on personal preferences. Such a community profile provides foundational knowledge relevant to the local community and use of water resources in moderately developed tourism communities.

In recent years, a combination of motorized, non-motorized and anchored on-water recreation activities has increased the frequency and concentration of users on freshwater resources like the Iowa Great Lakes Region to the point where residents have voiced concerns over social and environmental impacts (Wills, 2010). Conflict in on-water recreation experiences such as these lead to local concerns related to economic development, quality of life and the concern for potentially adverse environmental impacts. Results of this research not only can guide comprehensive sustainable tourism plans for the Iowa Great Lakes Region, but also serve as a resource for other locations where competing recreational activities lead to social, economic and environmental conflicts. The initial scope of this study

A. Mandić et al. (eds.), *Cultural Sustainable Tourism*, Advances in Science, Technology & Innovation,
https://doi.org/10.1007/978-3-031-10800-6_16

concentrated on answering the following questions related to recreation conflict in the region.

- To what extent do people recreating in the Iowa Great Lakes sense a change in their levels of on-water activity in recent years?
- What types of activity appear to occur the most/least in the region?
- What variation in attitudes and beliefs exist among long-term and/or short-term community members?

2 Literature Review

Unequal experiences within tourism communities remain a common phenomenon, particularly for water-based recreation. Since the mid- to late twentieth century, scholars have recognized the social, economic and environmental impacts of recreation on water resources (Jacob & Schreyer, 1980; Teclaff & Teclaff, 1972). While economic activity may thrive and draw visitors for short-term experiences, social and environmental impacts can result in an unsustainable outcome for local communities and stakeholders (Guaita Martinez et al., 2019; Ivars Baidal, 2004; Park et al., 2012). Sustainable development, which includes tourism as an economic driver, requires integrated consideration of environmental and social concerns along with economic prosperity (Bacher, 1987; Frank, 1987; Keeble, 1988). Globally, tourism and human development have spurred an environmental and social imbalance leading to broad-scale unsustainability for the sake of economic growth. Unfortunately, one of the greatest challenges in sustainable development and sustainable tourism has been the quest for appropriate metrics to assess perspectives of both short- and long-term stakeholders (Guaita Martinez et al., 2019; Toman, 1994).

As recreation activities grow in popularity, unintended consequences of resource exhaustion, degradation and a diminished individual experience can result. A positive public image and the prestige of recreation in a given area may increase, but property prices, security concerns, environmental damage and degradation due to resource pressure can polarize perspectives between long- and short-term visitors, particularly during peak seasonal visits (Almeida-García et al., 2016; Manning & Powers, 1984). The more popular an area becomes, the greater the risk of potential degradation of the very conditions tourists and visitors seek.

Non-compatible activities such as motorized and/or non-motorized activities can lead to what is considered recreation conflict where competition can not only lead to an exhaustion of resources, but also a diminished personal experience (Cahill, 2013). High-traffic, high-usage area managers such as state and national parks, forest lands and open water resources have implemented comprehensive recreation usage plans to mitigate such long-term effects (Kakoyannis, 2002; Spiers, 2012).

Asymmetrical conflict in recreation refers to when the competition for use of a given resource results in interference, change and often negative perceptions of competing actors and activities (Dawson, 2000). Goal interference in recreation tends to produce conflict in four major areas: impact based on personal meaning or identity; attachment to a given resource; the mode of an experience; and a generalized tolerance, acceptance or rejection for others (Jacob & Schreyer, 1980).

In the latter twentieth century, the concept of "sustainable tourism" evolved to describe, characterize and analyze development strategies in geographies reliant upon recreation and visitors for a thriving economy. This paradigm shift reconsiders tourism more as a philosophy than an industry, as a means of achieving socially worthy goals and integrating such approaches throughout the local economy for sustainable development as a whole (Bosak & McCool, 2016; McCool, 2016). In order for such endeavors to succeed, stakeholders must find commonality for the sake of communication and long-term cooperative agreements (Beritelli, 2011; Bramwell & Lane, 2003; Maiden, 2008).

The concept of ecotourism considers the positive and negative impacts of tourism falling within the same social, environmental and economic categories as the primary definition of sustainability (Bacher, 1987; Fennell, 2008; Frank, 1987; Keeble, 1988). However, ecotourism differs from the concept of sustainable tourism due to the significant ecological sensitivity of biological hotspots for the sake of nature-based tourism (Gill, 2018). Successful sustainable tourism endeavors can also occur in more developed areas. High-pressure tourist destinations within developed communities adjacent to natural systems rely on policies where resources are protected, but a level of common use is not only expected, but encouraged (Domínguez-Gómez & González-Gómez, 2017; McCool, 2016; Su et al., 2018). The fragility of ecotourism destinations emphasizes the need for careful planning that may or may not have been part of the process to establish more developed tourist destinations. Within more developed regions, the effort may be to mitigate further impacts beyond the current state.

Conflict can be from one or a combination of factors, and causality implies that if a factor is present at one point in time, then at a later point conflict will exist, assuming a social interaction has taken place. Recreation conflict results when intense participants must interact with casual ones. The importance of a specific recreation resource varies with one's range of experience which affects if viewed as unique or common; feelings of possession and role of a place in central life interest; and a connotation of status (Jacob & Schreyer, 1980). Differing views on technology can cause conflict, which may be generational. Furthermore, ethnic, racial and social distinction may also cause conflict including prejudice. Within an on-water recreation situation, such forms of conflict can lead to false perceptions of difference, when in reality a shared sense of value may exist.

In the context of recreation, conflict occurs as a function of satisfaction in relation to a given recreation activity, one's motivation to act, the level of disruption in relation to tolerance and the capacity for alternative action, as it occurs relative to other interpersonal experiences (Bartol & Locke, 2000; Ivy et al., 1992; Jacob & Schreyer, 1980; Locke, 1970; Locke et al., 1970; Maiolo et al., 1992; Vroom, 1964). If no equivalent alternative exists, conflict arises, as demonstrated in Fig. 1.

Stakeholders of tourism communities ranging from individual community members, local organizations as interest groups and private industry often maintain a shared sense of attachment and value to popular local recreation resources. Conflict can arise across personal, interpersonal and organizational planes as internal, external or interrelational challenges (Schneider, 2000). By identifying, if not embracing such challenges, the potential for growth across various aspects may exist for all involved.

Existing studies have assessed user experiences and levels of conflict related to commercial and recreational fishing, motorized versus non-motorized watercraft, interference by personal motorized watercraft (jetskis) in recreational fishing areas and the impact of snowmobiling trails on cross-country skiing experiences (Baas & Burns, 2016; Dawson, 2000; Marcouiller et al., 2005; Scott et al., 2005; Teclaff & Teclaff, 1972; Uleck, 1971; Vittersø et al., 2004). However, no peer-reviewed studies to date compared levels of satisfaction among those participating in on-water recreation in a developed, non-wilderness experience where the conflict occurs between established residents and day users. Despite an often shared perception of conflict, the need for cooperation between stakeholders in recreation and tourism communities abounds. While a distinct social divergence between temporary and permanent residence can likely be observed on the local scales, literature do not often indicate a statistically significant distinction between opinions and attitudes related to recreation experiences (Confer et al., 2005).

Existing literature have also considered quality of life impacts, emphasizing the individual experience as a means of assessing satisfaction and fulfillment based on a given experience (Andereck & Nyaupane, 2011; Jurowski & Brown, 2001; Perdue et al., 1999). A similar concept, quality of life (QOL), stems from research related to intellectual development as a holistic measure of individual emotional welfare, sense of safety, sense of fulfillment and satisfaction and inclusion (Brown & Brown, 2005; Su et al., 2018). Conflict within an on-water recreation experience can dramatically transform a positive activity to a negative experience, especially if personal safety is at risk.

In the United States, all navigable water bodies are accessible by any member of the public through common law. Beginning in 1812, "navigable waters" were defined by the United States Secretary of the Army as open waters for boat passage without obstruction, risk of harm to users, and accessible for public usage ("US Code Title 33, Chapter 1: Navigable Waters Generally," 1812). While the intent of original codification was to reduce interference with early development of United States watercourses, rulemaking related to navigable waters remains an active term in local policy. For the State of Iowa, the term refers to all lakes, rivers and streams, which can support a vessel capable of carrying one or more persons during a total of six months period in one out of every ten years (Hayward, 1912). Public access to navigable, sovereign waters such as the Iowa Great Lakes Region enables open access to users without paying a fee or regulation of the number of users at a given point in time ("Water Rights and Related Issues," 1994). As a result, fees for on-water recreation, boat launching, day usage and other activities in the public domain are prohibited due to common law. Local industry relies heavily upon what has been referenced as the "tourism machine" as the primary economic driver (Reynolds, 1993). With the inability to

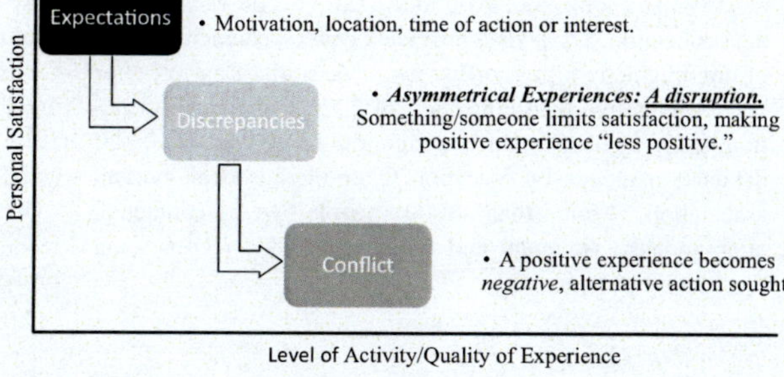

Fig. 1 A conceptual framework of recreation conflict related to personal experience and satisfaction

impose regulation or restricted access, such locations suffer from the vicious cycle of beginning as an attractive destination only to result in unattractive, unsustainable conditions (Russo, 2002). The question becomes whether private industry and market-driven corporate social responsibility (CSR) may play a stronger role in sustainable tourism initiatives in the future (Beritelli, 2011; Poudel et al., 2016; Su et al., 2018; Toman, 1994). If regulation and governmental strategies limit the capacity for enforced sustainable tourism measures, CSR may serve as a viable community-based alternative approach.

Tourism as an economic activity in the United States is based in the capitalist mode of production and in neoliberal economics (Bosak & McCool, 2016). The concept of sustainable tourism integrates social, economic and environmental factors as a means of mitigating the inevitable impacts of development through increased usage and visitors. Sustainable tourism research to date expresses the need for local stakeholder involvement in planning, particularly when adverse affects are more prevalent for local residents (Boley et al., 2017; Choi & Sirakaya, 2005; Panzer-Krause, 2019; Poudel et al., 2016). Similar to an importance-performance assessment, which can determine levels of satisfaction from a single user experience, a comparative "Mean Maximum Satisfaction" (MMS) calculation as an estimation of value can compare levels of satisfaction relative to multiple activities on a normalized scale (Bradford & D'Amato, 2012; Mimbs et al., 2020).

3 Approach to Research: Study Area and Methodology

The study was divided into two primary scopes to capture observational and qualitative data related to summer, on-water recreation activities. Observational data collection occurred from May 22 to September 7, 2020, assessing physical activities within a typical summer holiday season for the region. Every Wednesday, Friday, Saturday and Sunday (as well as Mondays of holiday weekends), the number of people and boats present in Miller's Bay were recorded at noon, 2:00, 4:00 and 6:00 p.m. on each day of data collection. These data provide a general characterization of ongoing activity in the Bay from Memorial Day to Labor Day. Qualitative data collection in 2020 gathered information from contacts over the age of 18 within an email database managed by Vacation Okoboji, the local tourism association. More than 3000 people were contacted, anonymously, by email and asked to provide feedback on the following topics:

- Beliefs About Other On-Water Recreation Users and Activities. Participants rate general beliefs about other on-water recreation users and related activities and provide an open-ended response explaining their answer. Survey Blocks consisted of the following categories and questions:
- Personal Preferences: Seven questions where individuals identify, rank personal on-water recreation activities and indicate preferred days and times for activities. Results determine most-preferred on-water recreational activities, times of recreation and preferred locations within the Iowa Great Lakes Region.
- Thoughts on Recreation Opportunities: Nine questions related to ideal recreation conditions, preferred locations (lakes) for such activities and whether on-water encounters with other people affect one's own recreation choices.
- Changes in Activity Over Time: Eleven questions comparing current levels of activity with two and five years prior to date, assessment of current policies and levels of personal safety regarding on-water recreation activities in the region. Questions also determine whether activities conducted by other people influence one's recreation choices.
- Demographics: Seven questions about property ownership status, on-water recreation equipment ownership and years of on-water recreation activities in the region, age, sex and ZIP code.

Chi-squared tests compare two categorical variables, including property owner and non-owner perspectives, while analysis of variance between age groups assesses relationship between categorical and numeric variables including recreation activities, time series and ranked values for personal preferences and Likert-based scales of measurement related to personal safety and satisfaction.

3.1 Study Area

Located in northwestern Iowa in the Midwestern United States of America, the Iowa Great Lakes Region is comprised of 12,687 acres of surface water available for fishing, boating, swimming, paddling and socialization in Dickinson County, Iowa. This chain of glacial, freshwater lakes includes West Lake Okoboji, East Lake Okoboji, Spirit Lake, Little Spirit Lake, Upper Gar Lake, Lower Gar Lake and Lake Minnewashta. For the sake of this study, Center Lake and Silver Lake were also included as recreation resources, as depicted in Fig. 2. This region provides broad-scale recreation opportunities in northwestern Iowa as high-quality freshwater resources. According to the Iowa State University Center for Agricultural and Rural Development, West Lake and East Lake Okoboji are two of Iowa's ten most visited lakes for recreation, within a region where intense modern agricultural activity dominates land use (Otto et al., 2012). Including Big Spirit Lake, these three water bodies annually generate approximately $120 million in direct spending for the surrounding community, and an estimated $190 million in total expenditure impacts.

With a 2021 population of 17,532 or 5.34% of the population of the State of Iowa, the population of Dickinson County ranks 41 out of 99 counties (US Census Bureau QuickFacts, 2019). However, in 2019, the county ranked ninth for direct expenditures associated with domestic travel in Iowa (The Economic Impact of Travel on Iowa Counties, 2019). As a result of unique, abundant opportunities for on-water recreation in this region, tourism dominates the local economy.

Given the economic significance of this glacial lake region, there is a critical need to understand the short- and long-term effects of both recreation conflict and potential for resource exhaustion as a result of increased activity (Otto et al., 2012). In 2008, the Iowa Natural Resource Commission adopted administrative rules to limit boat speed and distance zoning for areas within East and West Lake Okoboji, including Miller's Bay. An unintended consequence of this "no-wake zone" has been a dramatic increase in anchored boating, socializing, and subsequent conflict between users and adjacent dwellers along the shoreline, as depicted in Fig. 3.

3.2 Mean Maximum Satisfaction (MMS) Comparing Stakeholder Experiences

Analysis assumes stakeholders expect a certain amount of access and opportunity available for a preferred form of recreation, with dissatisfaction gained from the inability to meet the expected goal, by some form of disruption. 2020 survey responses prioritized specific recreation activities as ranked values, as well as categorical structures based on most- and least-preferred on-water activities. The magnitude of benefit B was calculated based on expectations (anticipated, A) and actual (observed, O) values for each surveyed activity A. A standardized ranked value was calculated based on Eq. 1 to generate a comparable value for each activity listed.

$$B = \frac{A_{\text{Max}} - A_{\text{MIN}}}{O_{\text{MAX}} - O_{\text{MIN}}} \tag{1}$$

For each activity option listed, magnitude of satisfaction for predicted and observed activities was based on the difference between average ranked values, with negative values indicating dissatisfaction and positive values assuming

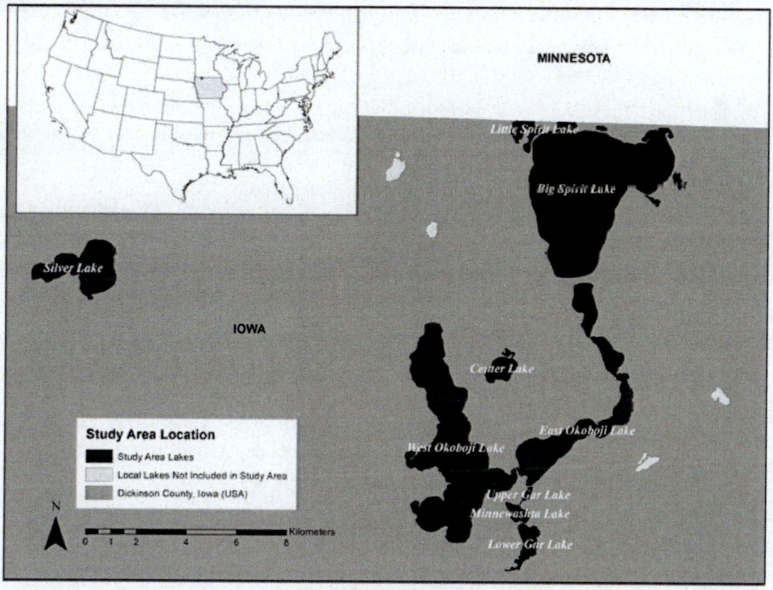

Fig. 2 Study area location

Fig. 3 July 2020 on-water recreation and anchored socializing in Miller's Bay, West Lake Okoboji

satisfaction. The result is a normalized value by which divergent perspectives may be compared across stakeholder groups. Perspectives from property owners and non-owners were compared based on ranked values for on-water recreation activities and a Likert-based scale assessing levels of satisfaction for each activity.

4 Results Summary

Results from both quantitative and qualitative studies indicate a high volume of activity for West Lake Okoboji for on-water recreation. As a subset, observations in Miller's Bay indicate a high volume of activity during holiday weekends, with afternoon activity throughout the season also consistent. Survey respondents were predominantly local property owners, comprising 82% of the total response rate. Despite ongoing local conditions of conflict among water resource users, results identify areas of agreement and shared values related to non-motorized recreation satisfaction levels, including anchored socializing on floatable equipment in calm waters across the region.

4.1 Quantitative Results: Observational Data

A total of 336 observations over the summer recreation season of Memorial Day to Labor Day 2020 track ongoing activity in Miller's Bay as anchored recreational activity occurs. Holiday weekends are when the greatest volume of anchored recreation activity occurs. For example, July 4 observations at 4:00 p.m. had approximately 3500 people in Miller's Bay. However, on the average, recreation in the Bay tends to attract fewer than 300 people, and the most frequent observation from the 2020 data was zero boats or people present. Table 1 lists the summary statistics for watercraft and people present from Memorial Day to Labor Day 2020 in Miller's Bay. Also included is a percentage representation of the number of days fewer than 100 of each present in the Bay. In 2020, 85% of observations counted fewer than 100

Table 1 Summary of 2020 observational data

Observations: 2020 season (N = 336)	Max	Min	Mean	Median	Mode	% Obs < 100
Watercraft present	790	0	59.42	28	0	85
People present	3512	0	279.16	119	0	46

watercraft both motorized and non-motorized, present in the Bay. Over the same sample period, 46% of observations counted fewer than 100 people present in the Bay. Mean and median watercraft present in the Bay over the 2020 season were 59.42 and 28, respectively, with the most common observation of zero watercraft present. Mean and median people present were 279.16 and 119, with zero people present most often during the 2020 season.

4.2 Qualitative Results: Email Survey Data

A total of 426 people participated in the email survey, with 46% completing the entire questionnaire. Forty-eight percent of respondents were female, and 50% were male, with 1.6% choosing not to disclose gender identity. A majority of survey participants were over the age of 50. See Table 2. Eighty-six percent of responses were collected from individuals who own property in the Iowa Great Lakes area, with 76% indicating ownership for more than 20 years.

A majority of respondents list West Lake and the adjacent East Lake Okoboji as the primary recreation resource, as displayed in Table 3.

Daytime recreation activity dominates the preferences of 2020 survey respondents. See Fig. 4, which indicates afternoon (12:00 p.m to 4:00 p.m.) is the prime recreation time for on-water activities for a majority (51%) of respondents. Morning activity is preferred for activities such as fishing and kayaking/paddle boarding. Afternoon is preferred for anchored socializing on water and power boating, which is also a preferred evening activity.

Non-motorized activity is preferred during morning hours, with kayaking/paddle boarding and fishing as the most common morning activities among all survey respondents. When asked specifically which activities are preferred during morning, afternoon and evening hours, 56.9% of respondents preferring fishing activity in the Iowa Great Lakes Region during morning hours. Kayaking and paddle boarding activities are preferred during morning hours for 55.32% of respondents. 2020 survey participants indicate a preference for motorized recreation activity during afternoon hours. This includes motorized boating (71.67%), as well as anchored socializing (82%). At the same time, afternoon hours are preferred for swimming by survey respondents, as 79.31% of responses indicated.

Swimming and fishing are viewed most positively among 81.12 and 79.87% of 2020 survey respondents, respectively. Activities viewed most negatively include jetskiing (54.42%) and indicating most negative association with jetski activity. Motorized "power" boating and anchored socializing were viewed negatively by 30.77% and 30.92% of respondents, respectively.

Sixty-seven percent of 2020 survey respondents claim adequate access to recreation in the Iowa Great Lakes Region. However, of the same sample, 59% claim other people's actions affect their own recreation experiences. Changes in activity based on comments indicate a shift from motorized to non-motorized activity, and reduced time fishing due to crowding, particularly on holidays and weekends. See Table 4.

Regarding a general sense of safety on the water in the Iowa Great Lakes Region, responses were relatively split, with 46.74% generally expressing a sense of safety (relatively safe and safe) and 46.19% feeling generally unsafe (relatively and unsafe). See Table 5. The number of boats in the general area, both anchored and moving, is expressed as reasons for a lost sense of safety. Conversely, those maintaining a general sense of safety cite personal responsibility,

Table 2 Age range of survey respondents (N = 189 of 426)

Age range	%	Count
18–25	0.00	0
26–31	2.65	5
32–39	5.82	11
40–49	12.70	24
50–59	23.81	45
60–69	35.45	67
70 or older	19.58	37
Total	100	189

Table 3 Area lakes ranked based on recreation activities

Lake name	%	Count
West Lake Okoboji	31.67	191
East Lake Okoboji	21.89	132
Spirit Lake	12.44	75
Little Spirit Lake	4.31	26
Upper Gar Lake	8.29	50
Lower Gar Lake	6.63	40
Lake Minnewashta	9.62	58
Center Lake	2.82	17
Silver Lake	2.32	14
Total	100	603

the existence of water patrol and boaters looking out for one another as a rationale.

A comparison between property owners and others, years of recreation experience in the area and age range indicate a strong, significant statistical relationship. Conversely, no relationship appears significantly significant regarding a general sense of safety, adverse effects of crowding, encounters with others impacting or hindering activity or a general sense of danger when recreating within the area relative to property ownership, age or sex of respondents. Levels of activity for recreation requiring specific equipment such as boats, kayaks or jetskis were highly dependent upon respondents' ownership, as most are not typically rented from local resources.

The age of survey respondents tends to indicate a likely divergence in activities and recreation preferences. A chi-squared test indicates a strong relationship between time of day for all recreation activities and the age of respondents, with $p = 0.0170$ and Cramér's V effect size of 0.218, $N = 187$. As depicted in Fig. 5, mid-day recreation activities are preferred by younger people, aged 26–31 as well as people aged 50–59.

A statistically significant relationship also exists between the preferred time of day for on-water recreation by all respondents and encounters with others on the water affecting recreation choices. A chi-squared test with $p = 0.032$ and effect size 0.196, $N = 181$ suggest this strong relationship. A ranked choice of 1 indicates the lowest level of influence and 5 as the highest level of others influencing recreation choices. Figure 6 indicates a similar pattern of impact during the noon to 4:00 p.m. time series containing a greater likelihood for altered recreation choices due to the influence of other people. A ranked analysis of variance (ANOVA) also indicates a relationship between preferred recreation times and the influence of other people on recreation choices, $p = 0.004$, with a Cohen's d effect of 0.285. Again, the noon to 4:00 p.m. time series is likely the most popular time for recreation, but also when potential conflict may arise.

When comparing the length of time respondents have spent recreating in the area, a strong statistical relationship exists between preferred times of day and years of experience in the region. A chi-square test with $p = 0.025$ and effect size 0.196, $N = 188$ suggest morning and mid-day recreation activities are strongly preferred by respondents with more than 20 years of experience recreating within the area. See Fig. 7.

Property ownership may also relate significantly to preferred times of day for on-water recreation. A chi-squared test of comparison at $p = 0.0135$ and Cramér's V effect size 0.227 for $N = 187$ suggest a relationship. However, with 82% of respondents identifying as property owners, more research is necessary to determine the strength of this relationship across variables.

4.3 Mean Maximum Satisfaction (MMS) of On-Water Recreation Activities

Overall mean values for all activities as expressed by property owners serve as the x-axis in Fig. 8 ($x = -0.057$). The y-axis represents overall mean values for non-property owners ($y = -0.011$) with a very slight but overall negative sense of satisfaction for both groups regarding non-motorized on-water recreation activities. Overall, property owners express a low level of dissatisfaction for their ability to conduct on-water recreation activity in the region beyond activities using a motorized boat on the water, including water skiing and general boating. Meanwhile, non-owners express a lower level of general satisfaction for their ability to conduct similar activities.

The lower left quadrant in Fig. 8 identifies a cluster of non-motorized activity where both non-owners and property owners share sentiments of general dissatisfaction with

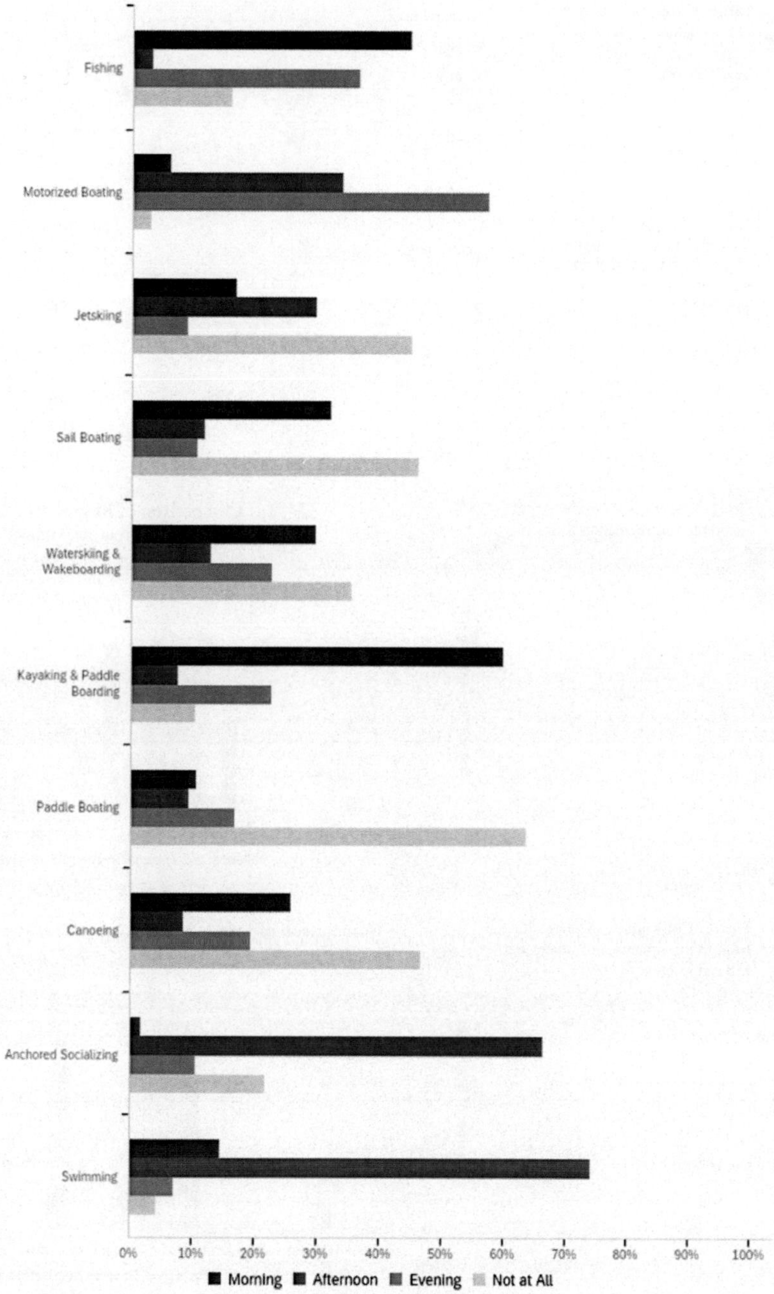

Fig. 4 Preferred time of day for specific on-water recreation activities

Table 4 Response to question: "I have adequate access to recreation in the Iowa Great Lakes Region"

Agreement	%	Count
Disagree	12.62	26
Neutral	23.79	49
Agree	63.59	131
Total	100	206

Table 5 Response to question: "I feel safe when recreating on water in the Iowa Great Lakes Region"

Scale range	%	Count	Sense of safety on water
1	16.30	30	Unsafe
2	29.89	55	Relatively unsafe
3	7.07	13	Neutral
4	33.15	61	Relatively safe
5	13.59	25	Safe
Total	100	184	

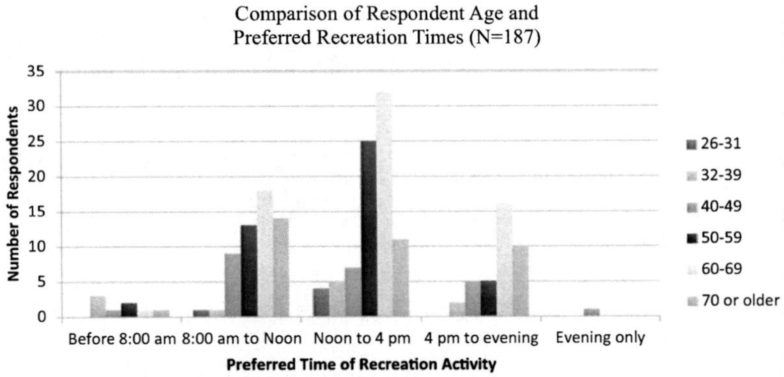

Fig. 5 Comparison of age ranges and preferred recreation times of day

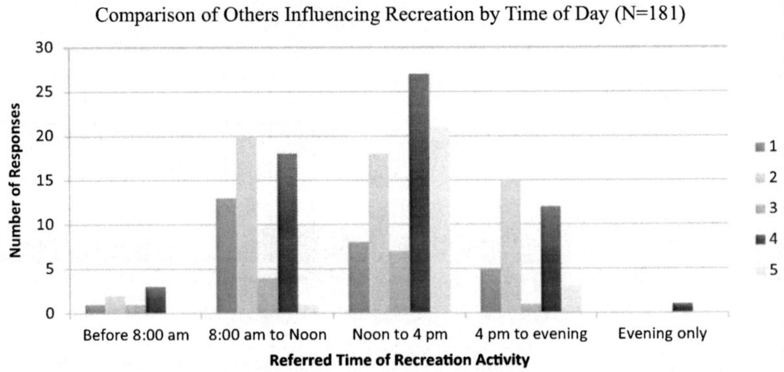

Fig. 6 Comparison of recreation time to the influence of others on recreation choices

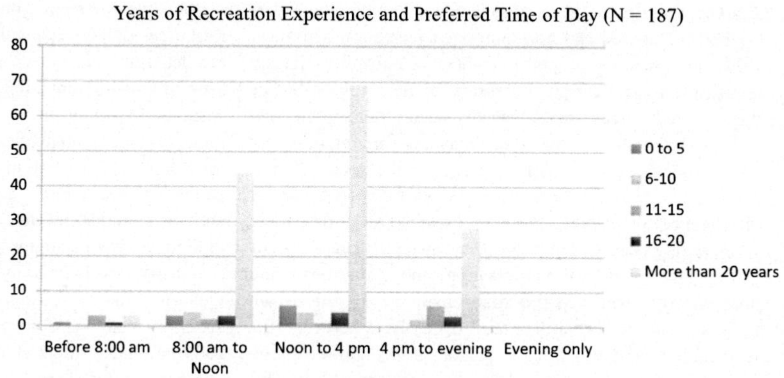

Fig. 7 Comparison with preferred on-water recreation times of day to years of recreation experience within the Iowa Great Lakes Region

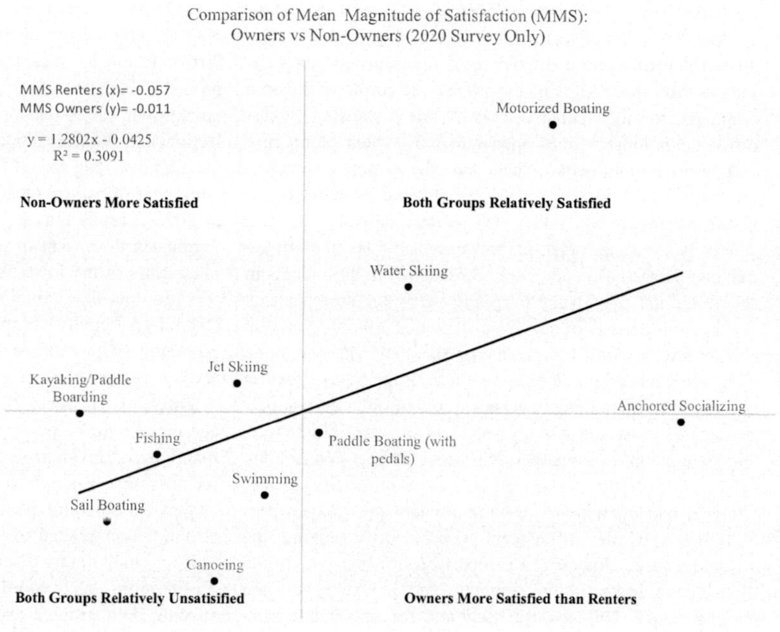

Fig. 8 Comparison of mean maximum satisfaction (MMS) for property owners and non-owners (2020 data only)

experiences. The upper right quadrant indicates shared satisfaction for motorized on-water activity, including motorized boating and water skiing. The red triangle indicates a relatively neutral, albeit positive level of satisfaction for anchored socializing on water, with property owners expressing a slight level of greater satisfaction than non-owners. A general slope of 1.2802 indicates an overall positive sense of satisfaction, and an intercept value of 0.043 indicates a relatively even split between owners and non-owners overall senses of satisfaction. However, the R^2 value of 0.3091 as a trendline suggests the data explain a relatively low proportion of overall variation in responses. Because 82% of 2020 survey respondents were property owners, it is necessary to gather additional feedback from non-owners to generate a statistically viable comparison.

5 Discussion and Conclusions

While areas of recreation conflict appear to exist, shared beliefs and values also tend to occur across community strata. Both property owners and non-owners surveyed in

2020 express similar levels of satisfaction and dissatisfaction for both motorized and non-motorized recreation activities. However, due to property owners dominating survey responses, more data are necessary to determine whether these shared values are statistically viable among the two stakeholder groups. A divergence may exist among on-water recreation community members regarding a sense of personal safety. When considering the strong perception of interference by others, further research may determine whether day users and non-property owners share this sentiment. Due to the low participation among day users and non-property owners in the 2020 email survey, efforts are underway to gather information from this subset of the recreation community. 2021 in-person survey administration concentrated on day users. 2022 data collection will also emphasize information from this community.

Statistically significant relationships appear to align conflict for on-water recreation in the area as both a generational disparity and a difference of opinion between individuals with more years of experience recreating in the area compared to more recent visitors. While a statistically significant relationship does appear to exist when comparing responses from property owners and others, more research is necessary in this area to draw a strong conclusion from this result. Relatively low effect size for test statistics suggests relatively weak association between variables within the existing dataset. More data are necessary from individuals in the 18–25 and 26–31 age ranges to verify the significance.

The non-normal frequency distribution of observational data prompt not only nonparametric statistical analysis, but a need for a large sample size. Extreme usage and recreation activity during holiday weekends juxtaposed with a percentage of most non-holiday weekend observations reflecting zero people or watercraft present during non-holiday observations result in a skewed data distribution. Statistically viable variation among property owners and non-owners indicates possible differences in perspective among the sample dataset. However, a comparison of Mean Maximum Satisfaction indicates both owners and non-owners share similar values. This is particularly true for satisfaction related to motorized and non-motorized recreation activity. Anchored socializing appears more favorable among property owners for 2020 data. This value may be skewed by the overwhelming response of property owners to the 2020 email survey. More data are necessary to adequately assess perspectives of non-property owners and day-user for on-water recreation. Additionally, more information is needed to assess perspectives of individuals in the 18–25 and 26–31 age ranges.

In Summer 2021, the same questions included in the email survey were administered in tandem with an "in-person intercept" targeting day users and others utilizing public boat ramps for on-water recreation access in the region. A partnership with the Iowa Department of Natural Resources (DNR) aquatic invasive species outreach program enabled Lakeside Lab interns to engage directly with lake users at five different public boat ramps over a ten-week time span. Each person the interns would engage with to determine awareness of aquatic invasive species was also asked to provide basic information on the age range and numbers of people in their parties; anticipated activities on the water; and destinations if any (including specific bays). These individuals were also invited to complete the same survey as was distributed in 2020 by emailing a link to an anonymized Qualtrics survey link. Future research plans include a cross-comparative analysis of responses to more dynamically compare beliefs, attitudes and action among property owners and non-owners as stakeholders within the community with a shared, public resource.

2021 data collection followed the same protocol as 2020, with an addition of survey activity at public boat ramps to further assess day users and people who do not own local property. Plans are to repeat the initial analysis for comparison with 2020 data. The addition of in-person data collection also enables comparison of activity across the region. Quantitative data from Miller's Bay can also be compared with the destinations cited by day users at public boat ramps in 2021. Results may help determine whether one location dominates the extent of activity, or if trends are similar in other parts of the Iowa Great Lakes Region.

Locations like Miller's Bay and the Iowa Great Lakes Region are heavily occupied for on-water recreation during weekends, with extreme usage occurring on holiday weekends. A general consensus is that conditions have changed in recent years. However, this perception may be a function of age and length of personal experience in the region. Additional data from short-term visitors and younger community members may strengthen the knowledge of this condition.

Despite ongoing conditions of recreation conflict, the shared values related to non-motorized recreation may provide a foundation for discourse among a mix of stakeholders. Furthermore, such commonalities indicate stakeholders of varying demographic groups may have more in common than previously perceived. Additional research is necessary to determine whether a strong, statistical relationship between property ownership and ongoing issue of conflict truly exists. In addition, exploration of CSR potential for tourism communities like the Iowa Great Lakes may enable a non-regulatory approach to sustainable on-water recreation which addresses ongoing issues of conflict.

Acknowledgements Special thanks to Dr. Martin St. Clair at Coe College for assistance with this project. 2020 and 2021 Iowa Lakeside Laboratory interns assist with ongoing data collection. The support from Mike Hawkins, Jed Siegwarth and Iowa Department of Natural Resources Fisheries and Lake Restoration Programs enabled data collection for this project. For their effort we are truly grateful. And thanks to the Friends of Lakeside Lab for financial support of this project.

References

Almeida-García, F., Pelaez-Fernandez, M. A., Balbuena-Vazquez, A., & Cortés-Macias, R. (2016). Residents' perceptions of tourism development in Benalmádena (Spain). *Tourism Management, 54*, 259–274.

Andereck, K. L., & Nyaupane, G. P. (2011). Exploring the nature of tourism and quality of life perceptions among residents. *Journal of Travel Research, 50*(3), 248–260. https://doi.org/10.1177/0047287510362918

Baas, J., & Burns, R. C. (2016). *Outdoor recreation planning*. Sagamore Publishing.

Bacher, J. (1987). The Brundtland report: Our common future. *Peace Magazine, 3*(4), 11.

Bartol, K., & Locke, E. (2000). Incentives and motivation. In S. Rynes, & B. Gerhart (Eds.), *Compensation in organisations*. Jossey Bass.

Beritelli, P. (2011). Cooperation among prominent actors in a tourist destination. *Annals of Tourism Research, 38*(2), 607–629. https://doi.org/10.1016/j.annals.2010.11.015

Boley, B. B., McGehee, N. G., & Tom Hammett, A. L. (2017). Importance-performance analysis (IPA) of sustainable tourism initiatives: The resident perspective. *Tourism Management, 58*, 66–77. https://doi.org/10.1016/j.tourman.2016.10.002

Bosak, K., & McCool, S. F. (2016). The way forward. In *Reframing sustainable tourism* (pp. 243–250). Springer.

Bradford, J. B., & D'Amato, A. W. (2012). Recognizing trade-offs in multi-objective land management. *Frontiers in Ecology and the Environment, 10*(4), 210–216.

Bramwell, B., & Lane, B. (2003). *Tourism collaboration and partnerships: Politics, practice and sustainability* (Vol. 2). Channel View Publications.

Brown, R. I., & Brown, I. (2005). The application of quality of life. *Journal of Intellectual Disability Research, 49*(10), 718–727. https://doi.org/10.1111/j.1365-2788.2005.00740.x

Cahill, K. (2013). *Recreation opportunities and conflicts*.

Choi, H.-S.C., & Sirakaya, E. (2005). Measuring residents' attitude toward sustainable tourism: development of sustainable tourism attitude scale. *Journal of Travel Research, 43*(4), 380–394. https://doi.org/10.1177/0047287505274651

Confer, J. J., Thapa, B., & Mendelsohn, J. L. (2005). Exploring a typology of recreation conflict in outdoor environments. *World Leisure Journal, 47*(1), 12–23. https://doi.org/10.1080/04419057.2005.9674382

Dawson, C. P. (2000, 2000). A comparison of recreation conflict factors for different water-based recreation activities.

Domínguez-Gómez, J. A., & González-Gómez, T. (2017). Analysing stakeholders' perceptions of golf-course-based tourism: A proposal for developing sustainable tourism projects. *Tourism Management, 63*, 135–143. https://doi.org/10.1016/j.tourman.2017.05.015

Fennell, D. A. (2008). *Ecotourism David A. Fennell* (3rd ed.). Routledge.

Frank, B. (1987). Our common future: The "Brundtland Commission" report. *Ambio, 16*(4), 217–218.

Gill, A. (2018). Understanding the evolving relationship between tourism and nature in an era of sustainability (pp. 79–104).

Guaita Martinez, J. M., Martin Martin, J. M., Salinas Fernandez, J. A., & Mogorron-Guerrero, H. (2019). An analysis of the stability of rural tourism as a desired condition for sustainable tourism. *Journal of Business Research, 100*, 165–174. https://doi.org/10.1016/j.jbusres.2019.03.033

Acts and Resolutions Passed at the Several Sessions of the Territorial Legislature of Iowa, 1840–1846 (1912).

Ivars Baidal, J. A. (2004). Tourism planning in Spain: Evolution and perspectives. *Annals of Tourism Research, 31*(2), 313–333. https://doi.org/10.1016/j.annals.2003.12.001

Ivy, M. I., Stewart, W. P., & Lue, C.-C. (1992). Exploring the role of tolerance in recreational conflict. *Journal of Leisure Research, 24*(4), 348–360.

Jacob, G. R., & Schreyer, R. (1980). Conflict in outdoor recreation: A theoretical perspective. *Journal of Leisure Research, 12*(4), 368–380. https://doi.org/10.1080/00222216.1980.11969462

Jurowski, C., & Brown, D. O. (2001). A comparison of the views of involved versus noninvolved citizens on quality of life and tourism development issues. *Journal of Hospitality & Tourism Research (Washington, D.C.), 25*(4), 355–370. https://doi.org/10.1177/109634800102500401

Kakoyannis, C. A. (2002). *Assessing and evaluating recreational uses of water resources: Implications for an integrated management framework* (Vol. 536). US Department of Agriculture, Forest Service, Pacific Northwest Research Station.

Keeble, B. R. (1988). The Brundtland report: 'Our common future.' *Medicine and War, 4*(1), 17–25. https://doi.org/10.1080/07488008808408783

Locke, E. A. (1970). Job satisfaction and job performance: A theoretical analysis. *Organizational Behavior and Human Performance, 5*(5), 484–500.

Locke, E. A., Cartledge, N., & Knerr, C. S. (1970). Studies of the relationship between satisfaction, goal-setting, and performance. *Organizational Behavior and Human Performance, 5*(2), 135–158.

Maiden, J. A. (2008). *Participation in sustainable tourism development: Stakeholders & partnership working*. ProQuest Dissertations Publishing.

Maiolo, J. R., Johnson, J., & Griffith, D. (1992). Applications of social science theory to fisheries management: Three examples. *Society & Natural Resources, 5*(4), 391–407.

Manning, R. E., & Powers, L. A. (1984). Peak and off-peak use: Redistributing the outdoor recreation/tourism load. *Journal of Travel Research, 23*(2), 25–31.

Marcouiller, D. W., Scott, I., & Prey, J. (2005). *Addressing recreation conflict: Providing a conceptual basis for management*.

McCool, S. F. (2016). Sustainable tourism in an emerging world of complexity and turbulence. In *Reframing sustainable tourism* (pp. 3–11). Springer.

Mimbs, B. P., Boley, B. B., Bowker, J. M., Woosnam, K. M., & Green, G. T. (2020). Importance-performance analysis of residents' and tourists' preferences for water-based recreation in the Southeastern United States. *Journal of Outdoor Recreation and Tourism, 31*, 100324.

Otto, D., Tylka, K., & Erickson, S. (2012). Economic value of outdoor recreation activities in Iowa. Saatavilla. http://www.card.iastate.edu/environment/items/DNR-AmenityRevised_9-25-12.pdf. Cited 24, 2018

Panzer-Krause, S. (2019). Networking towards sustainable tourism: Innovations between green growth and degrowth strategies. *Regional Studies, 53*(7), 927–938.

Park, D.-B., Lee, K.-W., Choi, H.-S., & Yoon, Y. (2012). Factors influencing social capital in rural tourism communities in South Korea. *Tourism Management, 33*(6), 1511–1520. https://doi.org/10.1016/j.tourman.2012.02.005

Perdue, R. R., Long, P. T., & Kang, Y. S. (1999). Boomtown tourism and resident quality of life: The marketing of gaming to host community residents. *Journal of Business Research, 44*(3), 165–177. https://doi.org/10.1016/S0148-2963(97)00198-7

Poudel, S., Nyaupane, G. P., & Budruk, M. (2016). Stakeholders' perspectives of sustainable tourism development: A new approach to measuring outcomes. *Journal of Travel Research, 55*(4), 465–480. https://doi.org/10.1177/0047287514563166

Reynolds, M. D. (1993). *Tourism committee proceedings: Final report.*

Russo, A. P. (2002). The "vicious circle" of tourism development in heritage cities. *Annals of Tourism Research, 29*(1), 165–182.

Schneider, I. E. (2000). Revisiting and revising recreation conflict research. *Journal of Leisure Research, 32*(1), 129–132.

Scott, I., Marcouiller, D., & Prey, J. (2005). Compatibility as a conceptual basis for outdoor recreation planning. *Work funded as part of the 2005–2010 Wisconsin Statewide Comprehensive Outdoor Recreation Planning process.*

Spiers, A. (2012). Achieving a better understanding of outdoor recreation conflict and its management in Canada's national parks.

Su, L., Huang, S., & Huang, J. (2018). Effects of destination social responsibility and tourism impacts on residents' support for tourism and perceived quality of life. *Journal of Hospitality & Tourism Research, 42*(7), 1039–1057.

Teclaff, L. A., & Teclaff, E. (1972). Saving the land-water edge from recreation, for recreation. *Arizona Law Review, 14*, 39.

The Economic Impact of Travel on Iowa Counties. (2019).

Toman, M. A. (1994). Economics and "sustainability": Balancing trade-offs and imperatives. *Land Economics, 70*(4), 399–413. https://doi.org/10.2307/3146637

Uleck, R. B. (1971). The challenge of recreation planning: Methodology and factors to consider. In *The Forest Recreation Symposium*

US Census Bureau QuickFacts. (2019). US Census Bureau. Retrieved August 31, from https://www.census.gov/quickfacts/fact/table/US/PST045219?

US Code Title 33, Chapter 1: Navigable Waters Generally. (1812).

Vittersø, J., Chipeniuk, R., Skår, M., & Vistad, O. I. (2004). Recreational conflict is affective: The case of cross-country skiers and snowmobiles. *Leisure Sciences, 26*(3), 227–243.

Vroom, V. H. (1964). Work and motivation.

Water Rights and Related Issues. (1994). In A. L. T. Association (Ed.).

Wills, J. (2010). *Iowa Great Lakes Watershed Management Plan.*